Ocean Instrumentation, Electronics, and Energy

Ocean Instrumentation, Electronics, and Energy

S.R. Vijayalakshmi, PhD
S. Muruganand, PhD

Mercury Learning and Information
Dulles, Virginia
Boston, Massachusetts
New Delhi

Publisher: David Pallai
MERCURY LEARNING AND INFORMATION
22841 Quicksilver Drive
Dulles, VA 20166
info@merclearning.com
www.merclearning.com
(800) 232-0223

S.R. Vijayalakshmi and S. Muruganand. *Ocean Instrumentation, Electronics, and Energy.*
ISBN: 978-1-944534-57-8

The publisher recognizes and respects all marks used by companies, manufacturers, and developers as a means to distinguish their products. All brand names and product names mentioned in this book are trademarks or service marks of their respective companies. Any omission or misuse (of any kind) of service marks or trademarks, etc. is not an attempt to infringe on the property of others.

Library of Congress Control Number: 2016935952

171819321   This book is printed on acid-free paper.

Our titles are available for adoption, license, or bulk purchase by institutions, corporations, etc. For additional information, please contact the Customer Service Dept. at 800-232-0223 (toll free).

All of our titles are available in digital format at authorcloudware.com and other digital vendors. The sole obligation of MERCURY LEARNING AND INFORMATION to the purchaser is to replace the book, based on defective materials or faulty workmanship, but not based on the operation or functionality of the product.

Contents

PREFACE

Ocean covers an area of about 140 million square miles or 362 million sq km. An estimated 50–80 % of all life on earth is found under the ocean surface and the oceans contain 99 % of the living space on the planet. Less than 10 % of that space has been explored by humans. The dark, cold environment called the deep sea constitutes 85 % of of the ocean's area and 90 % of its volume. The average depth of the ocean is 3,795 m. The average height of the land is 840 m. The oceans cover 71 % (and rising) of the Earth's surface and contain 97 % of the Earth water. Less than 1 % is fresh water, and 2–3 % is contained in glaciers and ice caps (and is decreasing). Oceans are the site of 90 % of all volcanic activity. Our oceans teem with life ranging from the blue whale, the biggest animal on Earth, to tiny microbes. But nobody knows exactly how many different species live in this environment. There is no data for around 20 % of the ocean's volume. It's a location whose details everyone is very much interested to know.

This book covers many topics encountered in the design and application of electronic instruments related to the ocean, both basic ideas and advanced technologies. It also covers requirements of ocean research.

Chapter 1 discusses oceanographic remote sensing, including applications, radar, radiometry, satellites, sonar, telemetry, laser and LIDAR, and research ideas. It also discusses the difference between satellite images and maps, remote sensing and GIS, remote sensing and aerial photography, remote sensing and sonar. It also describes oceanography, synthetic aperture radar, NADIR radar, microwave radiometry, infrared radiometers, imaging radiometer, microwave measurement of ocean wind, wind and wave height measurements, mapping the ocean floor with single beam echo sounding, multi-beam bathymetry, sound as underwater navigation, echo-sounders, applications of telemetry, data collection in LIDAR, discrete-return LIDAR, waveform recording devices, and applications of LIDAR remote sensing.

Chapter 2 discusses sensors for ocean monitoring and their measuring parameters. Sometimes satellites are called sensors, as well as the sensors they carry. It discusses the sensors, scanners, weather sensing, SAR sensors, marine observation sensors (MOS), ocean color monitoring sensor (OCM) and micro-sensors for ocean acidification monitoring. It also discusses the measurement of ocean parameters, such as ocean color, sediment monitoring, surface currents, surface wind, wave height, wind speed, sea surface temperature, upwelling, sampling, wave energy, and ocean floor. It also describes spatial resolution, pixel size, scale, spectral/radiometric resolution, temporal resolution, sensor design, sensor selection, and research on ocean phenomena.

Chapter 3 discusses underwater acoustics, including the interaction of sound with the seafloor, sound wave features, transmission of data underwater, wave height, wave velocities, bubbles study, water depth, sea temperature, global climate change, ocean current measurement using sound, fish finding, study of Earth history, and surf zone measurement using sound. It describes locating and identifying fish, methods of underwater communication, measurement of ocean temperature using acoustic tomography, inverted echo-sounders, acoustic doppler current profilers, RAFOS floats, and reciprocal transmission.

Chapter 4 discusses underwater wireless communication, including acoustic waves and acoustic communication, optical waves and optical communication, underwater acoustic communication, underwater optical communications, underwater mobile communication, types of modulation, Internet, and GPS. It describes the underwater communication **environment and propagation mediums,** limitations of underwater acoustic communication, using laser as optical communication above water and underwater, protocol stack for the underwater laser sensor network, wireless laser communication system description, **instrumentation system devices, the MEMS approach,** benefits of smart optical systems for underwater vehicles, systems and methods in underwater optical communication, frequency shift keying as applied to UAC, direct sequence spread spectrum (DSSS), multi-carrier modulation, multi-input multi-output techniques, Internet to ships using oceanographic tool (SeaNet), Internet access for voyagers, voice communication on ship, submarine communications cable, optical submarine cable repeaters, Antarctica, perfectly secure communication and environmental impact.

Chapter 5 discusses the wireless sensor network, oceanographic WSN, WSN architecture, WSN network topologies, WSN applications, and underwater network technology. It also discusses the wireless underwater acoustic sensor network, 2-dimensional underwater sensor network architecture, 3-dimensional underwater sensor network architecture, and sensor network architecture with AUV. It describes the difference between terrestrial WSNs and oceanographic WSNs, application areas of oceanographic WSN, general sensor node, sensing parameters of different sensors, challenges in oceanographic WSN, wireless communication technologies, oceanographic sensors protection, advanced buoy design, energy harvesting system design, system stability, reliability, wireless underwater sensor network, underwater sensing applications, underwater communication, physical layer, medium access control, resource sharing, the network layer, routing, transport, network services, sensing and application techniques, hardware platforms, test beds, simulators, models, the difference between underwater acoustic sensor networks and terrestrial networks, unique characteristics of underwater acoustic sensor networks, underwater acoustic sensor network architecture, network challenges for underwater acoustic sensor networks, and research challenges and opportunities in underwater WSN.

Chapter 6 deals with analog and digital images, EM spectrum, multi-layer images, spectral response patterns, multi-spectral images, multi-spectral remote sensing, superspectral images, hyperspectral images, hyperspectral remote sensing, sensor/platform systems, spatial resolution, pixel size, radiometric resolution, data volume, infrared remote sensing, black body radiation, microwave remote sensing, digital image processing and software for ocean color and algorithms. It describes image processing for the ocean, push broom scanning, interaction mechanisms, aerial photography, aerial videography, satellite-based scanning systems, interaction between microwaves and the Earth's surface, image preprocessing, image enhancement, image classification, and image transformation.

Chapter 7 discusses ocean energy as a renewable source of energy, wave energy, wave energy technologies, tidal power, ocean thermal energy conversion, ocean current energy, offshore wind energy, offshore wind energy technology, offshore solar energy, offshore solar energy technology and concentrating solar power technology. It also describes the operation of tidal barrages, ocean turbines, semi-submersible offshore wind turbines, offshore floating structures for mounting wind turbines, transmitting power ashore through subsea cables, and challenges in ocean power technologies.

Chapter 8 discusses the role of electronics in marine generator sets, marine instruments, wireless control stations, navigation equipment, autopilot systems, satellite phones, fire-fighting equipment, bubbler gauges, navigation instruments, AIS operation, electrical propulsion, and gas indicators. It also describes marine electronics, engine governors, timing considerations, fuel injection, marine instruments, salt assault, producing nautical charts, GPS-based instrument recovery, stray line buoys, weather monitoring systems, **routing and reporting,** integrated sail boat instruments, ship board level sensors, pressure-based level measurement, ultra sonic/microwave level sensors, capacitive level sensors, ship's bridge, controlling ship's speed and direction from the bridge, tracking ships using AIS, information transfer, boil-off for propulsion, TFDE propulsion layout, gas detection meters for ships, noncombustible gas indicators, and multi-gas analyzers.

Chapter 9 deals with instruments and their measured parameters, oceanographic instrumentation, Argo robots, measurements of hydrographic properties, measurement of dynamic properties, BIOMAPER, and many more instruments. It also discusses how to measure depth, temperature, salinity, oxygen, phosphate, silicate, nitrate, pH, water clarity, sound, current, waves and tides, seabed sampling, and bioluminescence. It describes research vessels, moorings, moored profilers, satellites, submersibles, towed vehicles, floats, and drifters, as well as reversing thermometers, CTDs, multiple water sample devices, thermosalinographs, remote sensors, current meters, wave measurements, tide gauges, shear probes, ferry box, glider, radar doppler current profiler (RDCP), X band radar, HF radar, satellite remote sensing, underwater nodes, zooplankton recorder, nucleic acid biosensor, sensors for pH and alkalinity, Air Sea Interaction METeorology (ASIMET), gravity corers, hydraulically damped gravity corers, marine magnetometers, ocean bottom seismometers, submersible incubation devices, deep ocean tsunami detection buoys, and oceanographic instruments.

Chapter 10 deals with the optical constituents of seawater, retrieval of oceanic constituents from ocean color measurements at sea level, ocean optics dip probe spectrometers, and interesting facts about ocean. It describes ocean optics, optical properties of the ocean, Inherent Optical Properties (IOP) variability, retrieval of oceanic constituents from ocean color measurements taken at top of atmosphere, the atmospheric correction problem, spectrometers, and ocean optics visible spectrophotometers.

The appendices present in-depth data about Indian satellites for ocean monitoring and acronyms used in the ocean electronics field. Almost every chapter can be read independently from the others; hence a flexible presentation of subjects can be realized. The consequence of this approach is that some of the details are repeated in different contexts, but this can only improve understanding. Furthermore, every chapter provides three levels of exercises for testing reader comprehension.

The authors would be glad to receive any suggestions for improving this textbook. Anyone requiring further information should contact the authors [email: srvijisiva@gmail.com]. The web links lead to relevant websites; appropriate material may be found in the web databases. S. R. Vijayalakshmi thanks the University Grants Commission, Government of India for the financial support in doing my postdoctoral research project work.

S. R. Vijayalakshmi
S. Muruganand

CHAPTER 1

REMOTE SENSING IN OCEANOGRAPHY

This chapter discusses remote sensing in oceanography; applications of remote sensing; radar, radiometry, satellites, sonar, telemetry, laser, and LIDAR in remote sensing; and research ideas in remote sensing.

1.1 INTRODUCTION TO REMOTE SENSING

The measurement of information about properties of an object by a recording device that is not in physical contact with the object under study (utilized at a distance, as from an aircraft, spacecraft, or ship) and the display of information important to the environment, such as measurements of force fields, electromagnetic radiation, or acoustic energy, is known as remote sensing. The technique employs such devices as cameras, lasers, and radio frequency receivers, radar systems, sonar, seismographs, gravimeters, magnetometers, and scintillation counters. As humans, we are intimately familiar with remote sensing in that we depend on visual perception to provide us with information about our surroundings. As sensors, however, our eyes are greatly limited by

1. Sensitivity to only the visible range of electromagnetic energy;

2. Viewing perspectives dictated by the location of our bodies; and

3. The inability to form a lasting record of what we view.

Because of these limitations, humans have continuously sought to develop the technological means to increase our ability to see and record the physical properties of our environment. Beginning with the early use of aerial photography, remote sensing has been recognized as a valuable tool for viewing, analyzing, characterizing, and making decisions about our environment. In the past few decades, remote sensing technology has advanced on three fronts:

1. From predominantly military use to a variety of environmental analysis applications that relate to land, ocean, and atmosphere issues;

2. From (analog) photographic systems to sensors that convert energy from many parts of the electromagnetic spectrum to electronic signals; and

3. From aircraft to satellite platforms.

Today, we define satellite remote sensing as the use of satellite-borne sensors to observe, measure, and record the electromagnetic radiation reflected or emitted by the Earth and its environment for analysis and extraction of information. What follows are the main points of similarity and difference between the field of remote sensing (analysis and images) and fields/products such as maps, satellite images, GIS, and sonar.

1.1.1 Satellite Images vs. Maps

A map is "a conventionalized image representing selected features or characteristics of geographical reality, designed for use when spatial relationships are of primary importance." A map shows us the world as we know it, and what we know is a very complex subject that comprises:

- The limits of matter, technology, and our measurement tools

- What we believe exists

- What we think to be important

- What we want and aspire to

Thus, a map is subjective, reflecting human decisions about what to put on it and how to represent these things. A remote sensing image, in contrast, is an objective recording of the electromagnetism reaching the sensor. Another important difference is that a map is a projection of the Earth on paper, without any relief displacement, while in a remote sensing image shows both relief displacement and geometrical distortion.

1.1.2 Remote Sensing vs. GIS

GIS (Geographic Information System) is a kind of software that enables:

- Collecting spatial data from different sources (remote sensing being one of them)

- Relating spatial and tabular data

- Performing tabular and spatial analysis

- Symbolizing and designing the layout of a map

GIS software can handle both vector and raster data (although some handle only one of them). Remote sensing data belongs to the raster type and usually requires special data manipulation procedures that regular GIS does not offer. However, after a remote sensing analysis has been done, its results are usually combined with GIS or put into a database of an area for further analysis (overlaying with other layers, etc.). In recent years, more and more vector capabilities are being added to remote sensing software, and some remote sensing functions are inserted into GIS modules.

1.1.3 Remote Sensing vs. Aerial Photography/Photogrammetry

Both remote sensing and aerial photography or photogrammetry gather data about the upper surface of the Earth by measuring electromagnetic radiation from airborne systems. The following major differences can be given:

Aerial photos are taken by an analog instrument; film from a (photogrammetric) camera is then scanned to be transformed to digital media. Remote sensing data is usually gathered by a digital Charge Coupled Device (CCD) camera. The advantage of film is its high resolution (granularity), while the advantage of CCD is that we measure quantitatively the radiation reaching the sensor (radiance values, instead of a gray-value scale bar). Thus, remote sensing data can be integrated into physical equations of, for example, energy balance.

- An aerial photograph is a central projection, with the whole picture taken at one instance. A remote sensing image is created line after line; therefore, the geometrical correction is much more complex, with each line (or even pixel) needing to be treated as a central projection.

- Aerial photos usually gather data only in the visible spectrum (although there are also special films sensitive to near infrared radiation), while remote sensing sensors can be designed to measure radiation all along the electromagnetic spectrum.

- Aerial photos are usually taken from planes, while remote sensing images may also be taken from satellites.

- Both systems are affected by atmospheric disturbances, aerial photos mainly by haze (that is, the scattering of light—the process which makes the sky blue), remote sensing images also by processes of absorption.

- Atmospheric corrections to aerial photos can be made while taking the picture (using a filter), or in post-processing, as done in remote sensing. Thermal remote sensing sensors can also operate at night and radar data is almost weather independent.

- Photogrammetry is mainly dedicated to the accurate creation of a 3D model, in order to plot with high accuracy the locations and boundaries of objects and to create a digital elevation model by applying sophisticated geometric corrections. Remote sensing is mainly dedicated to the analysis of the incoming electromagnetic spectrum, using atmospheric corrections, sophisticated statistical methods for classification of the pixels to different categories, and analyzing the data according to the known physical processes that affect light as it moves in space and interacts with objects.

- Remote sensing images are very useful for tracking phenomena on regional, continental, and even global scales, using the fact that satellites cover wide areas and take images all the time (whether fixed above a certain point or *revisiting* the same place every 15 days, for example).

- Remote sensing images have been available since the early 1970s. Aerial photos provide a longer time span for detecting landscape change (for example, with many aerial photos taken during World War I).

- Remote sensing images are more difficult to process and require trained personnel, while aerial photographs can be interpreted more easily.

1.1.4 Remote Sensing vs. Sonar

Sonar can also be considered a kind of remote sensing—that is, studying the surfaces of the sea (bathymetry and sea bed features) from a distance. Sonar is an active type of remote sensing (like radar; it is not dependent on an external source of waves, measuring the time between the transmission and reception of waves produced by our instruments and their intensity), but using sound waves rather than electromagnetic radiation. Both systems transmit waves through an interfering medium (water, air) that adds noise to the data we are looking for, and therefore corrections must be applied to the raw data collected. In remote sensing, however, radar is almost independent of weather, and atmospheric disturbances affect mainly passive remote sensing. To make these necessary corrections, both systems depend on calibration from field data (be it salinity, temperature and pressure measured by the ship while surveying, or measurements of the atmospheric profile parameters by a meteorological radiosonde, for example).

Sonar is mainly used to produce the bathymetry of the sea, while remote sensing techniques focus more on identification of a material's properties than its height. Echo-sounders (single- or multi-beam) can be compared to airborne laser scanning—both create point (vector) data containing X, Y, and Z that needs to be further post-processed in order to remove noise (spikes). An added complexity when dealing with bathymetry (as opposed to topography) is the need for tide corrections. Side scan sonar can be compared to side looking aperture radar, both creating images (raster) analyzing the surface. Another major difference is that in remote sensing the results of the analysis can be compared easily to the field (aerial photos, maps, field measurements), while in sonar the bottom of the sea is hidden from us, and we depend totally on the data gathered.

1.1.5 Applications of Remote Sensing

Each sensor is designed with a specific purpose. With optical sensors, the design focuses on the spectral bands to be collected. With radar imaging, the incidence angle and microwave band used plays an important role in defining which applications the sensor is best suited for. Each application itself has specific demands for spectral resolution, spatial resolution, and temporal resolution. There can be many applications for remote sensing, in different fields, as described below.

Agriculture: Satellite and airborne images are used as mapping tools to classify crops, examine their health and viability, and monitor farming practices. Agricultural applications of remote sensing include crop type classification, crop condition assessment, crop yield estimation, mapping of soil characteristics, mapping of soil management practices, and compliance monitoring (farming practices).

Forestry: Forests play an important role in balancing the Earth's CO_2 supply and exchange, acting as a key link between the atmosphere, geosphere, and hydrosphere. Forestry applications of remote sensing include forest cover updating, depletion monitoring, measuring biophysical properties of forest stands, collecting harvest information, updating of inventory information for timber supply, broad forest type, vegetation density, biomass measurements, and monitoring the quantity, health, and diversity of the Earth's forests.

Geology: Geology involves the study of landforms, structures, and subsurface to understand physical processes creating and modifying the Earth's crust. Geological applications of remote sensing include surficial deposit/bedrock mapping, lithological mapping, structural mapping, sand and gravel (aggregate) exploration/exploitation, mineral exploration, hydrocarbon exploration, environmental geology, geobotany, baseline infrastructure, sedimentation mapping and monitoring, event mapping and monitoring, geohazard mapping, and planetary mapping.

Hydrology: Hydrology is the study of water on the Earth's surface, whether flowing above ground, frozen in ice or snow, or retained by soil. Hydrological applications include wetlands mapping and monitoring, soil moisture estimation, snow pack monitoring/delineation of extent, measuring snow thickness, determining snow-water equivalent, river and lake ice monitoring, flood mapping and monitoring, glacier dynamics monitoring (surges, ablation), river/delta change detection, drainage basin mapping, and watershed modeling, irrigation canal leakage detection, and irrigation scheduling.

Sea ice: Ice covers a substantial part of the Earth's surface and is a major factor in the commercial shipping and fishing industries, Coast Guard and construction operations, and global climate change studies. Sea ice information and applications include ice concentration, ice type/age/motion, iceberg detection and tracking, surface topography, tactical identification of leads, navigation, safe shipping routes/rescue, ice condition (state of decay), historical ice and iceberg conditions and dynamics for planning purposes, wildlife habitat, pollution monitoring and meteorological/global change research.

Land cover and land use: Land cover refers to the surface cover on the ground, while land use refers to the purpose the land serves. Land use applications of remote sensing include natural resource management, wildlife habitat protection, baseline mapping for GIS input, urban expansion/encroachment, routing and logistics planning for seismic/exploration/resource extraction activities, damage delineation (tornadoes, flooding, volcanic, seismic, fire), legal boundaries for tax and property evaluation and target detection—identification of landing strips, roads, clearings, bridges, land/water interface.

Mapping: Mapping constitutes an integral component of the process of managing land resources, and mapped information is the common product of analysis of remotely sensed data. Mapping applications of remote sensing include the following:

- Planimetry: Land surveying techniques accompanied by GPS can be used to meet high accuracy requirements, but limitations include cost effectiveness and difficulties in attempting to map large or remote areas. Remote sensing provides a means of identifying and presenting planimetric data in convenient media and an efficient manner. Imagery is available in varying scales to meet the requirements of different users. Defense applications typify the scope of planimetry applications: extracting transportation route information, building and facilities locations, urban infrastructure, and general land cover.

- Digital Elevation Models (DEMs): Generating DEMs from remotely sensed data can be cost-effective and efficient. A variety of sensors and methodologies to generate such models are available and proven for mapping applications. Two primary methods of generating elevation data are stereogrammetry techniques using air photos (photogrammetry), VIR imagery, or radar data (radargrammetry), and radar interferometry.

- Baseline thematic mapping/topographic mapping: As a base map, imagery provides ancillary information to the extracted planimetric or thematic detail. Sensitivity to surface expression makes radar a useful tool for creating base maps and providing reconnaissance abilities for hydrocarbon and mineralogical companies involved in exploration. This is particularly true in remote northern regions, where vegetation cover does not mask the micro topography and, generally, information may be sparse. Multi-spectral imagery is excellent for providing ancillary land cover information, such as forest cover. Supplementing the optical data with the topographic relief and textural nuance inherent in radar imagery can create an extremely useful image composite product for interpretation.

Oceans and coastal monitoring: The oceans not only provide valuable food and biophysical resources, they also serve as transportation routes, are crucially important in weather system formation and CO_2 storage, and are an important link in the Earth's hydrological balance. Coastlines are environmentally sensitive interfaces between the ocean and land and respond to changes brought about by economic development and changing land-use patterns. Often coastlines are also biologically diverse intertidal zones and can also be highly urbanized. Ocean applications of remote sensing include ocean pattern identification, currents, regional circulation patterns, shears, frontal zones, internal waves, gravity waves, eddies, upwelling zones, shallow water bathymetry, storm forecasting, wind and wave retrieval, fish stock and marine mammal assessment, water temperature monitoring, water quality, ocean productivity, phytoplankton concentration and drift, aquaculture inventory and monitoring, oil spills, mapping and predicting oil spill extent and drift, strategic support for oil spill emergency response decisions, identification of natural oil seepage areas for exploration, shipping, navigation routing, traffic density studies, operational fisheries surveillance, near-shore bathymetry mapping, intertidal zones, tidal and storm effects, delineation of the land/water interface, mapping shoreline features/beach dynamics, coastal vegetation mapping, and human activity/impact.

1.1.6 Applications of Remote Sensing in Oceanography

Remote sensing is the science of obtaining information about objects or areas from a distance, typically from aircraft or satellites. Remote sensors collect data by detecting the energy that is reflected from the Earth. These sensors can be on satellites or mounted on aircraft. Remote sensors can be either passive or active. Passive sensors respond to external stimuli. They record radiation that is reflected from the Earth's surface, usually from the sun. Because of this, passive sensors can only be used to collect data during daylight hours. In contrast, active sensors use internal stimuli to collect data about the Earth. For example, a laser-beam remote sensing system projects a laser onto the surface of the Earth and measures the time that it takes for the laser to reflect back to its sensor. Remote sensing has a wide range of applications in many different fields of ocean.

Coastal applications:

1. To monitor shoreline changes: Remote sensing satellites images have been used effectively for coastal shore line change nitoring along the coast. Figure 1.1 shows the coastal shore line of Tamil Nadu state in India.
2. To track sediment transport: *Particle tracking*, or in the geological sciences *sediment tracing* or *sediment tracking*, offers a unique methodology for tracking the movement through space and time of environmental particulates. Utilizing this methodology, information can be garnered into source-sink relationships, the nature and location of the transport pathway(s), and the rate of transport.

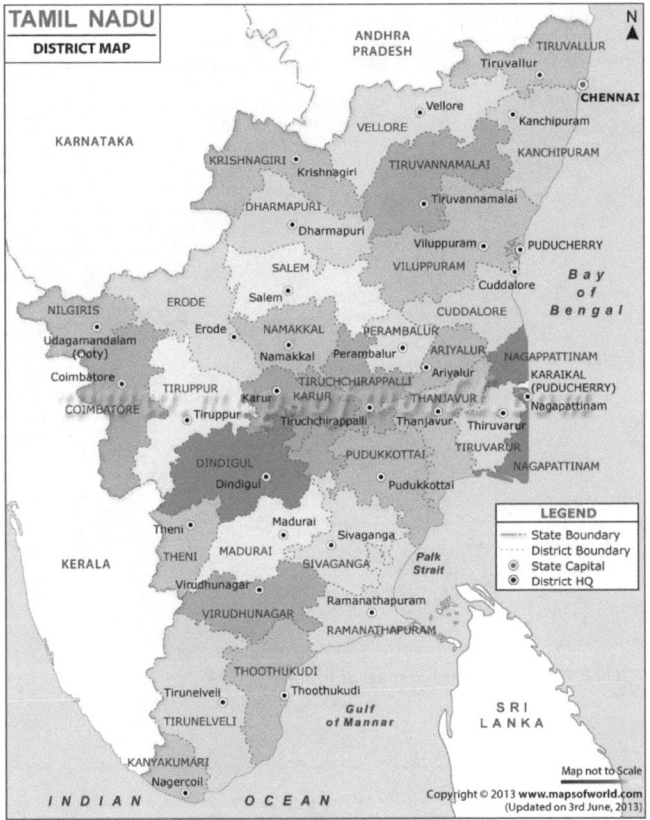

FIGURE 1.1 Tamil Nadu state coastal shore line.

3. To map coastal features: Geologic land features could be retrieved as seen in Figure 1.2, showing the physiographic map of India.

Coastal features can be also retrieved. For instance, in an area where the land meets the sea, its features vary depending on climate, wind, sea, and the type of rocks of which it is composed.

Common coastal features are natural arches, caves, stacks, sand islands, rocky islets, tombolos, spits, skerries, headlands, cliffs, dunes, river estuaries, lagoons, and beaches, as shown in Figure 1.3. A natural arch is the arch hollowed out of a headland by the sea. A cave is the natural underground cavity that results from the slow dissolution and erosion of rock by water. A stack is the needle-shaped column resulting from the collapse of an arch. A sand island is the exposed summit of a sand deposit formed near or occasionally far from a shoreline. A rocky islet is a small island made of rock.

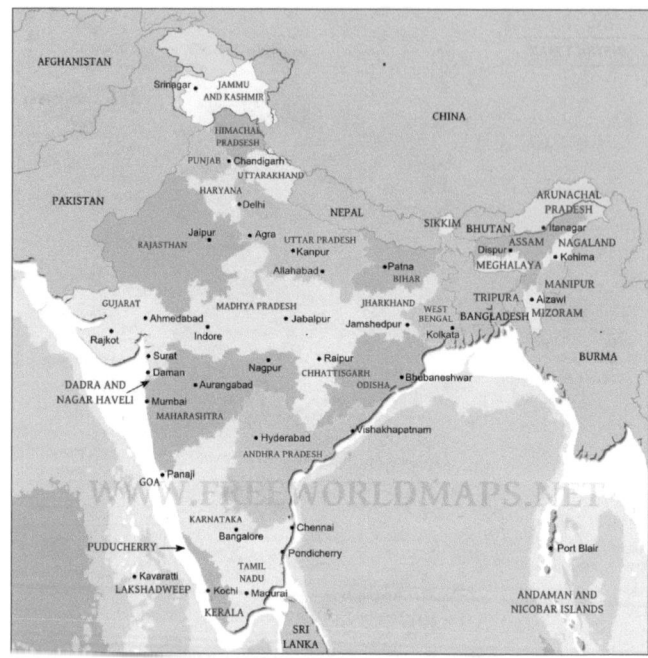

FIGURE 1.2 Physiographic map of India.

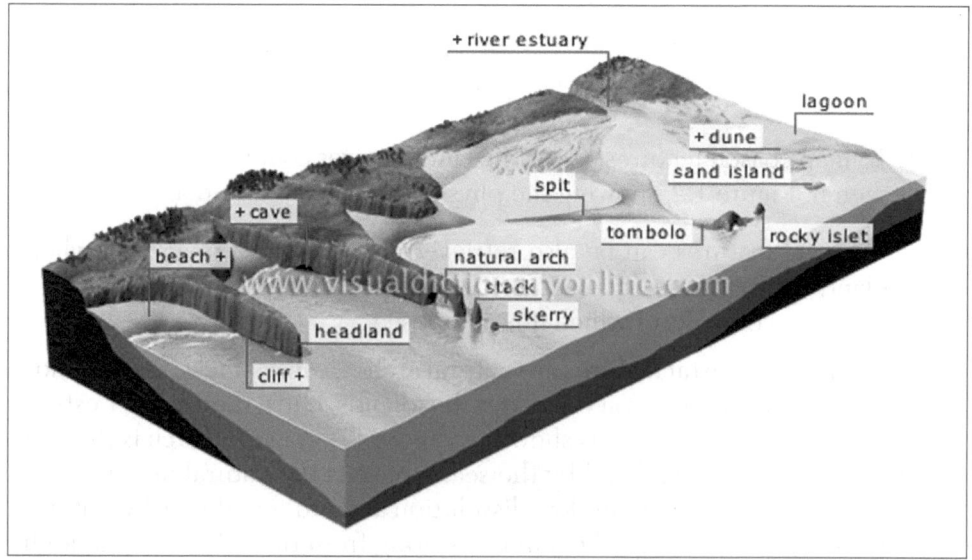

FIGURE 1.3 Coastal features.

A tombolo is the ridge of sand joining an island to the shoreline. A spit is the elongated ridge of sand or pebbles extending into the water. A skerry is a rock tip just above the surface of the water. A headland is the tapering strip of land jutting into the sea. A cliff is a steep rock face shaped by the sea. A dune is an accumulation of sand shaped by the wind. A river estuary is the mouth of a river that is influenced by the tides; it forms an indentation in the coastline that varies in width and depth. A lagoon is a shallow expanse of seawater separated from the sea by a ridge of sand or a barrier island. A beach is the accumulation of sand or pebbles along the coast.

4. Data can be used for coastal mapping and erosion prevention.

Ocean applications:

1. To monitor ocean circulation and current systems: An ocean current can be defined as a horizontal movement of seawater at the ocean's surface. Ocean currents are driven by the circulation of wind above surface waters. Frictional stress at the interface between the ocean and the wind causes the water to move in the direction of the wind.

Figure 1.4 describes the flow pattern of the major subsurface ocean currents. Near-surface warm currents are drawn in red. Blue depicts the deep, cold currents. Note how this system is continuously moving water from the surface to deep within the ocean and back to the top of the ocean.

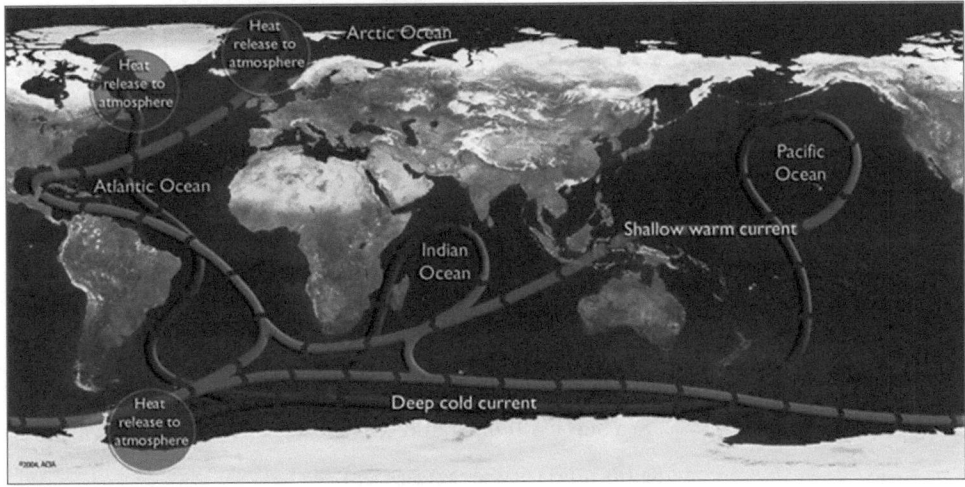

FIGURE 1.4 Ocean current flow pattern.

2. To measure ocean temperature: Temperature and density share an inverse relationship. As temperature increases, the space between water molecules increases—also known as density, which therefore decreases. If the temperature of water decreases, its density increases, but only to a point. At a temperature of 4°C, pure water reaches its maximum or peak density; cooled further, it expands and becomes less dense than the surrounding water, which is why when water freezes at 0°C it floats. Salinity and density share a positive relationship. As density increases, the amount of salts in the water (salinity) increases. Various events can contribute to change in the density of seawater. Salinity can decrease from the melting of polar ice or increase from the freezing of polar ice. Evaporation increases salinity and density while the addition of freshwater decreases salinity and density. Sea surface temperature (SST) is the water temperature close to the ocean's surface. The exact meaning of *surface* varies according to the measurement method used, but it is between 1 millimeter (0.04 in) and 20 meters (70 ft) below the sea surface. Figure 1.5 is the ocean weather picture for different places.

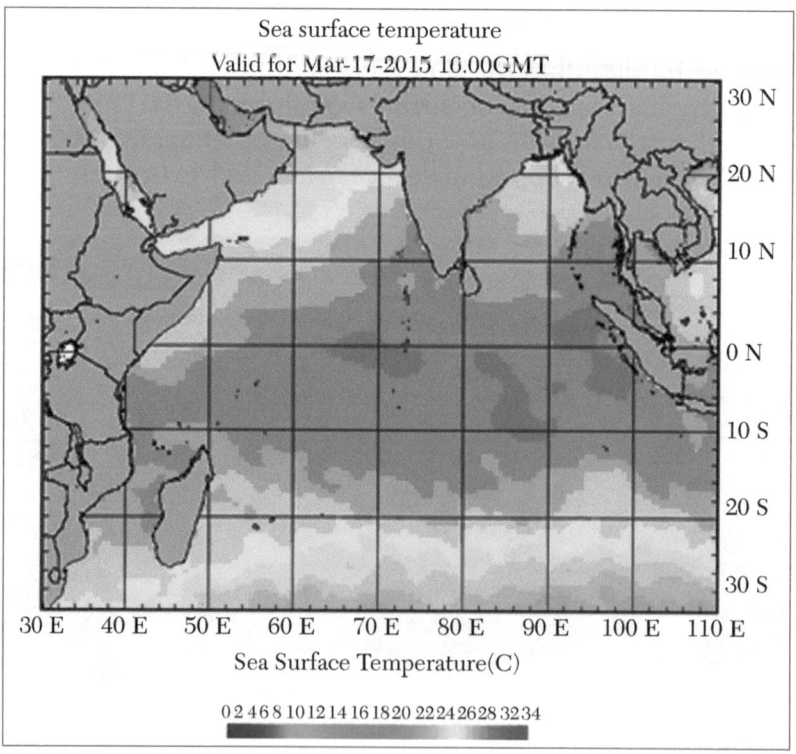

FIGURE 1.5 Ocean weather picture.

3. To measure wave heights:

As shown in Figure 1.6, wave height is defined as the height of the wave from the wave top (wave crest) to the bottom of the wave (wave trough). The wave length is

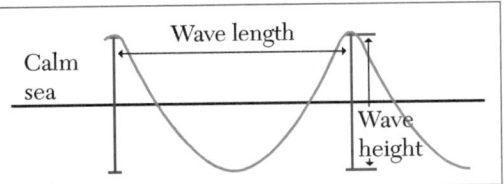

FIGURE 1.6 Wave height measurement.

defined as the horizontal distance between two successive crests or troughs. During storms, wave heights increase while the wave lengths decrease. Wave heights during storms may exceed 10 meters (33 ft). These waves are extremely dangerous on the water. Wave length during storms tends to decrease; some may be as small as 15 meters (50 ft). Figure 1.7 shows wave height with wave direction.

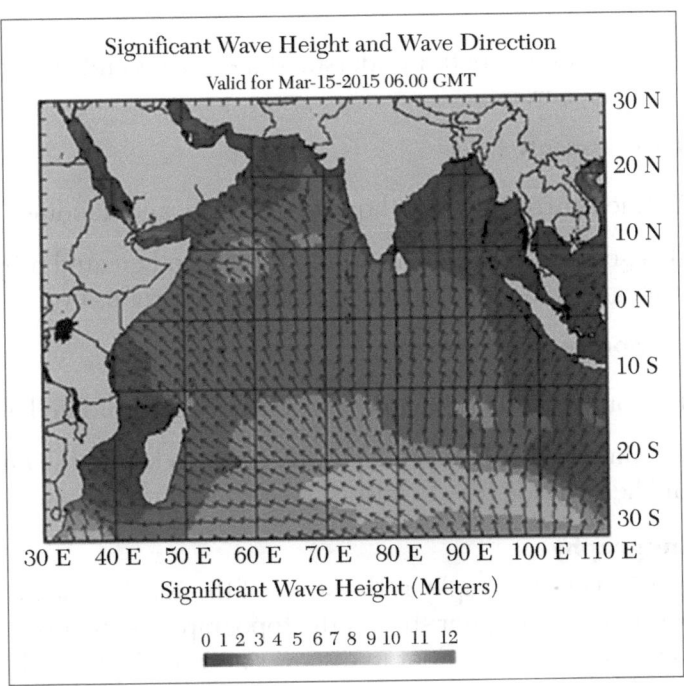

FIGURE 1.7 Wave height and wave direction.

4. To track sea ice: Arctic sea ice keeps the polar regions cool and helps moderate global climate. Sea ice has a bright surface; 80% of the sunlight that strikes it is reflected back into space. As sea ice melts in the summer, it exposes the dark ocean surface. Figure 1.8 shows the sea ice.

FIGURE 1.8 Sea ice.

5. Data can be used to better understand the oceans and how to best manage ocean resources.

Hazard assessment:

1. To track hurricanes, Earthquakes, erosion, and flooding

2. To assess the impact of a natural disaster and created preparedness strategies to be used before and after a hazardous event.

Natural resource management:

1. To monitor land use, map wetlands, and chart wildlife habitats.

2. To minimize the damage that urban growth has on the environment and help decide how to best protect natural resources.

1.1.7 Oceanography

Geological oceanography is the study of the Earth beneath the oceans. A geological oceanographer studies the topography, structure, and geological processes of the ocean floor to discover how the Earth and oceans were formed and how ongoing processes may change them in the future. Geological oceanography is one of the broadest fields in the Earth Sciences and contains many subdisciplines, including geophysics and plate tectonics, petrology and sedimentation processes, and micropaleontology and stratigraphy. Geological oceanographers study many ocean features, such as rises and ridges, trenches, seamounts, abyssal hills, the oceanic crust, sedimentation (chemical and biological), erosional processes, volcanism, and seismicity.

Many different tools are used by geological oceanographers. For example, the structure and topography of the ocean floor are studied through the use of satellite mapping, which measures the level of the ocean surface to estimate the shape of the ocean floor. Underwater mountains and valleys cause subtle variations in Earth's gravitational field. The stronger gravity near high massive formations attracts more water molecules, raising the level of the ocean slightly. Similarly, valleys on the ocean floor produce weaker areas of gravity, so the level of the ocean will be lower. Using microwave radiation techniques, a complete survey of the ocean floor has been accomplished.

Seismic techniques are used to measure the subsurface structure. This type of study is carried out by teams of two ships: one fires an explosive in the water and the other uses sensitive instruments to record the sound waves as they reach the second ship. Some waves travel directly to the second ship; others travel to the ocean floor, are refracted (bent) within the layers of sediment, and then travel to the second ship. By measuring the time it takes for the energy to arrive and the distance between the boats, the thickness of sediments and other features can be determined. Structures may also be analyzed by studying natural Earthquake waves that travel through deeper oceanic rocks and may be recorded at stations around the world. Geological oceanographers study the rocks and structure of the ocean floor, the ocean floor sediment that covers them, and the processes that formed them. Coastal geologists focus on these structures and processes in a coastal environment.

1.2 RADAR IN REMOTE SENSING

Radar stands for "RAdio Detection And Ranging." By sending out pulses of microwave electromagnetic radiation this type of instrument can be classified as an *active sensor*—it measures the time between pulses and their reflected components to determine distance. Different pulse intervals, different wavelengths, different geometry, and polarizations can be combined to roughness characteristics of the Earth's surface. Radar wavelengths range between less than 1mm to 1 meter. Radar uses relative long wavelengths, which allow these systems to "see" through clouds, smoke, and some vegetation. Also, being an active system, it can be operated day or night. Cameras capture reflected visible wavelengths. Radar captures emitted microwave wavelengths that are bounced back to the antenna. Figure 1.9 shows the difference between camera and radar antennas.

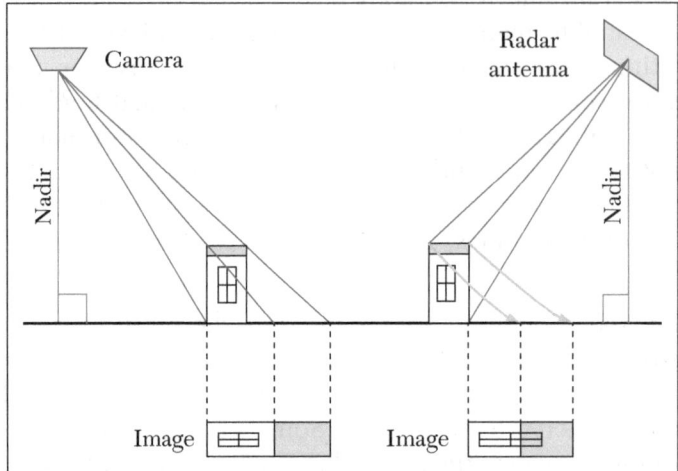

FIGURE 1.9 The difference between camera and radar antennas.

There are two types of radars.

1. SLAR (side-looking airborne radar) is airborne, with fixed antenna width, and sends one pulse at a time and measures what gets scattered back. Its resolution is determined by wavelength and antenna size (narrow antenna width = higher resolution).

2. SAR (synthetic aperture radar) was developed by those responsible for SLAR, but this configuration is not dependent on physical antenna size, although to achieve higher resolution the receiving antenna components and transmitter components need to be separated. It "synthesizes" a very broad antenna by sending multiple pulses.

1.2.1 Synthetic Aperture Radar (SAR)

Synthetic aperture radar is an instrument for producing microwave images of the Earth from space with a resolution comparable to optical systems. The frequency of the return signal from each scatterer is modulated linearly and, if the target is stationary, the modulation constant is a function of known parameters, namely the radar wavelength, the velocity of the platform, and its height. Broad-beam antennas give better resolution than narrow beam, since resolution is inversely proportional to the total time the target remains in the field of view. To reduce speckle, processors will add *multi-look* independent images of a target. Speckle—a type of random noise—is thereby reduced, but at the expense of resolution. Compared to land surfaces, the ocean is relatively homogeneous, with a low scattering cross section and low contrast. Images collected by SeaSat radar from the space shuttle have provided a wealth of information on surface features such as slicks, ships, currents, eddies, waves, and perhaps more surprisingly, on coastal bathymetry. Thus statistics on global wave climate are being built up. Figure 1.10 illustrates some of the common terms used to describe the geometry of a radar image. Most important are the *look angle*, the angle at which the radar pulse hits the surface, and the interval between pulses.

Using different combinations of wavelength and incidence angle, the characteristics of the recorded backscatter can be compared and interpreted. Acquiring a detailed dataset with the necessary calibration fieldwork and measurements is an ordeal to plan and execute, but with a good dataset, radar imagery can reveal characteristics of the landslide that is visible.

The assumption is that different vegetation types (e.g., desert, grasslands, forests, or frozen tundra) will all have different backscatter signatures. In addition, the basic reflectivity of the soil, called the *dielectric constant*, will change depending on the amount of water and organic matter that the soil contains. Dry soil has a low dielectric constant, so that little radar energy will be reflected. Saturated soil will have a high dielectric and will be a strong reflector. Moist and partially frozen soils will have intermediate values, as shown in Figure 1.11. High-resolution maps of topography and topographic change generated from SAR interferometry are also extremely valuable for studies of ice sheets and glaciers. Over 75% of the world's fresh water is presently locked up in ice and snow. While the general retreat of mountain glaciers globally is believed responsible for approximately one quarter to one half of the current 2 mm-per-year increase in sea level, the other sources are unknown. Radar provides a means of regionally monitoring the health of the ice sheets, which can be used to assess the threat of sea level rise, as in Figure 1.12.

1.2.2 NADIR Radar System

The NADIR radar system is an outdoor, ground based transportable system targeted to the dynamic measurements of full-scale ships in an inshore environment. NADIR radar is a flexible radar system for use in an inshore environment which is designed to perform dynamic measurements of full-scale ships, for radar cross section (RCS), high resolution range profiles (HRRP), hourglass plots, and inverse synthetic aperture radar (ISAR)

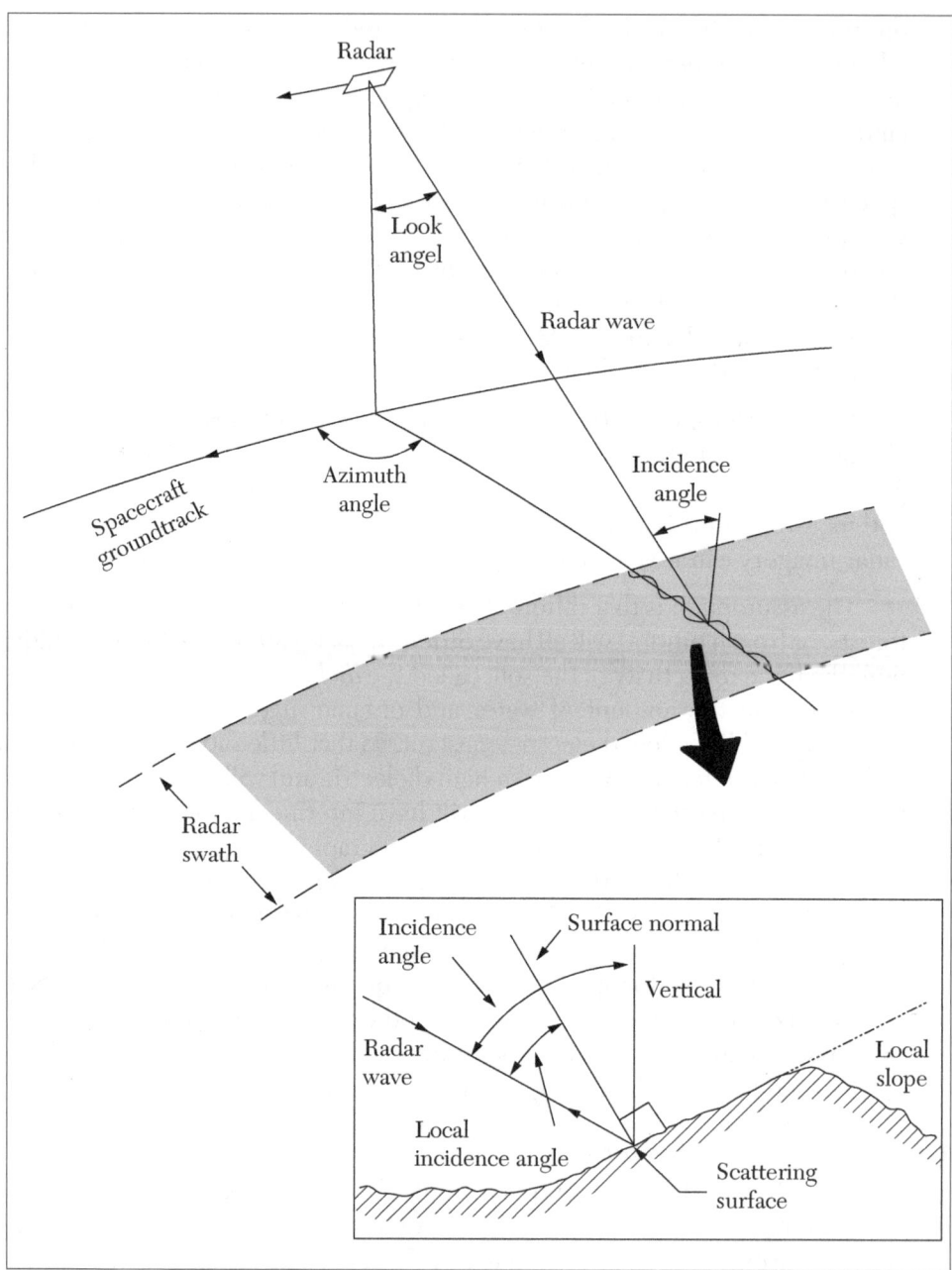

FIGURE 1.10 Radar look angle.

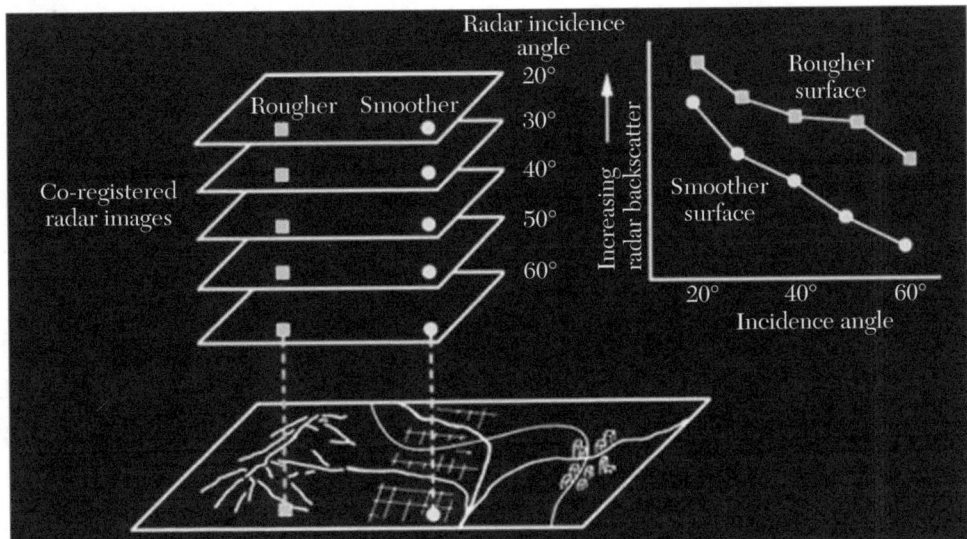

FIGURE 1.11 Radar image analysis.

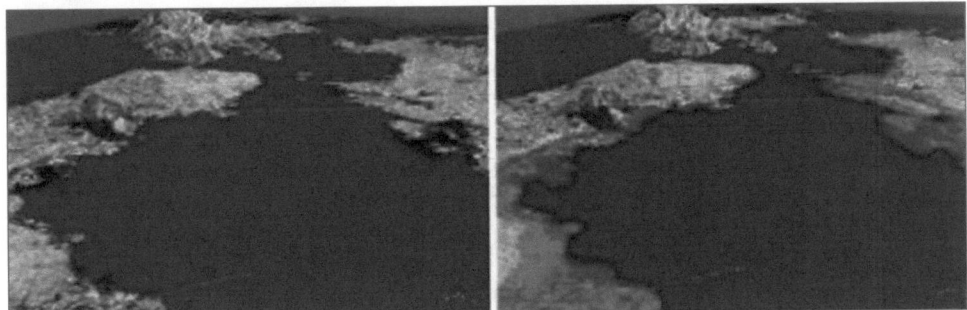

FIGURE 1.12 San Francisco sea level rise, map of radar image.

images. The system is customizable and expandable and only requires a short time to perform an accurate and comprehensive analysis, the acquisition time being limited only by the movement capability and speed of the ship itself. The radar, tracking equipment, and the cabin for its operators and control and monitoring equipment is all designed for outdoor use, is self-powered, and is fully transportable. The ship is tracked by NADIR radar from a range of 3 to 10 NM using its automatic target tracking capability. It allows very fast operation (5 KHz for 1 meter range resolution) and provides highly accurate data. It is fast to set up and easy to operate due to being designed specifically to measure ships at sea. It includes powerful post processing capabilities integrated in a unique software tool and database capability.

NADIR radar consists of three modules which are transportable by a single truck and trailer:

1. The antenna subsystem is mounted on an equipment trailer and includes an RX/TX radar antenna assembly and an elevation over azimuth positioner.

2. The operating shelter houses the positioning, processing, and control equipment, including the tracking equipment and post-processing computers.

3. The onboard subsystem which is carried on the target ship provides GPS and motion data and a radio data link.

Benefits:

1. Short time-to-measure

2. Accurate range profiling and imaging

3. Extremely fast waveform generation

4. Transportability and operative readiness

1.3 RADIOMETER IN REMOTE SENSING

The spectro-radiometer is used to analyze all the details and frequencies of an electromagnetic spectrum. Usually a radiometer is further identified by the portion of the spectrum it covers; for example, visible, infrared, or microwave. These instruments normally work by means of a sensitive element or detector that modulates the current passing through it in line with the electromagnetic energy that it receives. Different types of detector are used for the different wavelengths. Each machine is usually equipped with a single detector and thus takes readings on a certain wavelength interval, as in Figure 1.13.

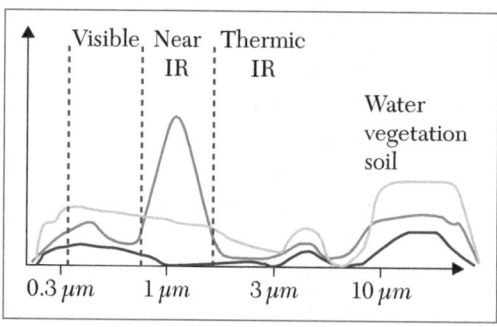

FIGURE 1.13 Radiometer to analyze EM spectrum.

1.3.1 Microwave Radiometer

Brightness temperature (T_B) is the fundamental parameter measured by passive microwave radiometers. Brightness temperatures, measured at different microwave frequencies, are used at remote sensing systems to derive wind, vapor, cloud, rain, and sea surface temperature (SST) products. Brightness temperature is a measurement of the radiance of the microwave radiation traveling upward from the top of the Earth's atmosphere to the satellite, expressed in units of the temperature of an equivalent black body.

Satellite passive microwave radiometers measure raw antenna counts from which we determine the antenna temperature and then calculate the brightness temperature of the Earth. Large antennas are used for the various channels of the radiometer, and during operation each antenna feed horn passes a hot and cold target in order to provide consistently calibrated raw counts. The conversion from radiometer counts to top-of-the-atmosphere T_B is called the calibration process. Several calibration processing steps are required to derive the T_B values. Microwave radiometer T_B are considered a fundamental climate data record and are the values from which we derive ocean measurements of wind speed, water vapor, cloud liquid water, rain rate, and sea surface temperature. Calculating T_B from raw radiometer counts is a complex, multi-step process in which many effects must be accurately characterized and adjustments made to account for them. These effects include radiometer nonlinearity, imperfections in the calibration targets, emission from the primary antenna, and antenna pattern adjustments. A rain-free ocean is used as the absolute calibration reference and the state-of-the-art radiative transfer model (RTM) of the ocean and intervening atmosphere in the absence of rain can predict the top-of-the-atmosphere T_B to a high degree of accuracy.

Space-borne passive microwave sensors observe radiation emitted from the Earth in the range 1–300 GHz but most ocean-related parameters are retrieved from observations below 40 GHZ. After a series of experiments at different microwave frequencies involving different polarizations and incidence angles, at different surface temperatures and wind speeds, a quasi-empirical set of relations was established.

Sea surface temperature is observed at low frequencies generally in the range of 5–10 GHZ. The great advantage of microwave sensors for measuring SST over the more commonly used infrared instruments is their ability to operate through cloud, but this must be offset against their resolution of around 150 km, which is too coarse to study mesoscale eddies. Higher resolution would require a much larger antenna than has been flown up to now. Another constraint is contamination by land masses, and in general reliable measurements must be made in the open ocean more than 600 km from a coast. Thus, again, interesting ocean features such as boundary currents and their associated eddy may not be capable of being studied with the microwave radiometer. Ocean surface emissivity is affected by surface winds through the generation of waves and foam. Measurements of ocean parameters by microwave radiometers are affected by atmospheric water vapor, clouds, and rainfall, and most sensors are therefore backed up by frequencies sensitive to water in the atmosphere.

1.3.2 Infrared Radiometers

The earliest repetitive set of sea surface observations taken from a satellite was of its temperature. The sensors have 3.7 and 11μm infrared channels. The total radiance at the sensor is the sum of the radiance from the ocean surface, from the atmosphere, and from reflected solar radiation. The along-track scanning radiometer (ATSR) flying employs a near-infrared channel (3.7 μm) and two infrared (10.8 and 12 μm) but seeks an increase in accuracy by viewing the surface at two angles—at nadir and at 47°—thus producing two independent measurements with different atmospheric path lengths. The ATSR aims at measuring sea surface temperature with an absolute accuracy of better than 0.5° K with a spatial resolution of 50 km in conditions up to 80% cloud cover, and a relative accuracy of 0.1° K with a 1 km resolution over a 500 km swath. The geometry of the ATSR view of the sea surface is illustrated in Figure 1.14.

It should be noted that what the radiometer measures is the *skin* temperature at the sea surface and this is not necessarily the same as the bulk or mixed layer temperature measured from ships and buoys.

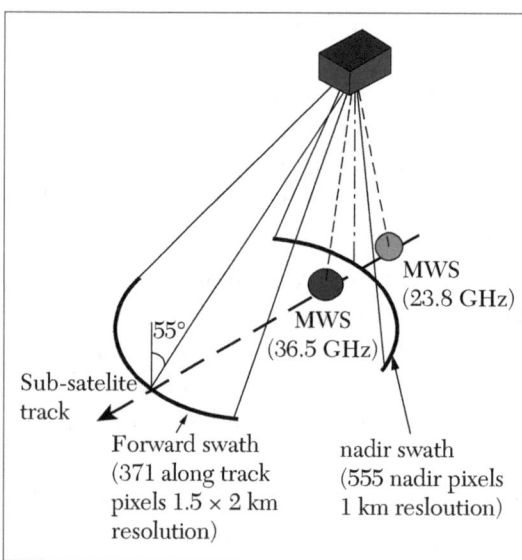

FIGURE 1.14 Viewing geometry of the along-track scanning radiometer.

Differences of a few degrees have been recorded in areas of flat calm and high solar radiation, so extreme care must be taken in calibrating the sensor against in situ observations. It is usually the bulk temperature which is required for oceanographic studies. Under most conditions the skin temperature is a good indicator of bulk temperature.

1.3.3 Imaging Radiometer

A radiometer that includes a scanning capability to provide a two-dimensional array of pixels from which an image may be produced is called an imaging radiometer. Scanning can be performed mechanically or electronically by using an array of detectors.

1.4 SATELLITE IN REMOTE SENSING

Space sensors monitor the surface of all oceans equally several times a day, 365 days a year. If there are many ocean phenomena within the volume of the sea which remain unobserved from an orbiting satellite, the continuous monitoring of its surface accumulates vital information on the processes which in the shorter term most affect human habitations. Satellite sensors have been developed to operate either in the optical/infrared part of the electromagnetic spectrum (λ = 0.4–12μm) or in the microwave part (λ = 0.3–30 cm). With these sensors, either from an aircraft or satellite, color, temperature, slope/height, and roughness properties of the ocean can be measured to useful accuracies. With the use of sensors all ocean features, such as physical, chemical, biological, or geological, can be monitored from space. Only ocean color cannot be detected by microwave sensors. The ocean color is monitored at five discrete wavebands in the visible, from blue to near infrared.

The first satellite is launched with the radar altimeter, microwave scatterometer, and scanning microwave multi-channel radiometer. The principle of extracting useful information on the state of the sea surface by analyzing the nature of the backscattered radiation is illustrated below. Scattering of electromagnetic waves from the sea surface depends on:

1. Surface roughness caused by winds, waves, currents and slicks

2. Radar parameters such as power, angle of incidence, frequency, polarization, and viewing angle

Scatterometer: The scatterometer is a radar designed to measure wind speed and direction at the sea surface. It transmits a fan beam of short pulses and measures the echo power backscattered from the surface at a variety of incidence angles. These wind observations have a wide variety of applications, including weather forecasting, marine safety, commercial fishing, and long-term climate studies. Wind is the movement of air and is caused by the difference in atmospheric pressure between high and low pressure systems. Over land, anemometers measure the surface wind speed and wind direction. These anemometers exist in high density for many areas and lower density in less populated regions. But over the oceans, measurement of surface wind characteristics is far more limited and is primarily obtained from anemometers located at small island weather stations, on ships, and on buoys floating in the ocean. Since the ocean regions are so large, especially the Pacific Ocean, knowledge of the wind characteristics over this vast space is important to weather forecasting, ocean navigation, and climate study. One of the first approaches was to use visible images to study cloud motion and indirectly determine wind speed and direction.

1.4.1 Microwave Measurement of Ocean Wind

Two types of microwave instruments measure ocean surface winds: the passive microwave radiometer and the active microwave scatterometer. The radiometer measures ocean surface roughness, which we correlate to wind speeds at 10 meters above the water's surface. WindSat is the first satellite microwave polarimetric radiometer, launched in 2003. The scatterometer is an active instrument and sends a signal to the Earth's surface, which reflects off the ocean Bragg waves (these are wind generated surface ripples—capillary waves) on the surface of the larger-scale ocean waves. The reflected energy measured by the scatterometer is translated using a geophysical model function into a 10 meter neutral wind speed and direction. Scatterometers typically operate at either C-band (~5GHz frequency) or Ku-band (~14 GHz frequency). With special processing techniques, one can obtain wind speeds and directions every 12 km over the oceans. Scatterometers can also be used to measure sea ice and land ice characteristics.

Scatterometer operation to measure wind speed and direction: A scatterometer is a microwave radar sensor used to measure the reflection or scattering effect produced while scanning the surface of the Earth from an aircraft or a satellite. The seawinds scatterometer is a microwave radar designed specifically to measure ocean near-surface wind speed and direction. The seawinds scatterometer consists of three major subsystems seen in Figure 1.15: the electronics subsystem (SES), the antenna subsystem (SAS), and the command and data subsystem (CDS).

The electronic subsystem is the heart of the scatterometer and contains a transmitter, receiver, and digital signal processor. It generates and sends high radio frequency (RF) waves to the antenna. The antenna transmits the signal to the Earth's surface as energy pulses. When the pulses hit the surface of the ocean, it causes a scattering effect referred to as backscatter. A rough ocean surface returns a stronger signal because the waves reflect more of the radar energy back toward the scatterometer antenna. A smooth ocean surface returns a weaker signal because less energy is reflected. The echo or backscatter is routed by the antenna to the SES through waveguides (rectangular metal pipes that guide RF energy waves from one point to another). The SES then converts the signals into digital form for data processing.

The CDS is essentially a computer housing the software that allows the instrument to operate. It provides the link between the command center on the ground, the spacecraft, and the scatterometer. It controls the overall operation of the instrument, including the timing of each transmitted pulse, and collects all the information necessary to transform the received echoes into wind measurements at a specific location on Earth. To locate the precise position on Earth at which the echo was taken, the CDS collects (for each pulse) the antenna rotational position, spacecraft time, and an estimate of the spacecraft position. The CDS also

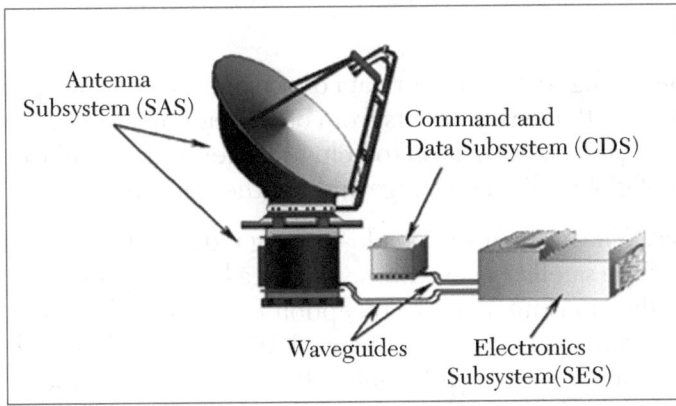

FIGURE 1.15 Schematic view of the SeaWinds scatterometer elements.

collects instrument temperature, operating voltages, and currents, so that the overall health of the instrument can be monitored. It is through the CDS that the other two subsystems receive the commands that control all their functions.

The SAS consists of a 1 meter parabolic reflector antenna mounted to a spin activator assembly, which causes the reflector to rotate at 18 rpms (revolutions per minute). The activator assembly provides very accurate spin control and precise position or pointing information to the CDS. Optical encodes, glass disks with small patterns printed on the surface, tell the CDS exactly where the antenna is pointing to about 10/1000 of a degree. The antenna spins at a very precise rate, and emits two beams about 6 degrees apart, each consisting of a continuous stream of pulses. The two beams are necessary to achieve accurate wind direction measurements. The pointing of these beams is precisely calibrated before launch so that the echoes may be accurately located on the ground from space.

Uses of scatterometry:

1. Data are vital in the study of air-sea interaction and ocean circulation, and their effects on weather patterns and global climate.

2. Data are useful in the study of unusual weather phenomena, the long-term effects of deforestation on our rain forests, and changes in the sea-ice masses around the polar regions. These all play a central role in regulating global climate.

3. Weather forecasting is important tool to meteorologists. Scatterometer data, with wide swath coverage, have been shown to significantly improve the forecast accuracy.

4. By combining scatterometer data of ocean-surface wind speed and direction with measurements from other scientific instruments, scientists can gather information to help us better understand the mechanisms of global climate change and weather patterns.

Altimeter: The altimeter is a nadir-looking radar that measures the precise altitude of the satellite above the sea surface by measuring the time interval between the transmission and reception of a stream of very short pulses. The technique (illustrated in Figure 1.16) is basically simple. The orbit height—which is the radial height of the satellite above the geocenter—is measured and calculated through satellite tracking—usually a combination of lasers, transponders, GPS, and onboard electronic systems. The height of the sea surface can then be determined by subtracting the altimetric measurement from the orbit height. Variations of this surface are measured with respect to a reference ellipsoid which approximates the Earth's surface and is defined by an internationally agreed formula. A precise knowledge of orbit and marine geoid are required. Changes in sea level are mostly determined by the Earth's gravity field. If the ocean were at rest, sea level would be a surface of constant gravity potential, referred to as the geoid. Across the surface of the globe the geoid varies by about 100 meters. By

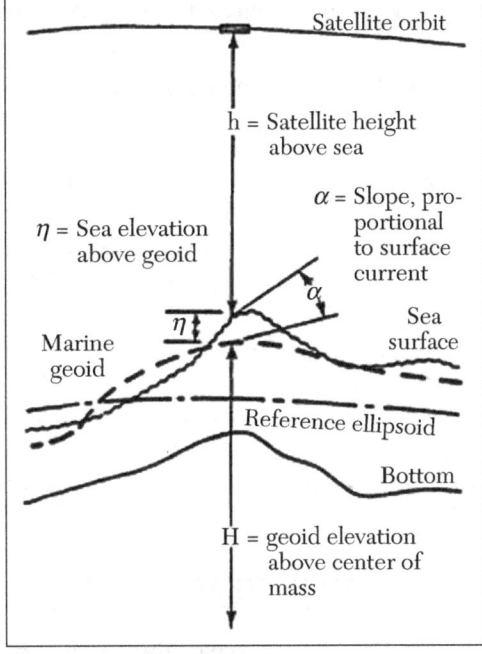

FIGURE 1.16 Measurement of sea surface elevation and sea surface slope from satellite altimeters.

contrast, the largest departures in sea level from the geoid brought about by tides, geostrophic currents, and other dynamic processes (referred to as ocean topography) are of the order of l meter. The difficulty is that only in a few parts of the world's oceans is the geoid known to accuracy sufficient to identify dynamic ocean changes. The altimeter has no way of differentiating gravity-induced from ocean-induced variability.

There are three main sources of error: instrument, orbit, and environment (at the sea surface and within the atmosphere through which the altimeter transmits its pulses). The latter source of error can and must be minimized by measuring the constituents of the atmosphere—its water vapor and liquid water content as well as variations in the electron density within the ionosphere—while at the sea surface the effects of waves (wave troughs tend to reflect more energy back than wave crests, causing a bias), tides (not always accurately known in the deep ocean), and atmospheric surface pressure (10 mbar is equivalent to 10 cm in sea level) must all be taken into account if the 5 cm precision of the instrument is to be fully exploited.

1.4.2 Wind and Wave Height Measurements

In addition to sea level measurements. the altimeter's return signals can be used to measure surface wind speed and wave height. The measurement of wind speed is based on radar backscatter cross section at nadir. As the wind increases, so the incident radiation is reflected away from a rougher sea surface, causing the backscatter cross section to decrease. The wave height measurement is based instead on the slope of the leading edge of the return pulse—the higher the waves, the wider the return pulse (Figure 1.17).

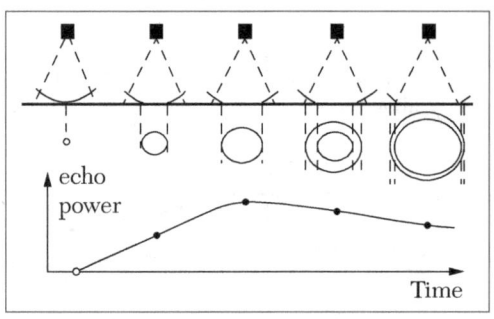

FIGURE 1.17 Intersections of an altimeter pulse with the sea surface.

1.5 SONAR IN REMOTE SENSING

The oceans have always played a big role in wars. Ships transported armies and supplies, blockaded harbors, besieged cities, and attacked enemy ships doing the same things. But the Civil War helped launch a stealthy new seagoing weapon that became common in 20th century warfare: submarines. To combat this new threat, naval leaders soon realized that they could detect submarines using sound transmitted through water. Huge efforts began to develop *sonar*, a word that is a combination of abbreviations for "sound," "navigation," and "ranging." (Interestingly, sonar was first developed to help avoid icebergs after the *Titanic* sank.) Data physics specializes in the design and manufacturing of low frequency acoustic projectors, and their associated electronic and handling systems, to simulate the acoustic signatures of ships, submarines, and underwater weapons.

For oceanographers, sonar provided a much easier way to measure the ocean depths accurately. The systems are even now used to aid in mine sweeping. Sonar allows scientists to use sound waves to measure the distance from the ocean surface to the seafloor. Ships' hulls are equipped with devices called transducers that transmit and receive sound waves. Echo-sounders were first used in oceanographic studies during the epic Germen expedition exploring the South Atlantic in the mid-1920s aboard the *Meteor*. Today echo-sounding remains the key method scientists use to make bathymetric maps of the seafloor. In recent years, marine scientists have used multi-beam sonar, which can automatically make very detailed contour maps of large area of seafloor as a research ship travels fast (about 12 knots) over the ocean surface. Today, there are many different types of sophisticated sonars. They can tell us not only about seafloor depths, but also about the structure of the ocean floor and even about currents and life in the ocean. The military also developed tools that proved useful to oceanographers, such as the magnetometer, which measures magnetic fields. The Navy uses it to detect the large metal hulls of submarines. Oceanographers use it to learn about magnetic properties of seafloor rocks. As it turned out, these properties provided key clues that completely changed our thinking about how our planet works.

1.5.1 Mapping the Ocean Floor with Single Beam Echo Sounding

Echo-sounding is the key method used to map the seafloor today. The technique, first used by German scientists in the early 20th century, uses sound waves bounced off the ocean bottom. Echo-sounders aboard ships have components called transducers that both transmit and receive sound waves. Transducers send a cone of sound down to the seafloor, which reflects back to the ship. Just like a flashlight beam, the cone of sound will focus on a relatively small area in places where the ocean is shallow, or spread out over the size of a football field when water depths reach 3,000 meters. The returned echo is received by the transducer, amplified electronically, and recorded on graphic recorders. The time taken for the sound to travel through the ocean and back is then used to calculate water depths. The faster the sound waves return, the smaller the water depths and the higher the elevation of the seafloor.

Echo sounders repeatedly *ping* the seafloor as a ship moves along the surface, producing a continuous line showing ocean depths directly beneath the ship. From the early days of ocean exploration until as recently as 20 years ago, marine geologists wrote down individual readings from recorders, plotted them on navigation charts showing their ship's position, and then drew contour lines joining points of equal depth. In this way, they produced *bathymetry maps* that displayed the ocean's changing water depths (and hence changes in seafloor elevation). These charts were accurate only within about 20–50 meters, but that was good enough for scientists to discover the mid-ocean ridge system in the Atlantic and Pacific Oceans. Echo-sounders use different frequencies of sound to find out different things about the seafloor. Scientists typically use echo-sounders that transmit sound at 12 kHz to determine how far down the seafloor lies. However, they use a lower frequency (3.5 kHz) sound, which penetrates the seafloor, if they want to "see" accumulated layers of sediments below it.

1.5.2 Multi-Beam Bathymetry

Multi-beam bathymetry is the successor to the single-beam echo-sounding. Multi-beam bathymetry is based on the fact that more beams are better than one. Instead of just one transducer pointing down, *multi-beam bathymetry systems* have arrays of 12 kHz transducers, sometimes up to 120 of them, arranged in a precise geometric pattern on ships' hulls. The swath of sound they send out covers a distance on either side of the ship that is equal to about two times the water depth. The sound bounces off the seafloor at different angles and is received by the ship at slightly different times. All the signals are then processed by computers on board the ship, converted into water depths, and automatically plotted as a bathymetric map with an accuracy of about 10 meters. In this way, ships traveling at speeds over 10 knots can produce a swath, rather than a line, of water-depth information. Multi-beam bathymetry systems are now routinely used during research cruises to map areas of seafloor as large as thousands of square kilometers.

1.5.3 Sound as Underwater Navigation

This technique is quite similar to satellite navigation. Instead of orbiting satellites transmitting radio signals, sound-transmitting transponders are sent overboard and anchored to the seafloor. The positions of the transponders are determined by the GPS system on board the ship ranging to them acoustically while the ship circles where the transponders were dropped. The positions of the transponders on the seafloor are known with an accuracy of about 10 meters. Transponders have accurate clocks to measure time very precisely. Each transponder is set to listen for sound signals transmitted either from the deep submergence vehicle or the ship at a specific frequency; in our case the frequency is 9 kHz. The clocks on the vehicle and ship are synchronized. When each transponder hears 9 kHz sound signals, it is programmed to "talk" or transmit a sound back to the vehicle and the ship. Each transponder "talks" at a different frequency (between 8 kHz and 15 kHz) so when we receive the signals at the vehicle or the ship, we can tell which transponder sent it. To calculate the vehicle's position, we use simple geometry and basic math. We know the speed of sound in water (about 1,500 meters per second) and the time it takes for signals from the transponders to reach the vehicle or the ship. Multiplying the travel times by the speed of sound in water gives us the distances between each transponder and the ship or sub. Using distance measurements from the ships and the transponders (triangulation again), computers can calculate the unique point in 3-dimensional space where all distances measured from all the transponders and the ship intersect. That is where the vehicle is. All of this happens very quickly because of computers and software so that we can constantly keep track of the deep submergence vehicle's position within about 5–10 meters during a dive.

1.5.4 Echo-Sounders

The term echo-sounder describes a way of using sound to measure distances underwater. Echo-sounders are a type of sonar (SOund Navigation And Ranging) device that can be used on ships or as part of an instrument placed underwater. Echo-sounders, or sonars, on research ships have two main uses:

- Looking for objects such as fish or bubbles from deep sea vents in the water column

- Locating the sea bed

Knowing the water depth is important for several reasons. Most important is to prevent the ship running aground. There are accurate charts for all the world's major ports, but our research ships sometimes visit poorly charted regions such as Antarctica and need to be able to measure how deep the water is so that they don't hit the sea bed. In the middle of the oceans where our ships often work the depths on the chart are very infrequent and so we use echo-sounders to measure exactly how deep the sea is. We often deploy remotely operated vehicles and sensors in the water and so we need to know how deep the water under the ship is. We also need to know the depth so that we can characterize the area of ocean that we are working in or so that we can hunt for features such as hydrothermal vents or wrecks.

Echo-sounder operation:

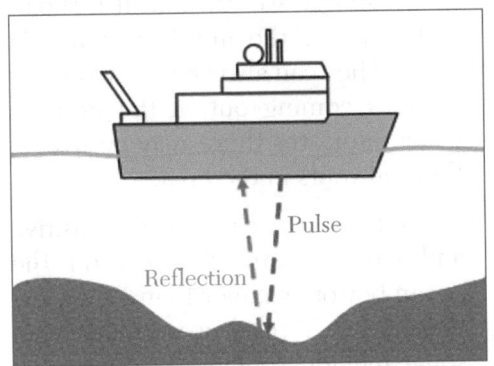
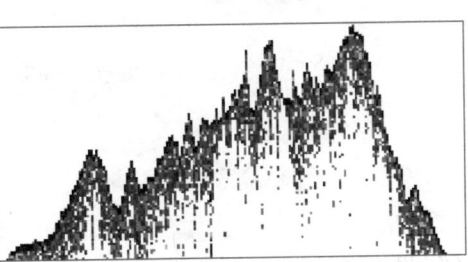

Pulse

Reflection

FIGURE 1.18 Echo-sounder operation and single-beam echo-sounder image.

An echo-sounder transmits a pulse of sound directly downwards from the bottom of the ship. The pulse of sound travels down through the water, bounces off the sea bed, and then travels upwards until the reflection is heard by the echo-sounder. The echo-sounder times how long the pulse of sound takes to travel to the sea bed and back up to the ship. The depth of the water can be calculated using the formula:

$$\text{Distance} = \text{time}/2 \times \text{speed of sound in water}$$

The speed of sound in water is sometimes assumed to be 1,500 ms^{-1}, or it can be measured using a sound velocity probe. Multi-beam echo-sounders allow us to map large areas of the sea bed from the ship. Multi-beam systems use an array of echo-sounder transducers and signal processing electronics to steer the echo-sounder beam across the sea bed, covering a large area of sea bed in each sweep. A single beam echo-sounder can be used for bathmetry, measuring the depth to the ocean floor directly underneath the ship. As the vessel travels forward, it builds up a profile of the sea bed it travels over. While this gives scientists the depth of an area, it is a narrow view, and it does not provide details about how ocean floor measurements relate to each other [Figure 1.18].

The multi-beam echo-sounder is used to build up an image of a large area of the sea bed. Here different colors represent different depths [Figure 1.19]. Blue is deep water and red is shallower water. The exact depth of each point is known and so a very accurate chart can be built from this data. As well as looking at the sea bed, sonar's sound pulses will reflect off items in the water column. Fishermen use this to look for fish and navies use it to look for submarines. Our research ships are fitted with fish-finding sonars so that we can find fish shoals and estimate the number of fish in them. They can also be used to detect bubbles coming out of the ground, which indicates there may be hydrothermal vents in the area.

Side scan sonar is very sensitive and can measure features on the ocean bottom smaller than 1 cm (less than .5 in). Typical uses of side scan sonar include: looking for objects on

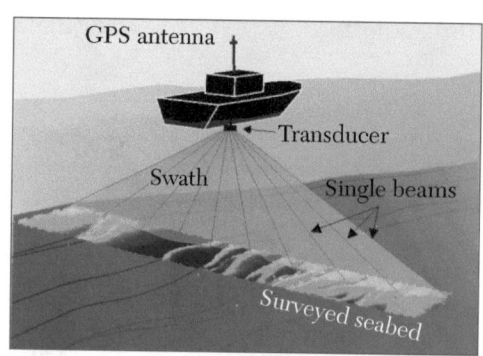

FIGURE 1.19 Multi-beam bathymetry.

the seafloor (sunken ships, pipelines, drowned aircraft, lost cargo), detailed mapping of the seafloor, investigation of seafloor properties (grain size, etc.) and looking at special fea-

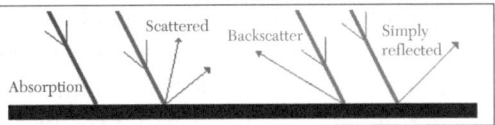

FIGURE 1.20 Sound properties.

tures on the seafloor like underwater volcanos. When the sound sent by sonar reaches the seafloor, several things can happen to that sound. Some of the sound may be absorbed by the seafloor. Some of the sound is almost always reflected. There are several different ways the sound can reflect. The sound can be directly back or it can be scattered in many different directions [Figure 1.20]. Sound that is scattered back toward the tow vehicle is called backscatter.

The amount of scattering, backscattering, and absorption depends on the properties of the seafloor. Hard materials, like rocks, will scatter more sound, while soft materials, like mud, will absorb more sound. Different amounts of scattering produce different amounts of sound returning to the tow vehicle and a different image of the seafloor is obtained. Figure 1.21 shows some examples of how one property of the sea floor (grain size) affects the side scan sonar image.

Figure 1.21 shows a side scan sonar image on the left. On the right are two samples taken from areas in the side scan sonar image. The fine sand is darker in the image because more energy is reflected from the uniform grain size. The gravel is lighter in the image because the gravel scatters more of the sound and less reflects back to the side scan instrument. Side scan sonar is often used to find objects like shipwrecks on the seafloor.

Echolocation using technology: Whales, dolphins, and bats have long possessed the ability to use sound to detect and map objects. Today, sonar is used to produce high-resolution maps of the seafloor. These maps show the location

FIGURE 1.21 Sea floor in a side scan sonar image.

of landscape features such as cliffs, ridges, and cracks in the seafloor. Sound travels through water at about 4,800 ft (1,400 m) per second, more than four times the speed through air. The exact speed through water depends on temperature: the warmer the water, the faster sound travels. Sonar maps can be used to track volcanic activity over time: lava flows and other traces of recent volcanic activity can be identified by comparing maps produced now with maps produced some time ago. The use of sonar to map the seafloor landscape is a means of bathymetry (from the Greek for "measuring depth").

Side scan sonar: Side scan sonar gives us information about the nature of the seafloor as well as its depth. An instrument towed behind the ship measures the intensity of reflected sound as well as the time taken for the sound to travel out and back. A strong signal means the seafloor is relatively hard (e.g., rock, hardened lava, or gravel). A weak signal indicates a soft or finer surface, such as silt or sand. Side scan surveys typically cover the seafloor in overlapping *swaths* or blocks 100–500 meters wide. At the end of a survey, the swaths are pieced together to form a comprehensive map of the seafloor.

Multi-beam sonar: Rather than sending out single pings like side scan sonar, multi-beam sonar equipment emits an array of sound in a fanlike pattern. The reflected sound waves can be used to determine information about sediment type as well as seafloor depth. Multi-beam sonar equipment is usually attached to the ship's hull rather than towed behind it.

1.6 TELEMETRY IN REMOTE SENSING

Telemetry is the highly automated communications process by which measurements are made and other data collected at remote or inaccessible points and transmitted to receiving equipment for monitoring. Remote sensing refers to the use of aerial sensor technologies to detect and classify objects on Earth (both on the surface and in the atmosphere and oceans) by means of propagated signals. This technology enables scientists to remotely monitor an aspect of the environment, like chlorophyll to measure water quality, or to monitor migration paths of sea turtles using satellite telemetry [Figure 1.22]. There are many other marine variables that can be remotely monitored, like sea surface temperature, wave height, and ocean currents.

Telemetry consists of sensors for pressure, temperature, and humidity and a wireless transmitter to return the captured data to an aircraft.

1.6.1 Applications of Telemetry

Oil and gas industry: Telemetry is used to transmit drilling mechanics and formation evaluation information uphole, in real time, as a well is drilled. These services

FIGURE 1.22 A salt water crocodile with a GPS-based satellite transmitter attached to its head for tracking.

are known as measurement and logging while drilling. The pressure wave is translated into useful information after DSP and noise filters. This information is used for formation evaluation, drilling optimization, and geosteering.

Water management: Telemetry is important in water management, including water quality and stream gauging functions. Major applications include AMR (automatic meter reading), groundwater monitoring, leak detection in distribution pipelines, and equipment surveillance. Having data available in almost real time allows quick reactions to events in the field. Telemetry control allows intervening with assets such as pumps and allows to remotely switching pumps on or off depending on the circumstances.

Marine animal tracking: Animals under study can be outfitted with instrumentation tags, which include sensors that measure temperature, diving depth and duration (for marine animals), speed and location (using GPS). Telemetry tags can give researchers information about animal behavior, functions, and their environment. This information is then either stored (with archival tags) or the tags can send (or transmit) their information to a satellite or handheld receiving device.

Satellite telemetry: Satellite Telemetry is a form of radio communication with satellites used to track objects globally. The technology is primarily used by biologists, environmental and other scientists to track the migratory patterns of animals for study. GPS and location tracking are a form of satellite telemetry especially useful for tracking animals, such as birds and fish. Because satellites have global coverage, they allow us to track animals that have been planted with transmitters to areas where radio transmitters or wired communication would not be feasible. Data acquired using location tracking satellite telemetry helps scientists follow mating and feeding patterns critical to species survival.

Data systems and telemetry: Ocean climate station (OCS) moorings are equipped with three different data collection systems, which send their data back to Seattle via satellite communications. These specialized computers talk to the various sensors, collect and average the raw measurements, and package it for transmission. Data is sent back in real-time, so scientists can study the observations on the same day it was taken thousands of miles away.

Mooring information: An OCS mooring is a surface float, loaded with scientific instruments—above, inside, and below—anchored in open ocean water depths of as much as 19,000 feet (5,700 m).

Surface float (buoy): The surface float has a tower, on which we mount sensors to measure the weather, and a frame just below the buoy, called a *bridle*, that holds instruments to measure water properties near the ocean surface [Figure 1.23]. Inside the buoy is a compartment that contains batteries and most of the data collection systems—specialized computers to collect data from the sensors and transmit the data back to our laboratory via satellite. The surface float used for OCS moorings is a 2.62 meter (8.5' ft) diameter fiberglass-over-foam discus buoy, with a central instrument well. It has an aluminum tower and a stainless steel bridle. Fully assembled, the system has an air weight of approximately 1,800 kg, a net buoyancy of nearly 2300 kg, and an overall extent (from the bridle bottom to the tallest instrument) of 6.37 meters (20.9 ft). OCS buoys are equipped with a radar reflector and can be seen on radar at 4–8 miles, depending on sea conditions. A flashing yellow light also helps alert ships to its location at night.

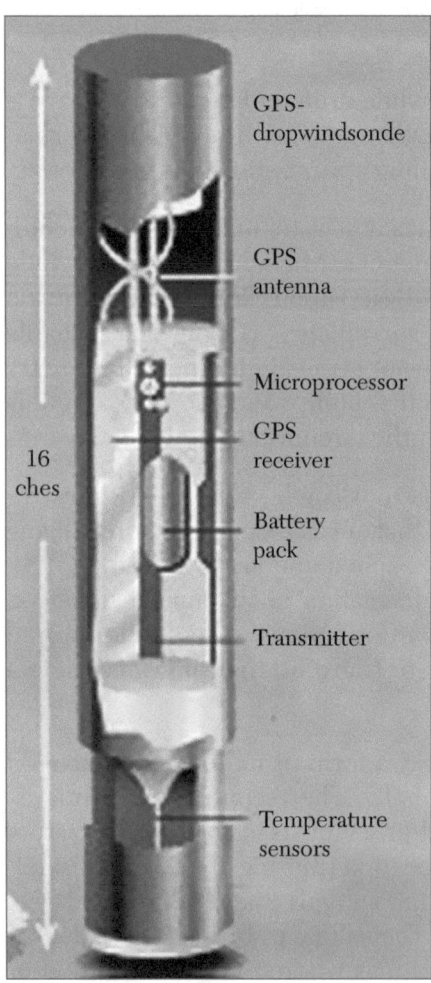

GPS-dropwindsonde

GPS antenna

Microprocessor

GPS receiver

Battery pack

Transmitter

Temperature sensors

16 ches

FIGURE 1.23 A dropsonde used to capture weather data.

Mooring line: Sensors are mounted on the mooring line that connects the buoy to the anchor. The line varies in length depending on the location, but we can use as much as 26,000 ft of line. The tower provides a mounting location for all the meteorological sensors, with the wind sensors above the other sensors. The white buoy well extends into the hull of the buoy.

1.7 LASER AND LIDAR IN REMOTE SENSING

LIDAR (Light Detection and Ranging) is a remote sensing method used to examine the surface of the Earth. LIDAR data supports activities such as inundation and storm surge modeling, hydrodynamic modeling, shoreline mapping, emergency response, hydrographic surveying, and coastal vulnerability analysis. LIDAR is a remote sensing method that uses light in the form of a pulsed laser to measure ranges (variable distances) to the Earth. These light pulses are combined with other data recorded by the airborne system to generate precise, 3-dimensional information about the shape of the Earth and its surface characteristics. A LIDAR instrument principally consists of a laser, a scanner, and a specialized GPS receiver. Airplanes and helicopters are the most commonly used platforms for acquiring LIDAR data over broad areas. Two types of LIDAR are topographic LIDAR, which typically uses a near-infrared laser to map the land, and bathymetric LIDAR, which uses water-penetrating green light to measure seafloor and riverbed elevations. LIDAR systems allow scientists and mapping professionals to examine both natural and manmade environments with accuracy, precision, and flexibility. LIDAR are used to produce more accurate shoreline maps, make digital elevation models for use in geographic information systems, to assist in emergency response operations, and in many other applications.

1.7.1 Data Collection in LIDAR

When an airborne laser is pointed at a targeted area on the ground, the beam of light is reflected by the surface it encounters. A sensor records this reflected light to measure a range. When laser ranges are combined with position and orientation data generated from integrated GPS and Inertial Measurement Unit systems, scan angles, and calibration data, the result is a dense, detail-rich group of elevation points, called a *point cloud*. Each point in the point cloud has 3-dimensional spatial coordinates (latitude, longitude, and height) that correspond to a particular point on the Earth's surface from which a laser pulse was reflected. The point clouds are used to generate other geospatial products, such as digital elevation models, canopy models, building models, and contours.

Laser altimetry, or LIDAR, promises to both increase the accuracy of biophysical measurements and extend spatial analysis into the third (z) dimension. LIDAR sensors directly measure the 3-dimensional distribution of plant canopies as well as subcanopy topography, thus providing high-resolution topographic maps and highly accurate estimates of vegetation height, cover, and canopy structure. In addition, LIDAR has been shown to accurately estimate aboveground biomass even in those high-biomass ecosystems where passive optical and active radar sensors typically fail to do so. The basic measurement made by a LIDAR device is the distance between the sensor and a target surface, obtained by determining the elapsed time between the emission of a short-duration laser pulse and the arrival of the reflection of that pulse (the return signal) at the sensor's receiver. Multiplying this time interval by the speed of light results in a measurement of the round-trip distance travelled. Dividing the results by two is equal to the distance between the sensor and the target. When the vertical distance between a sensor contained in a level-flying aircraft and the Earth's surface is repeatedly measured along transect, the result is an outline of both the ground surface and any vegetation obscuring it. Even in areas with high vegetation cover, where most measurements will be returned from plant canopies, some measurements will be returned from the underlying ground surface, resulting in a highly accurate map of canopy height.

LIDAR sensors are related to the laser's wavelength, power, pulse duration, and repetition rate, beam size and divergence angle, the specifics of the scanning mechanism (if any), and the information recorded for each reflected pulse. Lasers for terrestrial applications generally have wavelengths in the range of 900–1064 nm, where vegetation reflectance is high. In the visible wavelengths, vegetation absorbance is high and only a small amount of energy would be returned to the sensor. One drawback of working in this range of wavelengths is absorption by clouds, which impedes the use of these devices during overcast conditions. Bathymetric LIDAR systems (used to measure elevations under shallow water bodies) make use of wavelengths near 532 nm for better penetration of water. Early LIDAR sensors were profiling systems, recording observations along a single narrow transect. Later systems operate in a scanning mode, in which the orientation of the laser illumination and receiver field of view is directed from side to side by a rotating mirror, or mirrors, so that as the plane (or other platform) moves forward, the sampled points fall across a wide band or swath, which can be gridded into an image. The power of the laser and size of the receiver aperture determine the maximum flying height, which limits the width of the swath that can be collected in one pass. The intensity or power of the return signal depends on several factors: the total power of the transmitted pulse, the fraction of the laser pulse that is intercepted by a surface, the reflectance of the intercepted surface at the laser's wavelength, and the fraction of reflected illumination that travels in the direction of the sensor. The laser pulse returned after intercepting a morphologically complex surface, such as a vegetation canopy, will be a complex combination of energy returned from surfaces at numerous distances, the distant surfaces represented later in the reflected signal. The type of information collected from this return signal distinguishes two broad categories of sensors such as discrete LIDAR and waveform recording device.

1.7.2 Discrete-Return LIDAR and Waveform Recording Devices

Discrete-return LIDAR devices measure either one (single-return systems) or a small number (multiple-return systems) of heights by identifying, in the return signal, major peaks that represent discrete objects in the path of the laser illumination. The distance corresponding to the time elapsed before the leading edge of the peak(s), and sometimes the power of each peak, are typical values recorded by this type of system. Waveform-recording devices record the time-varying intensity of the returned energy from each laser pulse, providing a record of the height distribution of the surfaces illuminated by the laser pulse. By analogy to chromotography, the discrete-return systems identify, while receiving the return signal, the retention times and heights of major peaks; the waveform-recording systems capture the entire signal trace for later processing. Conceptual differences between the two major categories of LIDAR sensors are illustrated in Figure 1.24.

Both discrete-return and waveform sampling sensors are typically used in combination with instruments for locating the source of the return signal in three dimensions. These include global positioning system (GPS) receivers to obtain the position of the platform, inertial navigation systems (INS) to measure the attitude (roll, pitch, and yaw) of the LIDAR sensor, and angle encoders for the orientation of the scanning mirror(s). Combining this information with accurate time referencing of each source of data yields the absolute position of the reflecting surface, or surfaces, for each

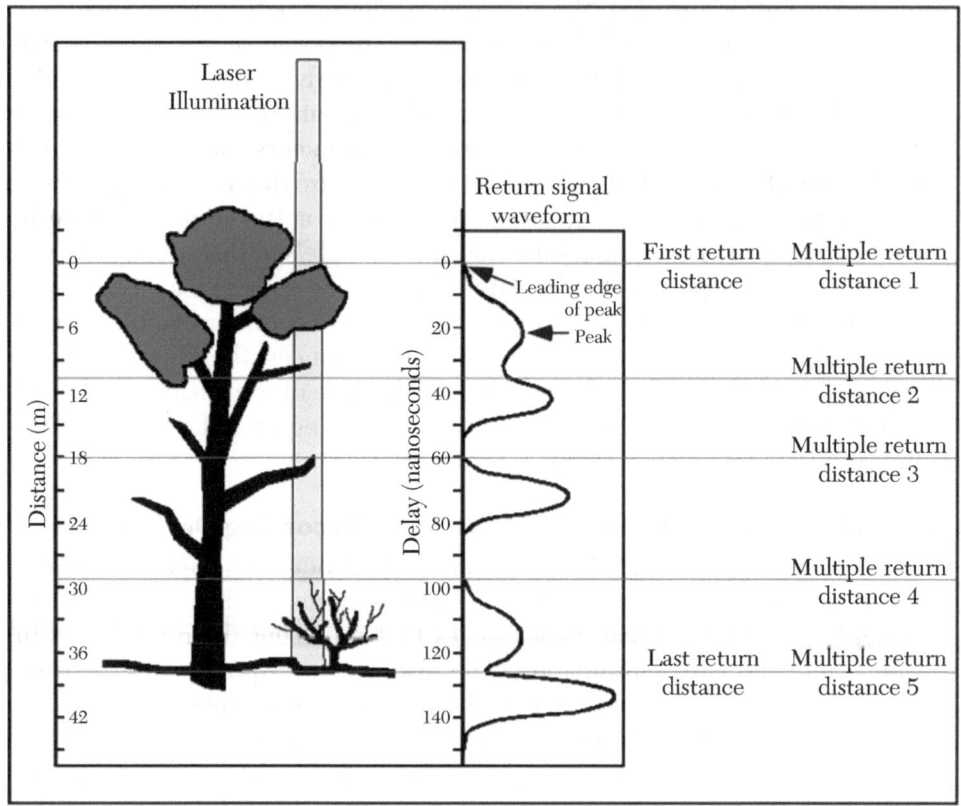

FIGURE 1.24 LIDAR on remote sensing.

laser pulse. There are advantages to both discrete-return and waveform-recording LIDAR sensors. For example, discrete-return systems feature high spatial resolution, made possible by the small diameter of their footprint and the high repetition rates of these systems (as high as 33,000 points per second), which together can yield dense distributions of sampled points. Thus, discrete-return systems are preferred for detailed mapping of ground and canopy surface topography, as in Figure 1.25.

An additional advantage made possible by this high spatial resolution is the ability to aggregate the data over areas and scales specified during data analysis, so that specific locations on the ground, such as a particular forest inventory plot or even a single tree crown, can be characterized. Finally, discrete-return systems are readily and widely available, with ongoing and rapid development, especially for surveying and photogrammetric

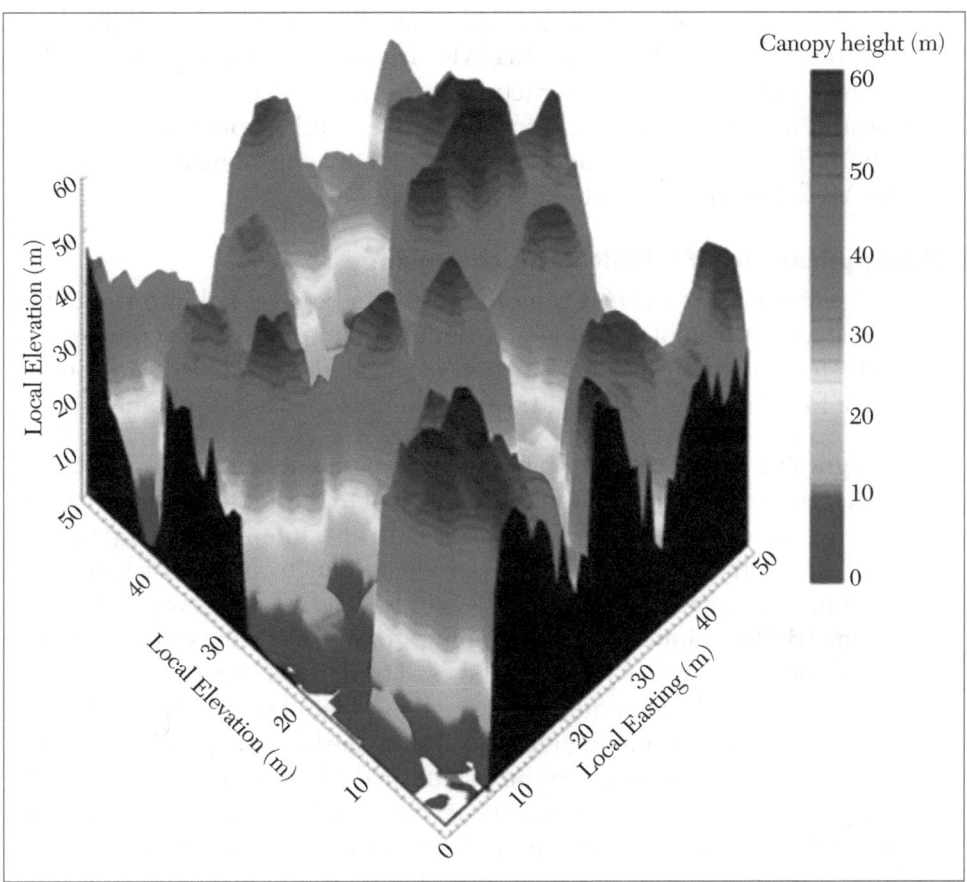

FIGURE 1.25 LIDAR topography.

applications. The primary users of these systems are surveyors serving public and private clients and natural resource managers seeking a cheaper source of high-resolution topographic maps and digital terrain models (DTMs). A potential drawback is that proprietary data-processing algorithms and established sensor configurations designed for commercial use may not coincide with scientific objectives. The advantages of waveform-recording LIDAR include an enhanced ability to characterize canopy structure, the ability to concisely describe canopy information over increasingly large areas, and the availability of global data sets (the extent of their coverage varies, however). One advantage of these waveform-recording LIDAR systems is that they record the entire time-varying power of the return signal from all illuminated surfaces and are therefore capable of collecting more information on canopy structure than all but the most spatially dense collections of small-footprint LIDAR. In addition, waveform-recording LIDAR integrates canopy structure information over a relatively large footprint and can store that information efficiently, from the perspective of both data storage and data analysis. Finally, only waveform-recording LIDAR will soon be collected globally from space.

1.7.3 Applications of LIDAR Remote Sensing

Applications of LIDAR remote sensing in ecology fall into three general categories: remote sensing of ground topography, measurement of the 3-dimensional structure and function of vegetation canopies, and prediction of forest stand structure attributes (such as aboveground biomass).

Topographic applications: Mapping of topographic features is the largest and fastest growing area of application for LIDAR remote sensing, because of its use in commercial land. Ecologists are also interested in topography (and bathymetry), which often has a strong influence on the structure, composition, and function of ecological systems. Traditional survey and photogrammetric techniques for determining ground elevations are limited in several ways. The primary disadvantages of traditional surveying are its substantial time and labor requirements and associated costs. Photogrammetric methods for determining elevations from aerial photographs or images collected by other sensors are an established alternative to field surveys. However, they are inaccurate in forested areas, where the ground is not visible, and in areas of low relief and texture, such as wetland areas and coastal dune systems. In these cases, airborne laser altimetry can be an accurate and cost-effective alternative.

Topographic applications most often use discrete-return data. When ranging information from the LIDAR is combined with position and pointing information, the result is a series of xyz data points, or *triplets*, describing the location of the observed surfaces in three-dimensional space. With adequate quality control, the accuracy of these points can achieve 50-cm root mean square error (RMSE) in the horizontal planes and 20-cm RMSE in the vertical. However, the elevations recorded in these triplets will be associated with myriad features, including the ground, human-made objects, clouds, vegetation, or anything else in the path of the laser pulse. To extract a topographic surface from these points, a series of filters must be applied to eliminate points not on the ground surface. Numerous methods exist for this process, but generally they combine highly automated processes with some manual correction. Examples of topographic applications of LIDAR include mapping of polar ice sheets for mass balance investigations, mapping of wetlands and shallow water, and high-resolution mapping of topography under forest for geomorphic investigations and hydrologic modeling. The mapping of dynamic features such as beaches and dunes is one application for which LIDAR is proving to be particularly well suited. The resulting data products are designed for accurate and cost-effective mapping of coastal erosion and could easily be applied to gain further understanding of the links between, for instance, geomorphologic and vegetation dynamics in coastal dune ecosystems.

Measuring vegetation canopy structure and function: In general, the single most important step in LIDAR mapping of topography involves the deletion of data points returned from vegetation and, in urban areas, buildings. However, for most ecological applications, it is the returns from the vegetation canopy that will be of primary interest. *Canopy structure*—"the organization in space and time, including the position, extent, quantity, type, and connectivity, of the aboveground components of vegetation"—contains a substantial amount of information about the state of development of plant communities and therefore about canopy function and vegetation-related habitat conditions for wildlife.

Prediction of forest stand structure: LIDAR data also have been used to predict biophysical characteristics of plant communities, most notably forests. Waveform-recording LIDAR use a set of indices describing the vertical distribution of the raw waveforms and the fraction of total power associated with the ground returns to predict field-measured quadratic mean stem diameter, basal area, and aboveground biomass, explaining up to 93%, 72%, and 93% of variance, respectively. LIDAR remote sensing is an extremely accurate tool for measuring topography, vegetation height, and cover, as well as more complex attributes of canopy structure and function. Another application of LIDAR data is the identification of forest areas with accumulations of fuels that make them particularly susceptible to large, especially damaging fires.

Laser altimeter: A laser altimeter uses a LIDAR to measure the height of the instrument platform above the surface. By independently knowing the height of the platform with respect to the mean Earth's surface, the topography of the underlying surface can be determined.

1.8 OCEANOGRAPHIC AND ATMOSPHERIC REMOTE SENSING RESEARCH

Oceanographic and atmospheric remote sensing is at a stage of rapid evolution, building upon vastly improved observing tools, increased computing power, and new analytical understanding. Major cooperative research efforts are dramatically changing our understanding of the dynamics of the ocean and atmosphere-ocean coupling. The results of these studies will contribute to solving a yet broader range of oceanic, geological, chemical, biological, engineering, and societal problems. The scope of our interests is global. The forefront of investigation areas are:

- The physics of rotating and stratified flow

- The dynamics of strong current systems (western boundary currents, meandering jets and fronts, equatorial current systems)

- Oceanic eddy phenomena

- Development and application of new tools for oceanic research

- Circulation and dynamics on the continental shelf

- Wind-driven and buoyancy-driven large-scale circulation

- Oceanic heat transport and storage, and effects upon global climate

- Physics at the air-sea interface

- Stirring and mixing processes in the coastal ocean

- Tropical cyclones

- Studying and predicting the behavior of the Loop Current and its eddies in the selective area

- Tracking and predicting the paths of hurricanes (real time data for emergency response planning

- Tracking oil spills

- Mapping sediment plume responses to various natural forcing events

- Studying circulation and biological impacts of river discharges along the selected path

- Mapping onshore flooded areas

EXERCISES: PART A (ANSWER IN A WORD OR A SENTENCE)

1. What is remote sensing?

2. What are the limitations of the eye as a sensor?

3. Define "map."

4. What are sonar, GIS, and LIDAR acronyms for?

5. Define "oceanography."

6. What are the two types of radars?

7. Define "GIS."

8. What is the definition of "sonar"?

9. _____ is the study of water on the Earth's surface.

10. What are radar and DEM acronyms for?

11. _____ is the science of obtaining information about objects from distance from satellites.

12. Define "ocean current."

13. _____ is the accumulation of sand or pebbles along the coast.

14. Define "wave height."

15. What is the wave length?

16. What is one difference between camera and radar antenna?

17. Define the look angle of the radar.

18. _____ is used to analyze all the details of an electromagnetic spectrum.

19. _____ is the fundamental parameter measured by passive microwave radiometers.

20. What are the products derived from the brightness temperature?

21. What is the imaging radiometer?

22. _____ is the radar designed to measure wind speed and direction at the sea surface.

23. Anemometers are used to measure _____ and _____.

24. List the uses of scatterometry.

25. Define the altimeter.

26. Define telemetry.

27. What is the use of LIDAR?

28. What is a laser altimeter?

EXERCISES: PART B (ANSWER IN A PAGE)

1. What are the similarities and difference between the field of remote sensing and satellite images, GIS, maps, and sonar?

2. List the applications of remote sensing.

3. List the applications of remote sensing in oceanography.

4. List the difference between the remote sensing vs. photogrammetry.

5. Describe one difference between sonar and remote sensing.

6. List the coastal applications of remote sensing in oceanography.

7. List the ocean applications in remote sensing.

8. Write a short note on synthetic aperture radar (SAR).

9. Write a note on NADIR radar system.

10. Write about the microwave radiometer.

11. Write a note on infrared radiometers.

12. Explain about the microwave measurement of ocean wind.

13. Discuss the working of scatterometer to measure wind speed and direction.

14. Write about the sea surface elevation and slope measurement using satellite altimeters.

15. Write the role altimeters in wind and wave height measurements.

16. Write about the echo sounders in mapping ocean floor.

17. Write in detail about echo-sounders and their operation.

18. How will you find the object on bottom of ocean using sonar?

19. Write the working difference between the side scan sonar and multi-beam sonar.

20. What are the applications of telemetry in oceanography?

21. Write about two different sensors of LIDAR devices.

EXERCISES: PART C (ANSWER WITHIN 2 OR 3 PAGES)

1. Write about radar in remote sensing.

2. What is role of radiometer in remote sensing?

3. Write about the importance of satellite work in remote sensing as applied to the ocean.

4. Discuss the use of sonar in ocean remote sensing.

5. Explain the use of telemetry in remote sensing.

6. Write a detailed study of laser and LIDAR in remote sensing.

WEB LINKS

http://www.oceanweather.com/

http://www.remss.com/missions/windsat

http://oceanservice.noaa.gov/facts/lidar.html

CHAPTER 2

SENSORS AND THEIR MEASUREMENTS FOR OCEAN MONITORING

This chapter discusses sensors for ocean monitoring and their measuring parameters. Sometimes satellites are called sensors, as well as the sensors they carry. It discusses the sensors, scanners, weather sensing, SAR sensors, marine observation sensors (MOS), ocean color monitoring sensor (OCM) and micro-sensors for ocean acidification monitoring. It also discusses the measurement of ocean parameters, such as ocean color, sediment monitoring, surface currents, surface wind, wave height, wind speed, sea surface temperature, upwelling, sampling, wave energy, and ocean floor. It also describes spatial resolution, pixel size, scale, spectral/radiometric resolution, temporal resolution, sensor design, sensor selection, and research on ocean phenomena.

2.1 INTRODUCTION TO SENSORS

A passive *remote sensing system* records the energy naturally radiated or reflected from an object. An active remote sensing system supplies its own source of energy, which is directed at the object to measure the returned energy [Figure 2.1]. Flash photography is active remote sensing, in contrast to available light photography, which is passive. Another common form of active remote sensing is *radar*, which provides its own source of electromagnetic energy in the microwave region. *Airborne laser scanning* is a relatively new form of active remote sensing, operating in the visible and near infrared wavelength bands.

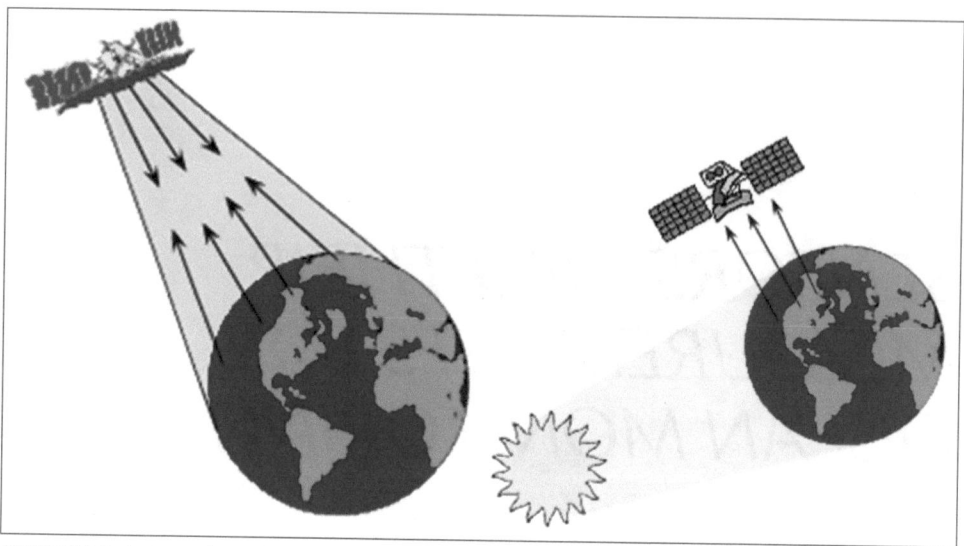

FIGURE 2.1 Active and passive remote sensing.

For a sensor to collect and record energy reflected or emitted from a target or surface, it must reside on a stable platform removed from the target or surface being observed. Platforms for remote sensors may be situated on the ground, on an aircraft or balloon (or some other platform within the Earth's atmosphere), or on a spacecraft or satellite outside of the Earth's atmosphere. Ground-based sensors are often used to record detailed information about the surface which is compared with information collected from aircraft or satellite sensors. In some cases, this can be used to better characterize the target that is being imaged by these other sensors, making it possible to better understand the information in the imagery. Sensors may be placed on a ladder, scaffolding, tall building, cherry-picker, crane, etc. Aerial platforms are primarily stable wing aircraft, although helicopters are occasionally used. Aircraft are often used to collect very detailed images and facilitate the collection of data over virtually any portion of the Earth's surface at any time. In space, remote sensing is sometimes conducted from the space shuttle or, more commonly, from satellites. *Satellites* are objects which revolve around another object—in this case, the Earth. For example, the moon is a natural satellite, whereas manmade satellites include those platforms launched for remote sensing, communication, and telemetry (location and navigation) purposes. Because of their orbits, satellites permit repetitive coverage of the Earth's surface on a continuing basis. Cost is often a significant factor in choosing among the various platform options.

2.1.1 History of Sensors

Since the early 1960s, numerous satellite sensors have been launched into orbit to observe and monitor the Earth and its environment. Most early satellite sensors acquired data for meteorological purposes. The advent of earth resources satellite sensors (those with a primary objective of mapping and monitoring land cover) occurred when the first LANDSAT satellite was launched in July 1972. Currently, more than a dozen orbiting satellites of various types provide data crucial to improving our knowledge of the Earth's atmosphere, oceans, ice and snow, and land. The path followed by a satellite is referred to as its *orbit*. Satellite orbits are matched to the capability and objective of the sensor(s) they carry. Orbit selection can vary in terms of altitude (their height above the Earth's surface) and their orientation and rotation relative to the Earth. Satellites at very high altitudes, which view the same portion of the Earth's surface at all times, have geostationary orbits [Figure 2.2]. These geostationary satellites, at altitudes of approximately 36,000 km, revolve at speeds which match the rotation of the Earth, so they seem stationary relative to the Earth's surface. This allows the satellites to observe and collect information continuously over specific areas. Weather and communications satellites commonly have these types of orbits. Due to their high altitude, some geostationary weather satellites can monitor weather and cloud patterns covering an entire hemisphere of the Earth.

Many remote sensing platforms are designed to follow an orbit (basically north-south) which, in conjunction with the Earth's rotation (west-east), allows them to cover most of the Earth's surface over a certain period.

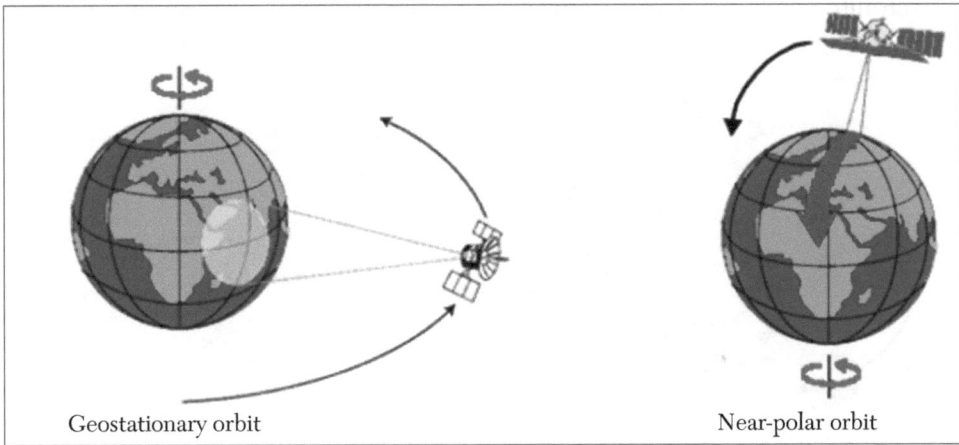

Geostationary orbit   Near-polar orbit

FIGURE 2.2 Satellite orbits.

These are near-polar orbits, so named for the inclination of the orbit relative to a line running between the North and South Poles. Many of these satellite orbits are also sun-synchronous such that they cover each area of the world at a constant local time of day called *local sun time*. At any given latitude, the position of the sun in the sky as the satellite passes overhead will be the same within the same season. This ensures consistent illumination conditions when acquiring images in a specific season over successive years, or over a particular area over a series of days. This is an important factor for monitoring changes between images as they do not have to be corrected for different illumination conditions.

Most of the remote sensing satellite platforms today are in *near-polar orbits*, which means that the satellite travels northwards on one side of the Earth and then toward the southern pole on the second half of its orbit. These are called ascending and descending passes, respectively. If the orbit is also sun-synchronous, the ascending pass is most likely on the shadowed side of the Earth while the descending pass is on the sunlit side. Sensors recording reflected solar energy only image the surface on a descending pass, when solar illumination is available.

Active sensors which provide their own illumination or passive sensors that record emitted (e.g., thermal) radiation can also image the surface on ascending passes. As a satellite revolves around the Earth, the sensor "sees" a certain portion of the Earth's surface. The area imaged on the surface is referred to as the swath [Figure 2.3]. Imaging swaths for space-borne sensors generally vary between tens and hundreds of kilometers wide. As the satellite orbits the Earth from pole to pole, its east-west position wouldn't

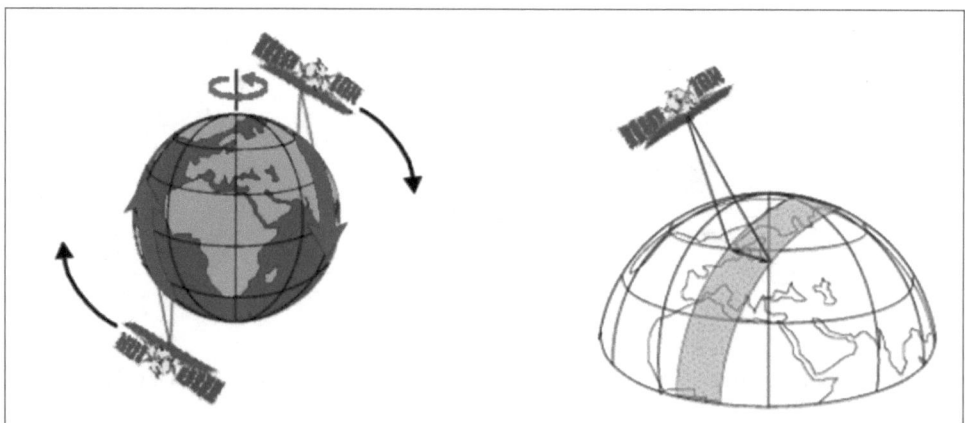

FIGURE 2.3 Ascending and descending pass and swath.

change if the Earth didn't rotate. However, as seen from the Earth, it seems that the satellite is shifting westward because the Earth is rotating (from west to east) beneath it. This apparent movement allows the satellite swath to cover a new area with each consecutive pass. The satellite's orbit and the rotation of the Earth work together to allow complete coverage of the Earth's surface, after it has completed one complete cycle of orbits.

If we start with any randomly selected pass in a satellite's orbit, an orbit cycle will be completed when the satellite retraces its path, passing over the same point on the Earth's surface directly below the satellite (called the nadir point) for a second time. The exact length of time of the orbital cycle will vary with each satellite. The interval of time required for the satellite to complete its orbit cycle is not the same as the *revisit period*. Using steerable sensors, a satellite-borne instrument can view an area (off-nadir) before and after the orbit passes over a target, thus making the "revisit" time less than the orbit cycle time. The revisit period is an important consideration for many monitoring applications, especially when frequent imaging is required (for example, to monitor the spread of an oil spill, or the extent of flooding). In near-polar orbits, areas at high latitudes will be imaged more frequently than the equatorial zone due to the increasing overlap in adjacent swaths as the orbit paths come closer together near the poles.

2.2 SCANNER SENSOR SYSTEMS

Electro-optical and spectral imaging scanners produce digital images with the use of detectors that measure the brightness of reflected electromagnetic energy. Scanners consist of one or more sensor detectors depending on type of sensor system used. One type of scanner is called a *whiskbroom scanner,* also referred to as across-track scanners (e.g., on a LANDSAT satellite). It uses rotating mirrors to scan the landscape below from side to side perpendicular to the direction of the sensor platform, like a whiskbroom. The width of the sweep is referred to as the sensor swath. The rotating mirrors redirect the reflected light to a point where a single or just a few sensor detectors are grouped together. Whiskbroom scanners with their moving mirrors tend to be large and complex to build. The moving mirrors create spatial distortions that must be corrected with preprocessing by the data provider before image data is delivered to the user. An advantage of whiskbroom scanners is that they have fewer sensor detectors to keep calibrated as compared to other types of sensors.

Another type of scanner, which does not use rotating mirrors, is the *pushbroom scanner*, also referred to as an along-track scanner (e.g., on SPOT). The sensor detectors in a pushbroom scanner are lined up in a row called a linear array. Instead of sweeping from side to side as the sensor system moves forward, the 1-dimensional sensor array captures the entire scan line at once, like a pushbroom would. *Step stare scanners* contain 2-dimensional arrays in rows and columns for each band. Pushbroom scanners are lighter, smaller, and less complex because of fewer moving parts than whiskbroom scanners, and they have better radiometric and spatial resolution. A major disadvantage of pushbroom scanners is the calibration required for the large number of detectors that make up the sensor system [Figure 2.4].

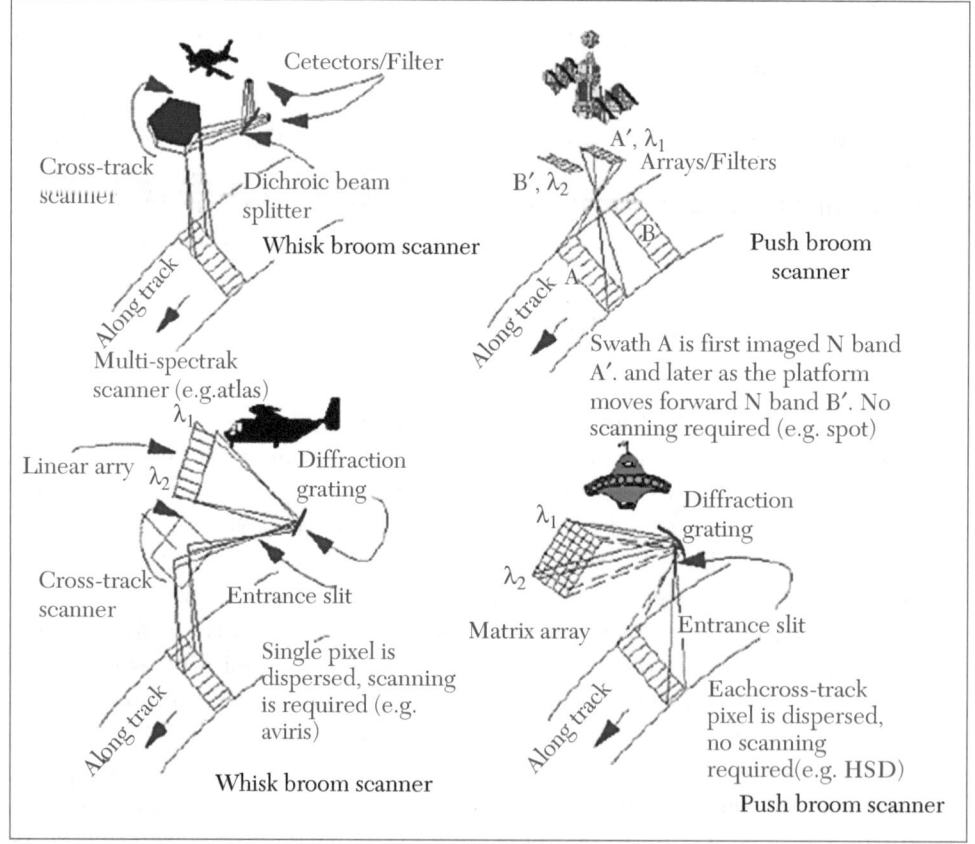

FIGURE 2.4 Scanners.

2.2.1 Spatial Resolution, Pixel Size, and Scale

For some remote sensing instruments, the distance between the target being imaged and the platform plays a large role in determining the detail of information obtained and the total area imaged by the sensor. Sensors on platforms far away from their targets typically view a larger area but cannot provide great detail. Compare what an astronaut onboard the Space Shuttle sees of the Earth to what you can see from an airplane. The astronaut might see your whole province or country in one glance, but couldn't distinguish individual houses. Flying over a city or town, you would be able to see individual buildings and cars, but you would be viewing a much smaller area than the astronaut. There is a similar difference between satellite images and air photos. The detail discernible in an image is dependent on the spatial resolution of the sensor and refers to the size of the smallest possible feature that can be detected.

Spatial resolution of passive sensors depends primarily on their *instantaneous field of view* (IFOV in Figure 2.5). The IFOV is the angular cone of visibility of the sensor (A) and determines the area on the Earth's surface that is "seen" from a given altitude at one particular moment in time (B). The size of the area viewed is determined by multiplying the IFOV by the distance from the ground to the sensor (C). This area on the ground is called the resolution cell and determines a sensor's maximum spatial resolution. For a homogeneous feature to be

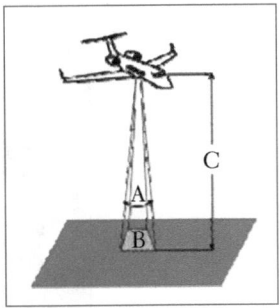

FIGURE 2.5 Instantaneous field of view (IFOV).

detected, its size generally has to be equal to or larger than the resolution cell. If the feature is smaller than this, it may not be detectable as the average brightness of all features in that resolution cell will be recorded. However, smaller features may sometimes be detectable if their reflectance dominates within an articular resolution cell, allowing subpixel or resolution cell detection.

Most remote sensing images are composed of a matrix of picture elements, or pixels, which are the smallest units of an image. Image pixels are normally square and represent a certain area on an image. It is important to distinguish between pixel size and spatial resolution—they are not interchangeable. If a sensor has a spatial resolution of 20 meters and an image from that sensor is displayed at full resolution, each pixel represents an area of 20 meters × 20 meters on the ground. In this case the pixel size and resolution are the same. However, it is possible to display an image with a pixel size different than the resolution. Many posters of satellite images of the Earth have their pixels averaged to represent larger areas, although the original spatial resolution of the sensor that collected the imagery remains the same.

A photograph can be represented and displayed in a digital format [Figure 2.6] by subdividing the image into small, equal-sized and -shaped areas, called picture elements or *pixels*, and representing the brightness of each area with a numeric value or digital number.

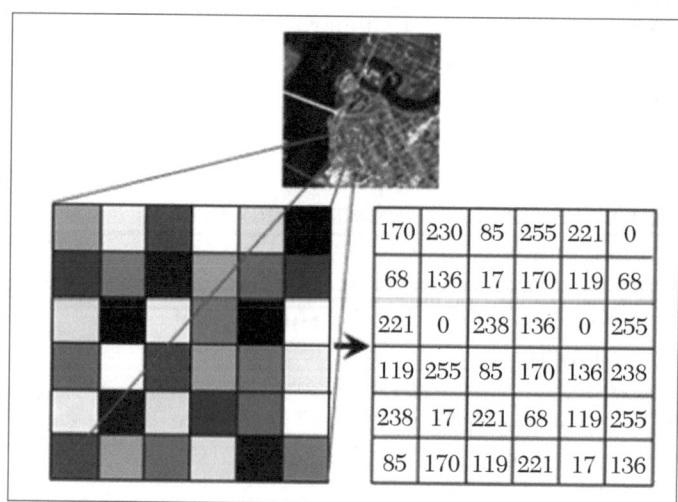

FIGURE 2.6 Satellite images.

Images where only large features are visible are said to have coarse or low resolution. In fine or high resolution images, small objects can be detected. Military sensors for example, are designed to view as much detail as possible, and therefore have very fine resolution. Commercial satellites provide imagery with resolutions varying from a few meters to several kilometers. Generally speaking, the finer the resolution, the less total ground area can be seen. The ratio of distance on an image or map, to actual ground distance is referred to as scale. If you had a map with a scale of 1:100,000, an object of 1 cm length on the map would actually be an object 100,000 cm (1 km) long on the ground. Maps or images with small *map-to-ground ratios* are referred to as small scale (e.g., 1:100,000), and those with larger ratios (e.g., 1:5,000) are called large scale.

2.2.2 Spectral/Radiometric Resolution

Spectral characteristics: While the arrangement of pixels describes the spatial structure of an image, the radiometric characteristics describe the actual information content in an image. Every time an image is acquired on film or by a sensor, its sensitivity to the magnitude of the electromagnetic energy determines the radiometric resolution. The radiometric resolution of an imaging system describes its ability to discriminate very slight differences in energy. The finer the radiometric resolution of a sensor, the more sensitive it is to detecting small differences in reflected or emitted energy.

Digital resolution is the number of bits comprising each digital sample. Imagery data are represented by positive digital numbers which vary from 0 to (one less than) a selected power of 2. This range corresponds to the number of bits used for coding numbers in binary format. Each bit records an exponent of power 2 (e.g., 1 bit = 2^1 = 2). The maximum number of brightness levels available depends on the number of bits used in representing the energy recorded. Thus, if a sensor used 8 bits to record the data, there would be 2^8 = 256 digital values available, ranging from 0 to 255—also called the dynamic range of the system. However, if only 4 bits were used, then only 2^4 = 16 values ranging from 0 to 15 would be available. Thus, the radiometric resolution would be much less. Image data are generally displayed in a range of grey tones, with black representing a digital number of 0 and white representing the maximum value (for example, 255 in 8-bit data). By comparing a 2-bit image with an 8-bit image, we can see that there is a large difference in the level of detail discernible depending on their radiometric resolutions.

The range of energy values expected from a system must *fit* within the range of values possible of the data format type, and yet the value must represent accurately the energy value of the signal relative to others. The cost of more bits per data point is longer acquisition times, the need for larger storage capacity, and longer processing time. Any signal outside the range is *clipped* and thus unrecoverable. On the other hand, if the dynamic range of the signal is widened too much to allow the recording of extremely high or low energy values, the true variability within the signal will be lost.

Many remote sensing systems record energy over several separate wavelength ranges at various spectral resolutions. These are referred to as *multi-spectral sensors*. Advanced multi-spectral sensors, called *hyperspectral sensors*, detect hundreds of very narrow spectral bands throughout the visible, near-infrared, and mid-infrared portions of the electromagnetic spectrum. Their very high spectral resolution facilitates fine discrimination between different targets based on their spectral response in each of the narrow bands. There are four general parameters that describe the capability of a spectrometer: 1) spectral range, 2) spectral bandwidth, 3) spectral sampling, and 4) signal-to-noise ratio (S/N).

Spectral range: Spectral range is important to cover enough diagnostic spectral absorption to solve a problem. There are general spectral ranges that are in common use, each to first order controlled by detector technology: a) ultraviolet (UV): 0.001 to 0.4 μm, b) visible: 0.4 to 0.7 μm, c) near-infrared (NIR): 0.7 to 3.0 μm, d) the mid-infrared (MIR): 3.0 to 30 μm, and d) the far infrared (FIR): 30 μm to 1 mm. The 0.4 to 1.0-μm wavelength range is sometimes referred to in the remote sensing literature as the VNIR (visible-near-infrared) and the 1.0 to 2.5-μm range is sometimes referred to as the SWIR (short-wave infrared).

Spectral bandwidth: Spectral bandwidth is the width of an individual spectral channel in the spectrometer. The narrower the spectral bandwidth, the narrower the absorption feature the spectrometer will accurately measure, if enough adjacent spectral samples are obtained. All the spectra are sampled at half Nyquist (critical sampling) except the near infrared mapping spectrometer (NIMS), which is at Nyquist sampling (named after H. Nyquist, who in his work published in 1928 stated that there must be at least two samplings per wavelength of the highest frequency in order to appropriately sample the waveform). Note, however, that the fine details of the absorption features are lost at the ~25 nm bandpass of NIMS. The visual and infrared mapping spectrometer (VIMS) and NIMS systems measure out to 5 μm and thus can see absorption bands not obtainable by the other systems.

Spectral sampling: Spectral sampling is the distance in wavelength between the spectral bandpass profiles for each channel in the spectrometer as a function of wavelength. The Nyquist theorem states that the maximum information is obtained by sampling at one-half the full width at half maximum (FWHM).

Signal-to-noise ratio: Finally, a spectrometer must measure the spectrum with enough precision to record details in the spectrum. The *signal-to-noise ratio* (S/N) required to solve a problem will depend on the strength of the spectral features under study. The S/N is dependent on the detector sensitivity, the spectral bandwidth, and intensity of the light reflected or emitted from the surface being measured. A few spectral features are quite strong and a S/N of only about 10 will be adequate to identify them, while others are weak, and a S/N of several hundred (and higher) are often needed.

2.2.3 Temporal Resolution

In addition to spatial, spectral, and radiometric resolution, the concept of temporal resolution is also important to consider in a remote sensing system. The revisit period of a satellite sensor is usually several days. Therefore, the absolute temporal resolution of a remote sensing system to image the exact same area at the same viewing angle a second time is equal to this period. However, the actual temporal resolution of a sensor depends on a variety of factors, including the satellite/sensor capabilities, the swath overlap, and latitude.

The ability to collect imagery of the same area of the Earth's surface at different periods of time is one of the most important elements for applying remote sensing data. Spectral characteristics of features may change over time and these changes can be detected by collecting and comparing multi-temporal imagery. For example, during the growing season, most species of vegetation are in a continual state of change and our ability to monitor those subtle changes using remote sensing is dependent on when and how frequently we collect imagery. By imaging on a continuing basis at different times we can monitor the changes that take place on the Earth's surface, whether they are naturally occurring (such as changes in natural vegetation cover or flooding) or induced by humans.

The time factor in imaging is important when:

- Persistent clouds offer limited clear views of the Earth's surface (often in the tropics)

- Short-lived phenomena (floods, oil slicks, etc.) need to be imaged

- Multi-temporal comparisons are required

- The changing appearance of a feature over time can be used to distinguish it from near-similar features

2.2.4 Sensor Design and Selection

Each remote sensing mission has unique requirements for spatial, spectral, radiometric, and temporal resolution. A number of practical considerations also arise in the design process, including system development and operational costs; the technical maturity of a particular design; and power, weight, volume, and data rate requirements. Because it is extremely expensive, or perhaps impossible, to gather data with all the characteristics a user might want, the selection of sensors or satellite subsystems for a mission involving several tasks generally involves compromises. Sensor performance may be measured by spatial and spectral resolution, geographical coverage, and repeat frequency. For example, sensors with very high spatial resolution are typically limited in geographical coverage. A video camera is one example of an instrument that employs an electro-optical sensor. The sun's angle with respect to the surface varies somewhat throughout the year, depending on the sun's apparent position with respect to the equator.

2.3 WEATHER SATELLITES/SENSORS

Weather monitoring and forecasting was one of the first civilian applications of satellite remote sensing. Today, several countries operate weather or meteorological satellites to monitor weather conditions around the globe. These satellites use sensors which have fairly coarse spatial resolution (when compared to systems for observing land) and provide large areal coverage. Their temporal resolutions are generally quite high, providing frequent observations of the Earth's surface, atmospheric moisture, and cloud cover, which allows for near-continuous monitoring of global weather conditions, and hence forecasting. These weather satellites carry sensors related to weather monitoring.

As India's first domestic dedicated Earth resources satellite program, the IRS series provides continuous coverage of the country. An indigenous ground system network handles data reception, data processing, and data dissemination. India's National Natural Resources Management System (NNRMS) uses IRS data to support a large number of applications projects.

India has orbited IRS satellites which carry two payloads employing *linear imaging selfscanning sensors* (LISS). The IRS-series have a 22-day repeat cycle. The LISS-I imaging sensor system consists of a camera operating in four spectral bands, compatible with the output from LANDSAT-series Thematic Mapper and SPOT HRV instruments. The LISS-IIA and

B comprise two cameras operating in visible and near-infrared wavelengths with a ground resolution of 36.5 meters and swath width of 74.25 km. As part of the National Remote Sensing Agency's international services, IRS data are available to all countries within the coverage zone of the Indian ground station located at Hyderabad. These countries can purchase the raw/processed data directly from NRSA Data Centre. India is designing sensors with resolutions of about 20 meters in multi-spectral bands and better than 10 meters in the panchromatic band. System designers intend to include a short-wave infrared band with spatial resolution of 70 meters. The system will also include a *wide field sensor* (WiFS) with 180 meter spatial resolution and larger swath of about 770 km for monitoring vegetation.

Operational uses of ocean satellites: The development and operation of SeaSat demonstrated the utility of continuous ocean observations, not only for scientific use, but also for those concerned with navigating the world's oceans and exploiting ocean resources. Its success convinced many that an operational ocean remote sensing satellite would provide significant benefits. The SAR, the scatterometer, and the altimeter all gathered data of considerable utility. Knowledge of currents, wind speeds, wave heights, and general wave conditions at a variety of ocean locations is crucial for enhancing the safety of ships at sea and for ocean platforms. Such data could also decrease costs by allowing ship owners to predict the shortest, safest sea routes.

SeaSat carried five major instruments—an altimeter, a microwave radiometer, a scatterometer, a visible and infrared radiometer, and synthetic aperture radar (SAR). Scientists used data from these instruments to measure the amplitude and direction of surface winds, absolute and relative surface temperature, the status of ocean features such as islands, shoals, and currents, and the extent and structure of sea ice. Data from the SeaWifs instrument aboard the privately developed SeaStar satellite will provide ocean color information, which could have considerable operational use.

Observations of sea ice: Because sea ice covers about 13% of the world's oceans, it has a marked effect on weather and climate. Thus, measurements of its thickness, extent, and composition help scientists understand and predict changes in weather and climate. Until satellite measurements were available, the difficulties of tracking these characteristics were a major impediment to understanding the behavior of sea ice, especially its seasonal and yearly variations. The AVHRR visible and infrared sensors aboard have been used to follow the large-scale variations in the Arctic and

Antarctic ice packs. Because they can "see through" clouds, synthetic aperture radar instruments are particularly useful in tracking the development and movement of ice packs, which pose threats to shipping, and in finding routes through the ice.

2.4 SYNTHETIC APERTURE RADAR SENSORS

Synthetic aperture radar (SAR) image data provide information different from that of optical sensors operating in the visible and infrared regions of the electromagnetic spectrum. SAR data consist of high-resolution reflected returns of radar-frequency energy from terrain that has been illuminated by a directed beam of pulses generated by the sensor. The radar returns from the terrain are mainly determined by the physical characteristics of the surface features (such as surface roughness, geometric structure, and orientation), the electrical characteristics (dielectric constant, moisture content, and conductivity), and the radar frequency of the sensor. By supplying its own source of illumination, the SAR sensor can acquire data day or night without regard to cloud cover.

2.5 MARINE OBSERVATION SATELLITES (MOS) SENSORS

The Earth's oceans cover more than two-thirds of the Earth's surface and play an important role in the global climate system. They also contain an abundance of living organisms and natural resources which are susceptible to pollution and other man-induced hazards. The Nimbus-7 satellite, launched in 1978, carried the first sensor, the *coastal zone color scanner* (CZCS), specifically intended for monitoring the Earth's oceans and water bodies. The primary objective of this sensor was to observe ocean color and temperature, particularly in coastal zones, with sufficient spatial and spectral resolution to detect pollutants in the upper levels of the ocean and to determine the nature of materials suspended in the water column. The Nimbus satellite was placed in a sun-synchronous, near-polar orbit at an altitude of 955 km. Equator crossing times were local noon for ascending passes and local midnight for descending passes. The repeat cycle of the satellite allowed for global coverage every six days, or every 83 orbits. The CZCS sensor consisted of six spectral bands in the visible, near-IR, and thermal portions of the spectrum each collecting data at a spatial resolution of 825 m at nadir over a 1566 km swath width. The accompanying table outlines the spectral ranges of each band and the primary parameter measured by each.

Channel	Wavelength Range (µm)	Primary Measured Parameter
1	0.43–0.45	Chlorophyll absorption
2	0.51–0.53	Chlorophyll absorption
3	0.54–0.56	Gelbstoffe (yellow substance)
4	0.66–0.68	Chlorophyll concentration
5	0.70–0.80	Surface vegetation
6	10.5–12.50	Surface temperature

TABLE 2.1 CZCS Spectral Bands

As can be seen from the Table 2.1, the first four bands of the CZCS sensor are very narrow. They were optimized to allow detailed discrimination of differences in water reflectance due to phytoplankton concentrations [Figure 2.7] and other suspended particulates in the water. In addition to detecting surface vegetation on the water, band 5 was used to discriminate water from land prior to processing the other bands of information. The CZCS sensor ceased operation in 1986.

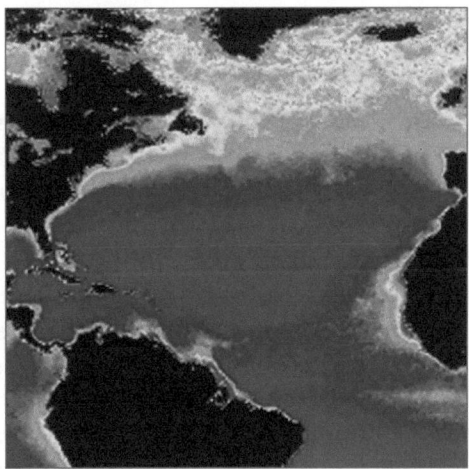

FIGURE 2.7 CZCS sensor image.

Marine observation satellite (MOS): The first Marine Observation Satellite carries three different sensors: a four-channel *multispectral electronic self-scanning radiometer* (MESSR), a four-channel *visible and thermal infrared radiometer* (VTIR), and a two-channel *microwave scanning radiometer* (MSR), in the microwave portion of the spectrum. The characteristics of the two sensors in the visible/infrared are described in Table 2.2.

The MESSR bands are thus useful for land applications in addition to observations of marine environments. The MOS systems orbit at altitudes around 900 km and have revisit periods of 17 days.

Sensor	Wavelength Ranges (µm)	Spatial Resolution	Swath Width
MESSR	0.51–0.59	50 m	100 km
	0.61–0.69	50 m	100 km
	0.72–0.80	50 m	100 km
	0.80–1.10	50 m	100 km
VTIR	0.50–0.70	900 m	1500 km
	6.0–7.0	2700 m	1500 km
	10.5–11.5	2700 m	1500 km
	11.5–12.5	2700 m	1500 km

TABLE 2.2 Characteristics of MESSR and VTIR Sensors

Sea-viewing wide-field-of view sensor (SeaWiFS): The SeaWiFS (*sea-viewing wide-field-of view sensor*) on board the SeaStar spacecraft is an advanced sensor designed for ocean monitoring. It consists of eight spectral bands of very narrow wavelength tailored for very specific detection and monitoring of various ocean phenomena including: ocean primary production and phytoplankton processes, ocean influences on climate processes (heat storage and aerosol formation), and monitoring of the cycles of carbon, sulfur, and nitrogen. The orbit altitude is 705 km with a local equatorial crossing time of 12 PM. Two combinations of spatial resolution and swath width are available for each band: a higher resolution mode of 1.1 km (at nadir) over a swath of 2,800 km, and a lower resolution mode of 4.5 km (at nadir) over a swath of 1,500 km. These ocean-observing satellite systems are important for global and regional scale monitoring of ocean pollution and health and assist scientists in understanding the influence and impact of the oceans on the global climate system.

Oceansat: Oceansat-2 is an Indian satellite designed to provide the ocean color monitor (OCM) instrument for users. It will also enhance the potential of applications in other areas. The main objectives of OceanSat are to study surface winds and ocean surface strata, observation of chlorophyll concentrations, monitoring of phytoplankton blooms, and study of atmospheric aerosols and suspended sediments in the water. (This is explained in detail in Appendix A: Indian Satellites for Ocean Monitoring.)

2.6 MICRO SENSORS FOR MONITORING OCEAN ACIDIFICATION

New technology that will measure pH levels in seawater with a cost-effective micro sensor for long-term monitoring of ocean acidification has been developed. Ocean acidification is occurring due to rising levels of atmospheric carbon dioxide (CO_2), which is absorbed by the oceans. When it dissolves in seawater, CO_2 forms a mild acid, which is decreasing ocean pH globally and could impact marine ecosystems. As well as monitoring global change, the sensors can be used to measure more localized human impact. The micro sensors could be deployed to detect leakages from carbon capture and storage sites—whereby CO_2 is artificially removed from the atmosphere and stored in subsea reservoirs—by measuring any proximal fluctuations in pH. The oil industry is also interested in this technology for monitoring seawater acidity around drilling sites.

The sensor works on the principles of litmus paper color changes depending on the acidity of the solution. The microfluidic chip within the sensor has great advantages because it is robust, small, reasonably cheap to produce, and uses small amounts of reagents—which is really key for in situ deployment where it may be collecting data out at sea for long periods of time. The sensor uses a dye which changes color with pH. The dye is added to the sample, then the color is measured using an LED light source and a device called a "spectrometer." The microfluidic element simply describes the component needed to mix the seawater sample with the dye and the cell to measure the color. The microfluidic chip used in the pH micro sensor with dimensions: 13×8 cm is shown in Figure 2.8.

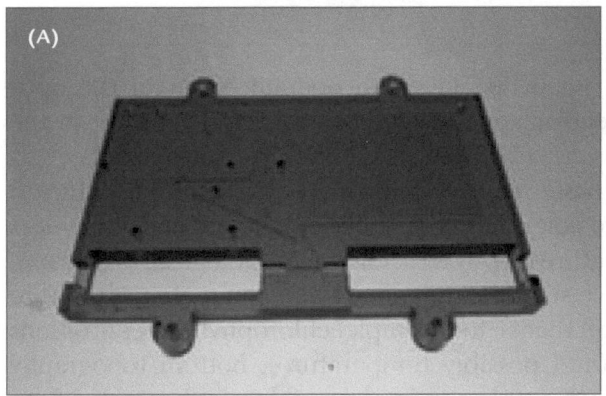

FIGURE 2.8 (a) pH micro sensor and (b) VULCANO buoy with atmospheric and communication sensors.

The VULCANO buoy with atmospheric and communication sensors is shown in Figure 2.8b. The increase in the production and emission of anthropogenic CO_2 and its absorption by the oceans leads to a reduction in oceanic pH, a process referred to as ocean acidification, which affects many other physicochemical processes. Atmospheric CO_2 is expected to continue its increase, and consequently the chemical changes will likely continue well into the future, affecting the ocean biogeochemical cycling. To characterize the ocean's chemical and ecosystem-related changes, examination of CO_2 system parameters over a wide range of temporal and spatial scales is necessary. Shipboard analyses in oceanic time series conducted during irregular ocean expeditions, which lasted between a fortnight and a month, have provided most of our understanding of recent trends in the oceanic CO_2 system. However, our ability to make frequent autonomous measurements over a broad range of spatial scales would greatly augment the current suite of open-ocean and coastal observations.

The carbon dioxide system in natural waters is defined by the measurement of two or more carbonate parameters: pH, carbon dioxide fugacity (fCO_2), total alkalinity (TA), and total dissolved inorganic carbon (DIC). The sensors for pH, pCO_2, DIC, and TA of seawater have been developed based in spectrophotometric techniques that in most cases require accurate dye additions and consideration of the effects of aging. A new family of rugged and extremely stable spectrophotometric pH sensors for both lab-based research and buoy monitoring, specifically designed for unattended operation independently of dye and aging effects in surface waters, is also developed.

2.7 MEASUREMENTS OF OCEAN MONITORING PARAMETERS

Satellite sensors have been developed to operate either in the optical infrared part of the electromagnetic spectrum (λ-0.4–12 µm) or in the microwave part (λ-0.3–30 µm). These sensors, either from an aircraft or satellite, have measured the basic sea surface properties to useful accuracies. These properties of sea are color, temperature, slope/height and roughness etc. At present all ocean features—physical, chemical, biological, or geological—must produce a surface signature in one of those parameters if they are to be monitored from space; that is, for example, chlorophyll concentrations must affect ocean color (and possibly temperature), bottom topography must be reflected in sea surface shapes (and possibly roughness), surface waves must modulate small-scale surface roughness patterns, and so on.

2.7.1 Ocean Color

Ocean color data is a vital resource for a wide variety of operational forecasting and oceanographic research, earth sciences, and related applications. Some of them are

- Mapping of chlorophyll concentrations

- Measurement of inherent optical properties such as absorption and backscatter

- Determination of phytoplankton physiology, phenology, and functional groups

- Studies of ocean carbon fixation and cycling

- Monitoring of ecosystem changes resulting from climate change

- Fisheries management

- Mapping of coral reefs, sea grass beds, and kelp forests

- Mapping of shallow-water bathymetry and bottom type for military operations

- Monitoring of water quality for recreation

Ocean color is the measurement of spectral distribution of radiance (or reflectance) upwelling from the ocean in the visible regime. Measurements of ocean color from space can provide quantitative maps of near-surface phytoplankton pigment concentration as well as identifying pollutants spilled into the sea by effluent discharges. The spectral composition of the radiation above the sea (that is, its *color*) is determined by the composition of solar irradiance plus the optical properties of the water column—in particular, the absorption, scattering, and to a lesser degree the fluorescence. The optical properties of the water column are, in turn, determined by those of pure sea water plus the concentrations of suspended and dissolved materials within it particulate organic matter (phytoplankton and its byproducts), inorganic suspended particles, and dissolved organic decay products of mainly terrestrial origin (*yellow substance*).

However, in the shorter term, fisheries research can be assisted enormously by a more precise knowledge of the elements in the food chain—larval recruitment, zooplankton production, and the patchiness of plankton. In this regard color (and temperature) information derived from orbiting satellites is

FIGURE 2.9 This MODIS image of blue water in the Caribbean Sea looks blue because the sunlight is scattered by the water molecules. Near the Bahama Islands, the lighter aqua colors are shallow water where the sunlight is reflecting off of the sand and reefs near the surface.

particularly useful in tropical and subtropical regions. The requirement to monitor the spread of pollutants may be stronger in some areas than others. In closed or semi-enclosed seas such as the North Sea or Mediterranean, the indiscriminate dumping of waste products into the sea has become a major source of concern and in many respects the ocean color signal can be detected and traced from satellites. Color imagery has also been successfully exploited in dynamical studies of the physics of shallow sea processes.

The "color" of the ocean is determined by the interactions of incident light with substances or particles present in the water. White light from the sun is made up of a combination of colors, which are broken apart by water droplets in a "rainbow" spectrum. When light hits the water surface, the different colors are absorbed, transmitted, scattered, or reflected in differing intensities by water molecules and other optically active constituents in the upper layer of the ocean.

If there are any particles suspended in the water, they will increase the scattering of light. For example, microscopic marine algae, called phytoplankton, have the capacity to absorb light in the blue and red region of the spectrum owing to specific pigments like chlorophyll. Accordingly, as the concentration of phytoplankton increases in the water, the color of the water shifts toward the green part of the spectrum. Fine mineral particles like sediment absorb light in the blue part of the spectrum, causing the water to turn brownish in case of massive sediment load.

The basic principle behind the remote sensing of ocean color from space is, the more phytoplankton in the water, the greener it is, and the less phytoplankton, the bluer it is. There are other substances that may be found dissolved in the water that can also absorb light. These substances are referred as colored dissolved organic matter (CDOM).

Ocean color sensors: Active remote sensing is a signal of known characteristics and sent from the sensor platform—an aircraft or satellite—to the ocean, and the return signal is then detected after a time delay determined by the distance from the platform to the ocean and by the speed of light. One example of active remote sensing at visible wavelengths is the use of laser-induced fluorescence to detect chlorophyll, yellow matter, or pollutants. In laser fluorosensing, a pulse of UV light is sent to the ocean surface, and the spectral character and strength of the induced fluorescence at UV and visible wavelengths gives information about the location, type, and concentration of fluorescing substances in the water body. Another example of active remote sensing is LIDAR bathymetry. This refers to the use of pulsed lasers to send a beam of short duration, typically about a nanosecond, toward the ocean. The laser light reflected from the sea surface and then slightly later from the bottom is used to deduce the bottom depth. The depth is simply $0.5(c/n)\,\Delta t$, where c is the speed of light in vacuum, n is the water index of refraction, Δt is the time between the arrival of the surface-reflected light and the light reflected by the bottom, and the 0.5 accounts for the light traveling from the surface to the bottom and back to the surface.

Passive remote sensing simply observes the light that is naturally emitted or reflected by the water body. The nighttime detection of bioluminescence from aircraft is an example of the use of emitted light at visible wavelengths. The most common example of passive remote sensing is the use of sunlight that has been backscattered within the water and returned to the sensor. This light can be used to deduce the concentrations of chlorophyll, CDOM, or mineral particles within the near-surface water; the bottom depth and type in shallow waters; and other ecosystem information such as net primary production, phytoplankton functional groups, or phytoplankton physiological state.

Passive ocean color remote sensing from satellites began with the coastal zone color scanner (CZCS), which was launched in 1978. CZCS was a multi-spectral sensor, meaning that it had only a few wavelength bands with bandwidths of 10 nm or more. After the phenomenal success of that "proof of principle" sensor, numerous other multi-spectral sensors have been developed and launched. Those later sensors generally had a few more bands with narrower bandwidths. Thus, the sea-viewing wide field-of-view sensor (SeaWiFS) added a band near 412 nm to improve the detection of CDOM. The near-IR bands are used for atmospheric correction.

There is today much interest in the use of hyperspectral sensors, which typically have 100 or more bands with nominal bandwidths of 5 nm or less. Figure 2.10 shows the wavelength bands for a few representative sensors. The MODIS (MODerate resolution Imaging Spectro radiometer) sensor has additional bands in the 400–900 nm range, which are used for detection of clouds, aerosols, and atmospheric water vapor. The bands shown are the ones used for remote sensing of water bodies. The compact airborne hyperspectral imager (CASI) is a commercially available hyperspectral sensor that is widely used in airborne remote sensing of coastal waters. It has 228 slightly overlapping bands, each with a nominal 1.9 nm bandwidth and covering the 400–1000 nm range. CASI users often select a subset of these bands as needed for an application. LIDAR bathymetry systems typically use either 488 nm in "blue" water or 532 nm in "green" water. Those wavelengths can be obtained from high-power lasers and give close to optimum water penetration for the respective water types.

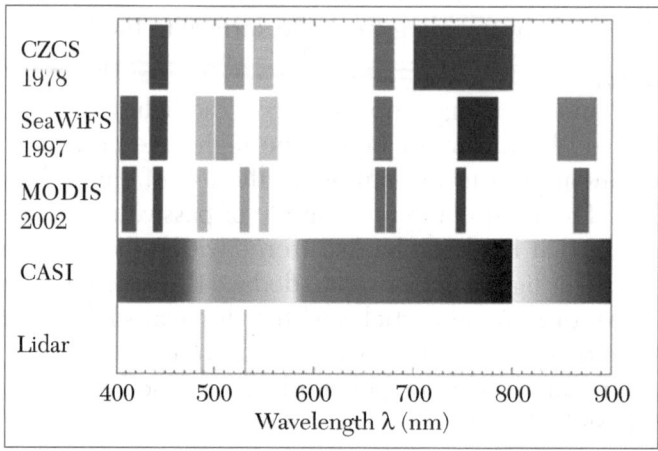

FIGURE 2.10 Wavelength bands used by various ocean color remote sensors.

Although remote sensing usually obtains information for one spatial point at a time, most applications combine measurements from many points to build up an image, i.e., a 2D spatial map of the ocean displaying the desired information at a given time. Imagery acquired at different times then gives temporal information. Satellite systems typically have spatial resolution (the size of one image pixel at the ocean surface) of 250 meters to 1 km. Those systems are useful for regional to global scale studies. Airborne systems can have resolutions as small as 1 meter, as required for applications such as mapping coral reefs.

Passive ocean-color remote sensing is conceptually simple. Sunlight, whose spectral properties are known, enters the water body. The spectral character of the sunlight is then altered, depending on the absorption and scattering properties of the water body, which of course depend on the types and concentrations of the various constituents of the water body. Part of the altered sunlight eventually makes its way back out of the water and is detected by the sensor on board an aircraft or satellite. If we know how different substances alter sunlight, for example by wavelength-dependent absorption, scattering, or fluorescence, then we can deduce from the altered sunlight what substances must have been present in the water and in what concentrations. This process of *working backwards* from the sensor to the ocean is an inverse problem that is fraught with difficulties. Nevertheless, these difficulties can be overcome, and ocean color remote sensing has completely revolutionized our understanding of the oceans at local to global spatial scales and daily to decadal temporal scales.

Ocean color radiometry: Ocean color radiometry is a technology, and a discipline of research, concerning the study of the interaction between the visible electromagnetic radiation coming from the sun and aquatic environments. In general, the term is used in the context of remote-sensing observations, often made from Earth-orbiting satellites. Using sensitive radiometers, one can measure carefully the wide array of colors emerging out of the ocean. These measurements can be used to infer important information such as phytoplankton biomass or concentrations of other living and nonliving material [Figure 2.11] that modify the characteristics of the incoming radiation.

Start time of the remote sensing of color: Remote sensing of ocean color from space began in 1978 with the successful launch of NASA's coastal zone color scanner (CZCS). Ten years passed before other sources of ocean-color data became available with the launch of other sensors, and in particular the sea-viewing wide field-of-view sensor (SeaWiFS) in 1997 on board the NASA SeaStar satellite. Subsequent sensors have included NASA's moderate-resolution imaging spectroradiometer (MODIS) on board the Aqua and Tearra satellites and ESA's medium resolution imaging spectrometer (MERIS) on board its environmental satellite Envisat. Several new ocean-color sensors have recently been launched, including

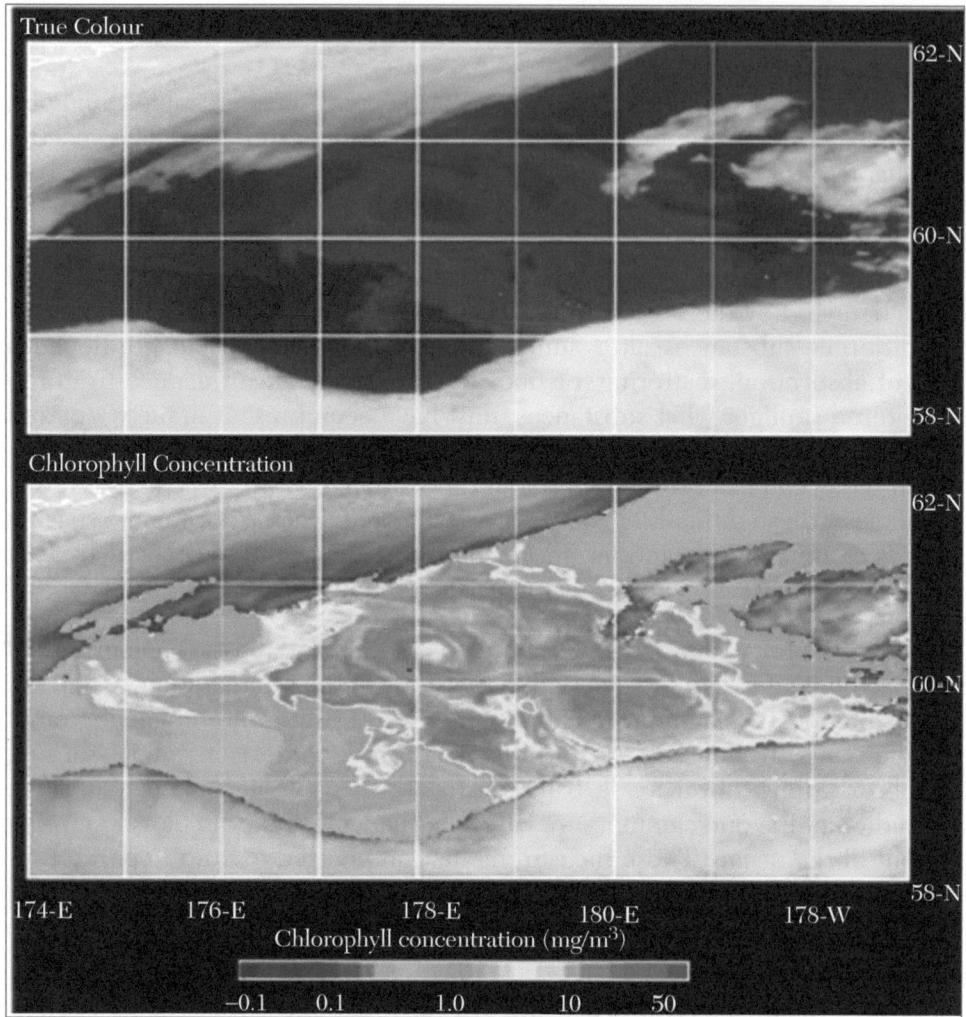

FIGURE 2.11 Phytoplankton bloom in the sea.

the Indian ocean color monitor (OCM-2) on board ISRO's Oceansat-2 satellite, the Korean geostationary ocean color imager (GOCI), which is the first ocean color sensor to be launched on a geostationary satellite, and the visible infrared imager radiometer suite (VIIRS) aboard NASA's Suomi NPP. More ocean color sensors are planned over the next decade by various space agencies. Table 2.4 gives information about the sensing of ocean color sensors launched by various agencies and Table 2.5 gives information about scheduled color sensors.

Sensor/Data Link	Agency	Satellite	Launch date	swath (km)	Spatial resolution (m)	Band	spectral coverage (nm)	Orbit
COCTS CZI	CNSA (China)	HY-1B (China)	11 April 2007	2400 500	1100 250	10 4	402–12,500 433–695	Polar
GOCI	KARI/KIOST (South Korea)	COMS	26 June 2010	2500	500	8	400–865	Geostationary
MERSI	CNSA (China)	FY-3A (China)	27 May 2008	2400	250/1000	20	402–2155	Polar
MERSI	CNSA (China)	FY-3B (China)	5 November 2010	2400	250/1000	20	402–2155	Polar
MERSI	CNSA (China)	FY-3C (China)	23 Sept. 2013	2400	250/1000	20	402–2155	Polar
MODIS-Aqua	NASA (USA)	Aqua (EOS-PM1)	4 May 2002	2330	250/500/1000	36	405–14,385	Polar
MODIS-Terra	NASA (USA)	Terra (EOS-AM1)	18 Dec. 1999	2330	250/500/1000	36	405–14,385	Polar
OCM-2	ISRO (India)	Oceansat-2 (India)	23 Sept. 2009	1420	360/4000	8	400–900	Polar
VIIRS	NOAA (USA)	Suomi NPP	28 Oct. 2011	3000	375/750	22	402–11,800	Polar

TABLE 2.4 Ocean Color Sensors Launched 1999–2013

Sensor	Agency	Satellite	Schedule launch	swath (km)	Spatial resolution (m)	bands	spectral coverage (nm)	orbit
OLCI	ESA/EUMETSAT	Sentinel 3A	2015	1270300/1200		21	400–1020	Polar
COCTSCZI	CNSA(China)	HY-1C/D (China)	2015	2900 1000	1100 250	10 10	402–12500 433–885	Polar
SGLI	JAXA(Japan)	GCOM-C	2016	1150-1400250/1000		19	375–12500	Polar
COCTSCZI	CNSA(China)	HY-1E/F (China)	2017	2900 1000	1100 250	10 4	402–12500 433–885	Polar
HSI	DLR (Germany)	EnMAP	2017	30	30	242	420–2450	Polar
OCM-3	ISRO(India)	OCEANSAT-3	2018	1400	360/1	13	400–1010	Polar
OLCI	ESA/EUMETSAT	Sentinel-3B	2017	1265	260	21	390–1040	Polar
VIIRS	NOAA / NASA(USA)	JPSS-1	2017	3000	370 / 740	22	402–11800	Polar
Multi-spectral Optical Camera	INPE / CONAE	SABIA-MAR	2019	200/2200	200/1100	16	380–11800	Polar
GOCI-II	KARI/KIOST (South Korea)	Geo Kompsat 2B	2019	1200 × 1500	250/1000	13	412–1240	Geostationary
HYSI-VNIR	ISRO (India)	GISAT-1	*(planned)	250	320	60	400–870	Geostationary
OES	NASA	ACE	>2020	TBD	1000	26	350–2135	Polar
Coastal Ocean Color Imaging Spec	NASA	GEO-CAPE	>2022	TBD	250 – 375	155 TBD	340–2160	Geostationary
VSWIR instrument	NASA	HyspIRI	>2022	145	60	10 nm	380–2500	LEO, Sun Sync.

TABLE 2.5 Scheduled Ocean Color Sensors

How satellites "see" ocean color: Satellite instruments or sensors "see" variations in the color of the ocean by detecting different wavelengths or bands of reflected light. The colors are recorded as numerical values that can be downloaded and converted back into images. Scientists can then compare the color in satellite images at specific times and locations to actual measurements of chlorophyll or other suspended matter in samples of ocean water collected at the same time and location. The comparison, called ground truthing, enables scientists to develop mathematical formulas or algorithms for analyzing the same parameters in other satellite images. This technique has enabled scientists to track the timing and spread of offshore phytoplankton blooms around the oceans, which has helped explain the timing and success of year classes—fish in a stock that hatched in the same year.

Satellite imagery and ground truthing: Collecting ground truth data is essential to interpreting what satellite instruments "see" in the ocean. Ground truthing involves comparing pixels on a satellite image with direct observations and measurements on the ground or, in this case, the ocean. This enables scientists to verify what they are seeing in the image and to convert the colors to tangible quantities, such as biomass of plankton or sediment concentration. Satellite and ground truth data must be collected at the same time and location.

It is easier to interpret offshore satellite images because the only thing that changes the color of the ocean, other than the water itself, is the presence of phytoplankton. Interpreting near shore satellite imagery is more complicated because in addition to phytoplankton, other suspended matter including sediment and colored dissolved organic matter (CDOM) from terrestrial matter such as decomposing leaves can change the ocean color and we don't necessarily know which factor has a more significant effect. Since the main source of CDOM is from the land, concentrations vary with river outflow and rainfall. This means that an algorithm for the near shore that is valid for today may not be valid tomorrow and we will probably have to develop separate ones for different seasons, conditions, and locations.

Coastal zone color scanner: The CZCS measured ocean color in four discrete bands with a further low sensitivity band designed for coast and cloud identification. A further band in the infrared operated intermittently. The four color bands are shown in Table 2.6.

Band	Wavelength (nm)	Properties
1	433–453	Blue radiance shows effect of chlorophyll absorption
2	510–530	Blue-green radiances are increased by particles in water. Some chlorophyll absorption
3	540–560	Green radiances increased due to particles in the water
4	660–680	Red radiances due almost entirely to scattering in the atmosphere for case 1 waters

TABLE 2.6 IR Band for Ocean Color Monitoring

To understand the link between the color of the sea and the concentration of suspended matter within its surface layer, models of radiative transfer were constructed from a study of the spectral characteristics of a number of substances. Much of the approach was necessarily empirical. The total radiance observed at the sensor can effectively be divided into two components:

1. *Water-leaving* radiance, which is that part of the signal that has penetrated the sea surface and been reflected back, times the diffuse transmission between the sea surface and the sensor, and

2. Radiance that has not penetrated the sea surface but has been reflected or scattered from other sources into the sensor.

Whereas the effects of the ocean form part of the first, atmospheric effects dominate the second and make up the unwanted noise. The basic task of processing the CZCS record is to identify and remove this noise and then, from the water leaving part of the signal, to make the best estimate of phytoplankton pigment concentrations. Much effort has gone into the development of the best processing *algorithms*. Although the CZCS included reference signals for calibration, it became apparent at a comparatively early stage of the mission that the blue channel was losing sensitivity. The largest effort has been directed to correcting for atmospheric effects which have two main components: radiance resulting from molecular scattering (Raleigh), and scattering due to aerosols (Mie scattering) [also see Figure 10.1].

2.7.2 Sediment Monitoring Methods

Sediment lies at the bottom of the ocean floor; it is made from many items, including:

- Tiny particles of rock, sand, silt, and clay

- Marine snow (clumps of living and dead microscopic organisms, faecal pellets, and dust)

- Materials vented out from the Earth's surface

Over time the sediment forms layer upon layer building up a record of the ocean floor at that moment. Scientists examine these layers to find out about the past. For example, the size of particles shows how close to the shore the ocean floor used to be, because large particles sink faster and therefore nearer to land.

Rock samples are collected because they can tell about:

- What the rocks are made of

- How the rock was formed

- Mineralization—minerals can be an important resource

- The mantle—the layer underneath the Earth's crust

- Earth's evolution—how tectonic plates used to be arranged

- Future geohazards—earthquakes, volcanoes, tsunamis

The goal of sediment monitoring is to quantify changes in sedimentation and soil characteristics that relate the changes in elevation, vegetation, and invertebrates. The rate of sediment accretion or erosion is a determining factor of tidal wetland sustainability with sea level rise and a primary driver of habitat evolution over time. Accretion rates vary spatially due to many factors including elevation, vegetation type and productivity, distance to channels, wave climate, and salinity dynamics. Elevation of the mud flat and the associated inundation frequency and duration are critical to understanding the potential vegetation community that will colonize post-restoration actions. In the context of climate change, long-term adaptability and persistence of tidal marsh depends in part on sediment sources, quantities, and distribution patterns. Basin scale restoration and enhancement projects, including the blocking of large naturally eroding bluffs, are projected to significantly reduce historical sediment delivery rates. In contrast, changes in the flow regime (e.g., dike removal) and in storm frequency and intensity may increase sediment delivery. Multiple methods are used to measure sedimentation in tidal marshes and should be chosen based on restoration and monitoring objectives and site-specific considerations.

The following methods can be used for repeated measures of sedimentation at localized spots over time.

1. Sediment pins

Description: Poles installed within a study site. Height of pole is measured through time to show sediment gain or loss.

Benefits: Inexpensive, easy to install and measure underwater or on land.

Limitations: Sedimentation rates limited to where pin is installed, cm resolution.

2. Sediment plates

Description: In areas of soft sediment, a hard plate is placed below the sediment surface. Measure the sediment accumulation on top of plate.

Benefits: Easy, inexpensive, possible to measure accumulation and erosion, reduces error of rod penetrating into soft sediments, mm resolution, can be used to calculate sediment volume.

Limitations: Plates can be undercut due to hydrologic scour.

3. Marker horizons

Description: A thick marker layer (usually white in color, i.e., feldspar clay) placed on top sediment surface. Sediment cores are later taken to measure sediment accumulation. It can be paired with surface elevation tables (SET) to explain processes behind elevation increases or decreases (i.e., sedimentation, shallow subsidence, etc.).

Benefits: Easy, inexpensive, mm resolution.

Limitations: Repeated measures can deplete marker horizon layer, can be affected by invertebrate bioturbation, does not measure erosion, can be eroded/washed away (typically in unvegetated areas, in this case, use of plastic grid or sediment plate is recommended), can be difficult to measure in areas of standing water (may need to freeze sediment core using liquid nitrogen).

4. Surface elevation table (SET):

Description: Portable mechanical leveling device for measuring relative sediment elevation changes. Is often paired with marker horizon to explain processes behind elevation increases or decreases (i.e., sedimentation, shallow subsidence, etc.).

Benefits: Accurate and precise as measurements are always taken in the exact location, mm resolution.

Limitations: Expensive to install, sedimentation rates limited to where poles are installed, poles can sometimes cause erosion in unvegetated areas.

Mapping of sea floor: Geoscientists are interested in the terrain of an area: Is it flat or rocky? Are there any canyons, mountains, or volcanoes? To find out, scientists need to map the area. On dry land, satellites can be used to measure vast areas, but in water, satellite signals (microwaves) can be absorbed. The further a signal travels through water, the less likely it is to bounce back, so another method is needed to map the ocean floor. Sound waves use pressure to move through gases, liquids, and solids. In air, sound moves at around 340 meters per second but in seawater it zooms along at around 1,500 meters per second. Light cannot be used as it is absorbed by water very quickly; usually lighting no further than 30 meters. So by using sound, scientists can find out different properties about the sea floor, as shown in Table 2.7.

Method used	Property discovered	How it's done
Single-beam echo-sounder	Bathymetry—the measurement of depth to the bottom. Can also be used to gain information about the subsurface, i.e., deep or shallow sediment.	Measured using the time it takes for the sound to be sent and returned.
Multi-beam echo-sounder	Swath bathymetry—by taking lots of depth measurements from a single place the shape of the sea bed is revealed.	Measured using the time it takes for the sound to be sent and returned.
Backscatter	Measures the reflectivity of the floor—this will show what the floor is made of, e.g., rocks, sand, mud.	The change to the strength of the returned sound wave.
Sound Velocity Profilers	Measures the speed sound moves through the water, as it varies depending on the water properties at different depths. Important to know otherwise depth calculations will be wrong.	Measures the temperature, and sometimes the salinity, of the water to determine the speed.

TABLE 2.7 Properties of Sea Floor for Different Methods

2.7.3 Surface Currents

The ocean is maintained in a state of continual motion through a combination of solar radiation and the earth's rotation. Currents are produced by the equilibrium established between a number of forces which include gravity, pressure gradients within the volume of the sea, the Coriolis force due to the Earth's rotation, and frictional forces mostly due to wind. The pressure gradients within the water column are produced by differences in density at fixed depths brought about by different values of temperature and salinity. Because the earth rotates from west to east, an observer in the Northern Hemisphere would observe that a moving object was deflected to the right (that is, in a clockwise direction). In the Southern Hemisphere the deflection is to the left, while at the equator there is no deflection. The strength of the horizontal component of the Coriolis force depends on the latitude and on the speed of a moving object such as a water particle. If there were no Coriolis force acting (that is, if the earth were not in rotation) the pressure gradient would cause the water to move directly from high to low pressure. On a rotating Earth, the Coriolis force deflects the motion; the acceleration on the water will reduce to zero when the speed of the current at given latitude is fast enough to produce a Coriolis force in exact balance with the horizontal pressure gradient. In the absence of frictional forces this balance is referred to as geostrophic flow. From the simple geostrophic relationship linking current speed and latitude to the Earth's rate of rotation and to the acceleration due to gravity, the surface slope is easily calculated. Surface slopes produced by currents are measured against a reference equipotential mean sea level referred to as the geoid. Over the surface of the Earth, variations in the gravity field cause the geoid itself to undulate with respect to a reference ellipsoid by about 200 meters. (The minimum is found in the Indian Ocean just south of Sri Lanka while the maximum is relatively close in the Eastern Pacific off Borneo).

A method that has been used to compute patterns of sea level changes is to calculate a mean geoid over a selected area from (say) a year's altimeter's observations, subtract this out—making corrections to the calculated height of the altimeter by minimizing the differences at track intersections—and record the differences [Figure 2.12]. At middle and high latitudes, the cross-track difference between the Geosat ground tracks was approximately 100 km, making the altimeter observations useful for mapping the tracks of mesoscale eddies. In areas of extremely high sea level variability, the Geosat observations were used to compute a frequency-wave number spectrum of sea level variability which identified significant eddy energy at time scales longer than 34 days and spatial scales longer than 200 km.

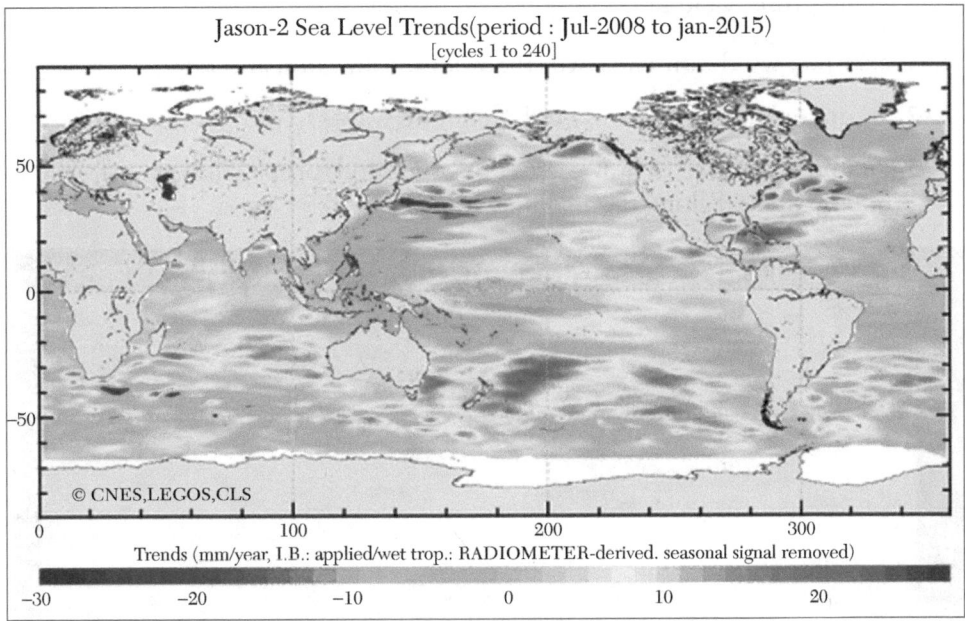

FIGURE 2.12 Sea level.

2.7.4 Surface Wind and Waves

Ever since sailors first put to sea, efforts have been made to improve the forecasts of the conditions they may find there. Sudden storms still take their toll in terms of ships and men lost at sea. Each month sees two ships disappear without trace off the face of the globe. There are strong arguments therefore for improving the collection of reliable measurements of wind and wave conditions over the oceans. One obviously successful area of international cooperation is the meteorological global network in which nations share their observations with each other. Routine measurements of surface pressure, humidity, precipitation, wind velocity, temperature, and other parameters form the basis of the weather charts issued up to four times daily around the globe. Ships play their own part in reporting conditions at sea. It remains true, however, that their observations are unevenly distributed over large tracts of open ocean, especially in the Southern Hemisphere where comparatively few are reported. It is clear that satellites will play a primary role in helping to achieve its ultimate potential. In general terms ocean waves are the result of winds blowing over the surface for a certain time (duration) and over a certain area (fetch).

The primary elements of good forecasting are:

1. accurate wind estimates over the relevant duration and fetch

2. an understanding of how winds generate waves

3. an understanding of how waves are generated, propagated, transformed, and dissipated along their route

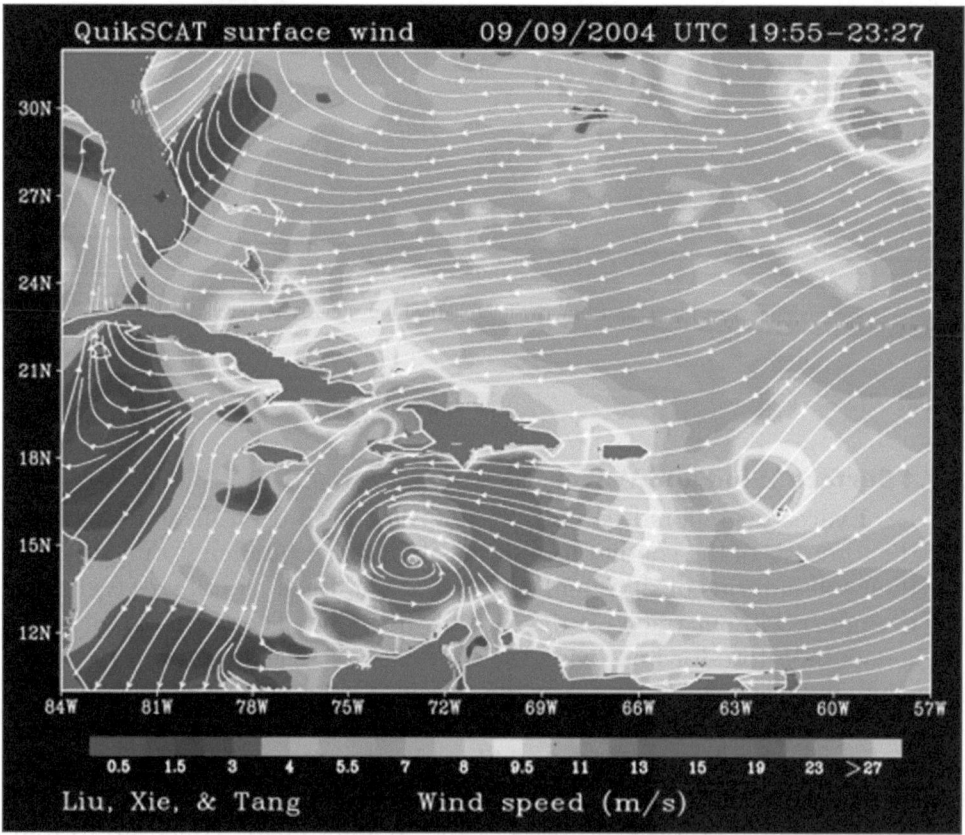

FIGURE 2.13 Surface winds during Hurricane Ivan estimated from QuikScat scatterometer measurements.

2.7.5 Wave Height and Wind Speed

The largest forces on any offshore or coastal structure, from ships to offshore rigs to coastal defenses, generally result from surface waves, which can cause destruction and devastation, in association with forces from winds, currents, and sea level surges. Ships—even 100,000-ton carriers—routinely disappear in storms; offshore structures have been severely damaged and millions of dollars of damage have been inflicted upon breakwaters in recent years. Knowledge of ocean waves is essential for any activity connected with the seas. Forecasts of wave conditions are required for operational planning, both at specified locations and across oceans for ship routing. Estimates of wave climate, such as monthly average wave height or 50-year wave height, are needed for design purposes. Forecasts are prepared using numerical wave models with forecast winds as input and using the physics of wave growth, transmission, and decay developed in recent years, but forecasts are considerably improved if the models are initiated and updated with observations of wave height and period.

2.7.6 Sea Surface Temperature

The ocean-atmosphere system is a heat engine powered by solar radiation. The average daily amount of incoming radiation decreases from the equator to the poles. Low latitudes receive relatively large amounts of radiation each year while winter darkness and the obliqueness of the sun's rays reduce the amount of radiation received at the higher latitudes. The Earth also reemits radiation from the sun at slightly longer wavelengths, most of which is absorbed by natural greenhouse gases in the atmosphere such as carbon dioxide, water vapor, and cloud droplets. There is a net gain of radiation energy at low latitudes and a net loss at higher latitudes. But since there is no net gain of heat at low latitudes (or loss at higher latitudes) there must be a net transfer of heat to maintain a balance. This is brought about mainly through winds in the atmosphere and currents in the oceans. In the tropics, it is the oceans which contribute more to the poleward transfer of heat while the atmosphere contributes more at the higher latitudes. Across the equator, the oceans provide the mechanism for a southward net transport from the Northern to the Southern Hemisphere. The two principal components of the ocean contribution are wind-driven surface currents and density-driven (thermohaline) deep circulation. Of the total amount of energy received from the sun by the oceans, about 41% is lost to the atmosphere as long-wave radiation and about 54% as latent heat by evaporation from the sea surface. Temperature is a measure of the thermal energy possessed by the oceans and if the average temperature is to remain constant then the gains and losses must even out—that is, the heat budget must balance.

Sea surface temperature (SST) is observed at low frequencies generally in the range of 5–10 GHz. The great advantage of microwave sensors for measuring SST over the more commonly used infrared instruments is their ability to operate through cloud but this must be offset against their resolution of around 150 km, which is too coarse to study mesoscale eddies. Higher resolution would require a much larger antenna than has been flown up to now. Another constraint is contamination by land masses, and in general reliable measurements must be made in the open ocean more than 600 km from a coast. Thus, again, interesting ocean features such as boundary current and their associated eddy may not be capable of being studied with the microwave radiometer.

Measurements of ocean parameters by microwave radiometers are affected by atmospheric water vapor, clouds, and rainfall and most sensors are therefore backed up by frequencies sensitive to water in the atmosphere. This information is also of considerable importance where precise altimetry is required to measure very small changes in ocean topography. It is the role of the ocean in redistributing heat by currents and mixing which ultimately will be one of the key factors in determining whether or not there will be a net warming over the globe due to a manmade increase in greenhouse gases. Attempts to measure the oceanic heat transport directly by traditional observations have been less than successful in providing credible estimates.

This is particularly true in the Southern Hemisphere, where observations are so sparse that neither the magnitude nor even the direction of heat transport estimated directly agree with those inferred from the apparent global radiation budget. Patterns of surface temperature revealed by satellite have been used to study the flow of currents by following certain distinguishable features, shown in Figure 2.14.

In a similar way, sea surface temperature has been used to infer other variables in the upper ocean. Upwelling around the coasts of many tropical and subtropical countries carries a strong temperature signal which can be identified and studied from space-borne infrared devices. By their very nature many such areas also exhibit a strong increase in phytoplankton, as revealed in the record of the coastal zone color scanner. Thus, in some areas sea surface temperature may be used as a useful tracer of nutrient concentrations. Sea surface temperature (SST) is an important geophysical parameter, providing the boundary condition used in the estimation of heat

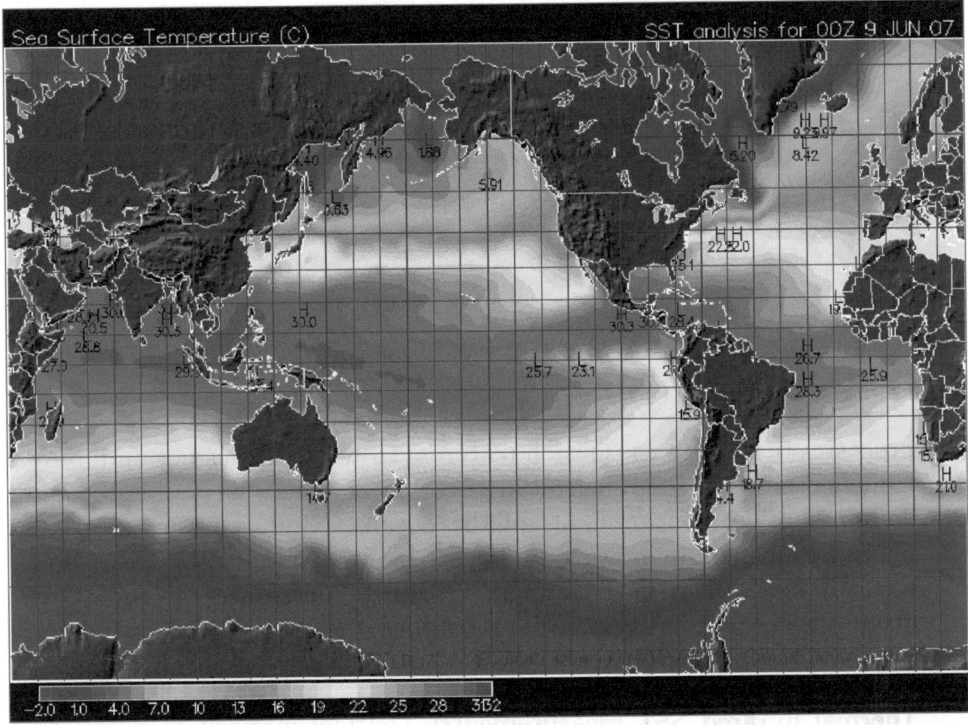

FIGURE 2.14 Sea surface temperature.

flux at the air-sea interface. On the global scale this is important for climate modeling, study of the earth's heat balance, and insight into atmospheric and oceanic circulation patterns. On a more local scale, SST can be used operationally to assess eddies, fronts, and upwellings for marine navigation and to track biological productivity. Satellite technology has improved upon our ability to measure SST by allowing frequent and global coverage. In the past, SST could only be measured by ships and buoys, whose ranges were limited. Figure 2.15 shows two maps illustrating this point.

Methods for determining SST from satellite remote sensing include thermal infrared and passive microwave radiometry. Interest in using satellites to measure ocean phenomena began in the 1960s. In 1978, the polar-orbiting TIROS satellites began to gather data on sea surface temperatures using the AVHRR and microwave sensors. The maps of sea surface temperatures produced from these data demonstrate complex surface temperature patterns that have led to considerable speculation about the physical

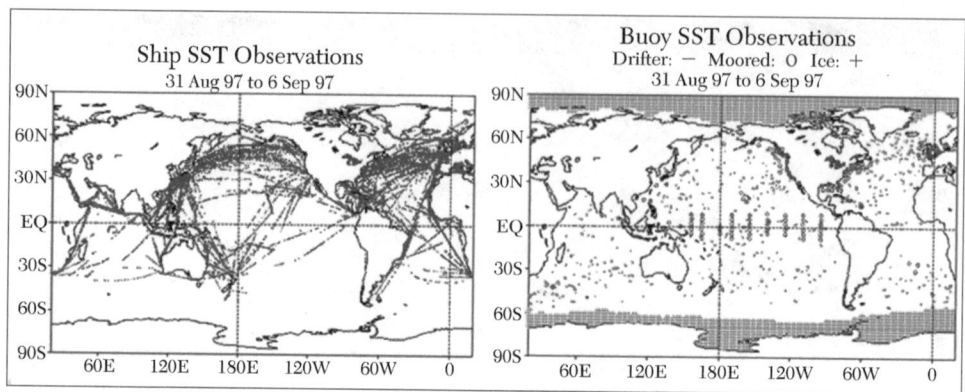

FIGURE 2.15 SST measurement by ship and buoy.

processes that might cause such patterns. However, it was not until NASA launched Nimbus 7 and SeaSat in 1978 that scientists were able to gather comprehensive measurements of the oceans. Nimbus-7 carried a scanning multichannel microwave radiometer (SMMR) that provided accurate measurements of sea surface temperatures. By measuring the color of the ocean surface, its coastal zone color scanner (CZCS) provided estimates of ocean biological productivity.

Thermal infrared SST measurements: Thermal infrared SST measurements have a long heritage (~20 years). They are derived from radiometric observations at wavelengths of ~3.7 μm and/or near 10 μm. Though the 3.7 μm channel is more sensitive to SST, it is primarily used only for nighttime measurements because of relatively strong reflection of solar irradiation in this wavelength region, which contaminates the retrieved radiation. Both bands are sensitive to the presence of clouds and scattering by aerosols and atmospheric water vapor. For this reason, thermal infrared measurements of SST first require atmospheric correction of the retrieved signal and can only be made for cloud-free pixels. Thus, maps of SST compiled from thermal infrared measurements are often weekly or monthly composites which allow enough time to capture cloud-free pixels over a region. Thermal infrared instruments that have been used for deriving SST include advanced very high resolution radiometer (AVHRR) on NOAA polar-orbiting operational environmental satellites (POES), along-track scanning radiometer (ATSR) aboard the European remote sensing satellite (ERS-2), the geostationary operational environmental satellite (GOES) imager, and moderate resolution imaging spectroradiometer (MODIS) aboard NASA earth observing system (EOS) Terra and Aqua satellites.

Strengths: Good resolution and accuracy and long heritage (~ 20 years)

Weaknesses: Obscured by clouds and atmospheric corrections required

Passive microwave SST measurements: Due to lower signal strength of the Earth's Planck radiation curve in the microwave region, accuracy and resolution is poorer for SST derived from passive microwave measurements compared to SST derived from thermal infrared measurements. However, the advantage gained with passive microwave is that radiation at these longer wavelengths is largely unaffected by clouds and generally easier to correct for atmospheric effects. This is well illustrated in the two SST images in Figure 2.16. Though the two images cover the same period, the thermal infrared composite (AVHRR) has lots of white patches where cloud-free pixels could not be obtained over such a short period of time.

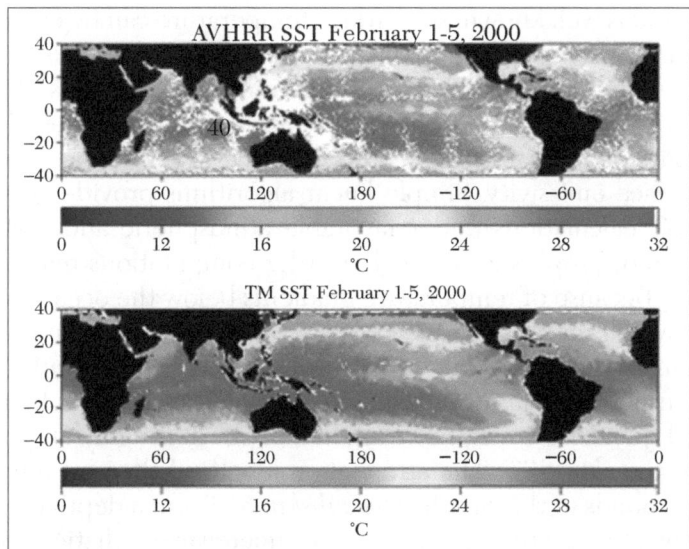

FIGURE 2.16 Comparison between thermal infrared and passive microwave.

Phenomena which do effect passive microwave signal return, however, are wind-generated roughness at the ocean's surface and precipitation. These can usually be corrected for, however, using multiple frequencies. SST measurements are primarily made at a channel near 7 GHz with a water vapor correction enabled by observation at 21 GHz. Other frequencies used for correction of surface roughness (including foam), precipitation, and what little effect clouds do have on microwave radiation are 11, 18, and 37 GHz. Passive microwave instruments that have been used for deriving SST include the scanning multichannel microwave radiometer (SMMR) carried on Nimbus-7 and SeaSat satellites, the tropical rainfall measuring mission (TRMM) microwave imager (TMI), and upcoming data from the advanced microwave scanning radiometer (AMSR) instrument on the NASA EOS Aqua satellite and on the Japanese advanced earth observing satellite (ADEOS II).

Strengths: Clouds are mostly transparent and relatively insensitive to atmospheric effects

Weaknesses: Poorer accuracy and resolution; sensitive to surface roughness and precipitation

Derivation of SST: Radiation emitted by a surface is the Planck emission times the surface emissivity. Since the Planck function is dependent on temperature and is well known, sea surface temperature can be estimated if the surface emissivity can be sufficiently estimated using models or regression techniques that employ independent in situ measurements. After atmospheric corrections, then, coefficients are applied to the retrieved brightness temperature signals in the derivation of SST which factor in estimations of the surface emissivity. Simple linear algorithms provide reasonably accurate SST calculations under favorable atmospheric and surface conditions, but more sophisticated higher-order computations may be required otherwise. Because of temperature gradients below the ocean's surface, the depth at which measurements are made will significantly impact the SST. Measurements made at only a depth of one or two molecules below the ocean's surface are considered the *interface SST* and cannot be realistically measured. Just below this, however, at a depth of roughly 10 µm is what is known as the *skin SST*. The attenuation length of thermal infrared radiation corresponds to this depth. The *subskin SST* is at a depth of ~1 mm and corresponds to the attenuation length of microwave radiation. Beyond this depth is what is commonly referred to as the *bulk SST, near-surface SST*, or SST_{depth}. Figure 2.17 is an illustration of these different depths of SST, showing two different temperature gradients:

As can be discerned from Figure 2.17, the bulk SST (or SST_{depth}) may vary greatly from the skin and subskin SSTs depending on the temperature gradient. The skin temperature may also vary from the subskin temperature for the same reason. Diurnal heating will cause these differences to be greatest during the afternoon and least right before dawn. Since SST

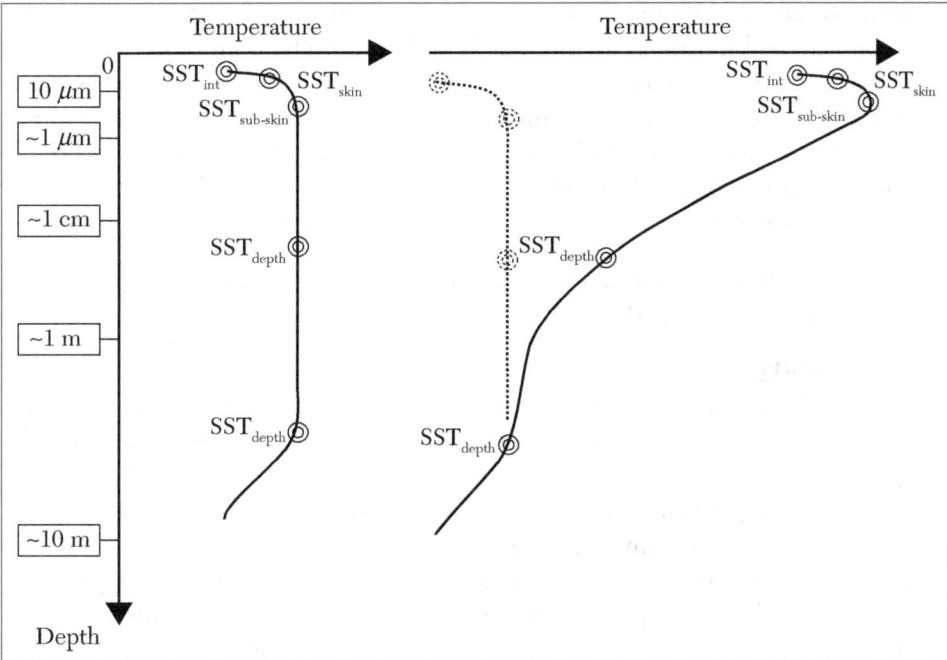

FIGURE 2.17 Different depth SST with different temperature gradients.

measurements made from buoys and ships are usually bulk temperature measurements, temperature gradients must be taken into consideration when comparing them to SST measurements made by either thermal infrared or passive microwave remote sensing observations.

Since thermal infrared instruments measure the skin temperature and passive microwave instruments measure the subskin temperature, one must also consider differences due to evaporative cooling at the sea surface when comparing measurements derived from these methods. The difference can be as great as 1 Kelvin in combination with diurnal heating effects, and so both properties must be properly accounted for when comparing or blending thermal infrared and microwave products.

Weakness: Diurnal heating and evaporative cooling make comparison of SSTs at different depths difficult; special care must be taken to correct for their effects

Blending thermal infrared and passive microwave SST: Given the desire to combine the high accuracy and resolution of the thermal infrared SST measurements with the better temporal and spatial coverage of passive microwave SST measurements (due to cloud transparency), efforts are being made to create a blended product which combines these strengths. In the effort to combine these two kinds of SST products, careful consideration must be made to correct for differences due to diurnal heating and evaporative cooling as well as biases introduced by high wind speeds, water vapor, and other atmospheric conditions. Models are being tested for each of these considerations. Algorithms which incorporate these models still use in situ measurements, as well, to quality assure and to adjust the final product.

Strength: Helps scientists better model climate change with improved SST product

2.7.7 Upwelling

Winds blowing across the ocean surface often push water away from an area. When this occurs, water rises from beneath the surface to replace the diverging surface water. This process is known as *upwelling*. Figure 2.18 highlights major upwelling areas along the world's coasts. These subsurface waters are typically colder, rich in nutrients, and biologically productive. Therefore, good fishing grounds typically are found where upwelling is common. For example, the rich fishing grounds along the west coasts of Africa and South America are supported by year-round coastal upwelling.

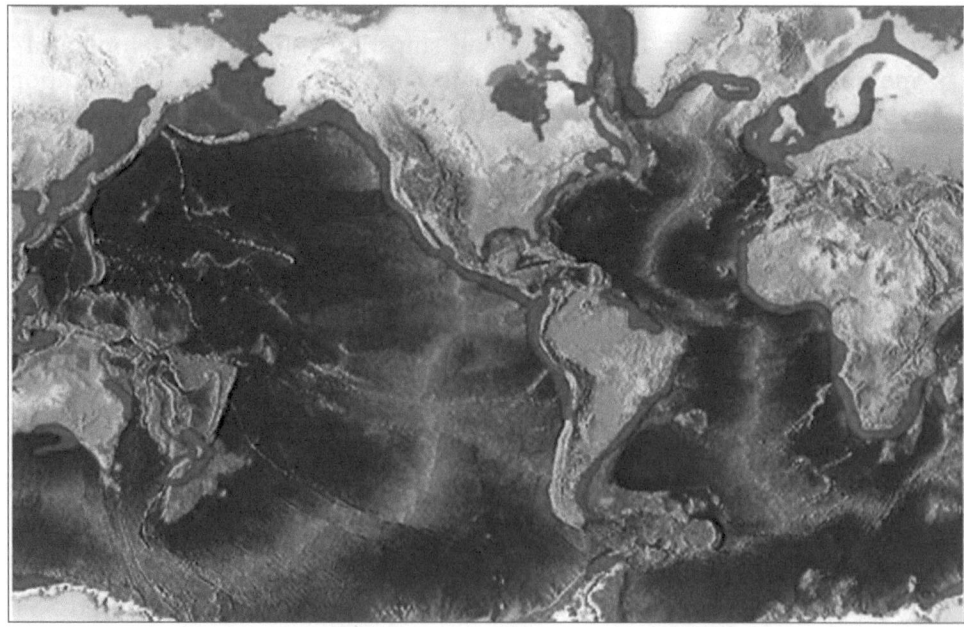

FIGURE 2.18 Upwelling.

2.7.8 Sampling

Properties like satellite speed, satellite ground speed, satellite period, number of orbits/day, separation of ascending track on equator, track separation, if evenly distributed for H months, and satellite precession rate determine the sampling and sampling rates. Table 2.8 describes the sampling scheme for the measurements type and notes differences between the systems.

2.7. 9 Wave Power

Wave power is the transport of energy by ocean surface waves, and the capture of that energy to do useful work. For example, electricity generation, water desalination, or the pumping of water (into reservoirs) is done using wave power. A machine able to exploit wave power is generally known as a wave energy converter (WEC).

2.7.10 Ocean Floor

The bottom of a body of water is known as the benthic zone, regardless of how deep it occurs. In coastal waters the sea floor sits upon the continental shelf and is generally less than 200m deep. Most of the ocean floor lies upon the ocean crust and is between 4,000—6,000

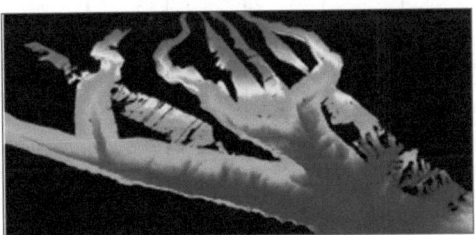

FIGURE 2.19 The ocean floor was mapped using an EM120 echo-sounder.

meters deep. Working at such depths requires specialist tools and skills to obtain the data needed [Figure 2.19].

The reasons for study the benthic zone are just as varied but can include:

- Understanding—exploration of areas, learning about ecosystems and habitats

- Resources—locations and quantities of minerals, oil, gases, and food supplies

- Geohazards—landslides such as the one that caused the 2004 Tsunami, turbidity currents. It is not safe to dump waste at sea.

Ocean sensing and the ice caps: Because the oceans cover about 70% of Earth's surface, they make a significant contribution to Earth's weather and

Measurement	Data System	Sample rate	Sample period	Sample time	Stored data interval	Real-time Transmitted data
Wind – UV components, scalar speed, direction Air temperature	ATLAS	2-hz	2 min	2359–0001, 0009–0011...	10 min	Daily mean and 2-min mean on the hour, 16 x per day
Relative humidity	FLEX	2-hz	2 min	2359–0001, 0009–0011...	10 min	Hourly mean and standard deviation - 24 x per day
Barometric Pressure	FLEX	2-hz	2 min	2359–0001, 0009–0011...	10 min	Hourly mean and standard deviation—24 x per day
Shortwave radiation	ATLAS	1-hz	2 min	2359–0001, 0001–0003...	2 min	Daily mean and standard deviation
	FLEX	1-hz	1 min	2359–0000, 0000–0001...	1 min	Hourly mean and standard deviation—24 x per day
Long wave radiation (thermopile, case and dome temperatures)	ATLAS	1- hz	2 min	2359–0001, 0001–0003...	2 min	Daily mean
	FLEX	1-hz	1 min	2359–0000, 0000–0001...	1 min	Hourly mean and standard deviation—24 x per day
Rain rate	ATLAS	1-hz	1-hz	0000–0001, 0001–0002...	1 min	Daily mean, standard deviation, and percent time raining
	FLEX	1-hz	1-hz	0000–0001, 0001–0002.	1 min	Hourly mean and percent time raining, 24 x per day
Temperature & Conductivity (sea surface and subsurface)	ATLAS	1 per 10 min	instantaneous	0000, 0010...	10 min	Daily mean
	FLEX	1 per 10 min	instantaneous	0000, 0010.	10 min	Hourly reading—24 x per day
Currents	ATLAS	1 per 20 min	instantaneous	0000, 0010...	20 min	Daily mean
	FLEX	1 per 20 min	instantaneous	0000, 0020.	20 min	Hourly reading —24 x per day
Air and sea surface water pCO_2, O_2, pH, turbidity, gas tension	MAP-CO2	2-hz	30 sec	0000, 0300...	3 hours	8 data points/day

TABLE 2.8 Sampling Time

climate. The oceans interact constantly with the atmosphere above them and the land and ice that bound them. Yet scientists know far too little about the details of the oceans effects on weather and climate, in part because the oceans are monitored only coarsely by ships and buoys. Improving the safety of people at sea and managing the seas vast natural resources also depend on receiving better and more timely data on ocean phenomena. Satellite remote sensing is one of the principal means of gathering data about the oceans.

2.8 RESEARCH ON OCEAN PHENOMENA

To understand the behavior of the oceans and to make more accurate predictions of their future behavior, scientists need to gather data about sea temperature, surface color, wave height, the distribution of wave patterns, surface winds, surface topography, and currents. Fluctuations in ocean temperatures and currents lead to fluctuations in the atmosphere and therefore play a major part in determining weather and climate. Understanding and predicting the interactions are major goals of climatologists. The study of other ocean phenomena would enhance scientists understanding of the structure and dynamics of the ocean. For example, observations of wave conditions are important for modeling ocean dynamics. Because winds create waves, measurements of wind speed and direction over wide areas can lead to estimates of wave height and condition. Closely observing the color of the ocean surface provides a powerful means of determining ocean productivity. Variations in ocean color are determined primarily by variations in the concentrations of algae and phytoplankton, which are the basis of the marine food chain. Because these microscopic plants absorb blue and red light more readily than green light, regions of high phytoplankton concentration appear greener than those with low concentration. Because fish feed on the photoplankton, regions of high concentration indicate the possibility of greater fish population.

EXERCISES: PART A (ANSWER IN A WORD OR A SENTENCE)

1. What is passive remote sensing?

2. Give examples for active remote sensing.

3. Define "satellite orbit."

4. What are the geostationary satellites?

5. What is the use of having remote sensing satellites in near polar orbits?

6. Define "swath."

7. What is the other name of whiskbroom scanner?

8. Give the definition of instantaneous field of view (IFOV).

9. _____ describes the ability of the image to discriminate very slight differences in energy.

10. Define "spectral range."

11. What is spectral bandwidth?

12. What is spectral sampling?

13. What is the importance of signal-to-noise ratio?

14. What is the significance of ocean color?

15. Define "ocean color radiometry."

16. What is upwelling?

17. Define "wave power."

18. What is the benthic zone?

EXERCISES: PART B (ANSWER IN A PAGE)

1. Write a note on remote sensor platforms.

2. Differentiate between passive and active remote sensing.

3. Write about scanner sensor systems.

4. Compare the whiskbroom scanner with the push broom scanner.

5. Write a note on temporal resolution.

6. What factors determine the selection and design of sensors?

7. Write a short note on synthetic aperture radar sensors.

8. How do you measure the surface current?

9. Write a note on surface wind and waves.

10. Write about ocean color sensors.

11. How do satellites see the ocean color?

12. Write about the different methods to measure sedimentation.

EXERCISES: PART C (ANSWER IN 2 OR 3 PAGES)

1. Write in detail about weather satellites.

2. Explain MOS satellites.

3. Write about micro-sensors for ocean monitoring.

4. How do you measure ocean color using sensors?

5. Write in detail about sediment monitoring methods.

6. What are different methods to measure the sea surface temperature?

WEB LINKS

http://modis.gsfc.nasa.gov

http://www.planetdiary.com

https://cimss.ssec.wisc.edu/sage/oceanography

https://stevengoddard.wordpress.com

http://weather.unisys.com/

CHAPTER 3

UNDERWATER ACOUSTICS

This chapter discusses underwater acoustics, including the interaction of sound with the seafloor, sound wave features, transmission of data underwater, wave height, wave velocities, bubbles study, water depth, sea temperature, global climate change, ocean current measurement using sound, fish finding, study of Earth history, and surf zone measurement using sound. It describes locating and identifying fish, methods of underwater communication, measurement of ocean temperature using acoustic tomography, inverted echo-sounders, acoustic Doppler current profilers, RAFOS floats, and reciprocal transmission.

3.1 INTERACTION OF SOUND WITH SEAFLOOR

The seafloor is a reflecting and scattering boundary. It is often layered with a density and sound speed that may change gradually or abruptly with depth or even over short ranges. The seafloor is highly variable in its acoustic properties since its composition may vary from hard rock to soft mud. Because of the variable stratification of the bottom sediments in many areas, sound is often transmitted into the bottom where it is refracted and internally reflected further down. Further, the small-scale roughness of the seafloor causes scattering and attenuation of sound. Thus the ocean bottom is a complicated propagating medium.

A typical bottom-structure section is presented in Figure 3.1. This section consists of 5 km of water, about 0.5 km of unconsolidated sediments,

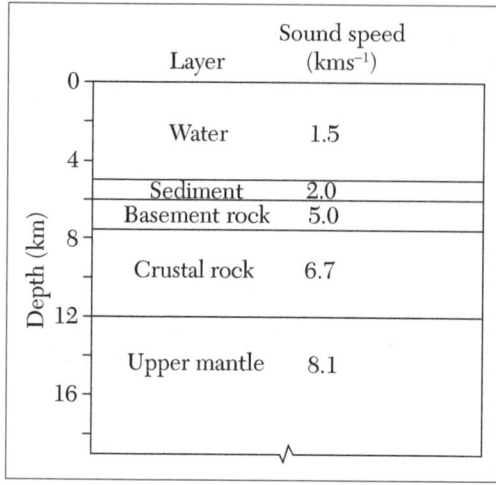

FIGURE 3.1 Bottom structures of sea and sound speed.

1–2 km of basement rock and 4–6 km of crustal rock overlying the upper mantle. Also, typical sound speed values are indicated for each layer.

The crustal rock, or oceanic crust, consists largely of basalt, and is usually covered by sedimentary rock (denoted "basement rock" in Figure 3.1). The continental crust largely consists of granite and is much thicker than the oceanic crust. It is covered by a wider range of sediments, but usually coarser than those in the deep ocean.

3.1.1 Underwater Acoustics

Underwater acoustics is the science of utilizing sound waves underwater as a method of navigating, communicating, or detecting. This technology is also used to determine profiles of the earth's layer immediately below the ocean. Underwater sound waves can be generated in the frequency range from as low as 1 KHz (one thousand cycles or waves per second) to as high as 500 KHz (five hundred thousand cycles per second). Most underwater acoustic equipment operates in the frequency range of 10 KHz to 100 KHz. As a rule, the lower the frequency of an underwater acoustic signal, the farther it will travel through water. For example, a 12 KHz acoustic signal will travel farther than a 50 KHz signal if both are transmitted with the same amount of power. Another determining factor in how far a sound wave will travel is the power with which it is driven. As a rule, the higher the power of the acoustic signal, the greater the distance it will travel through the water. The strength of an underwater acoustic signal is measured in "watts" or decibels, also referred to as dB.

The sound wave or acoustic signal is created by a *transducer*, which is a technical term for an underwater antenna. The transducer converts electrical energy into mechanical energy (vibrations) which in turn create the sound waves. A transducer, depending upon the application, can transmit as well as receive. The heart of the transducer is a piezo-electric ceramic element, usually in the shape of a tube, which is then encapsulated in polyurethane. Applying an electric signal (voltage) to this piezo-electric ceramic element will cause it to alternately contract and expand thereby creating a pressure or sound wave. Thus a sound wave or acoustic signal is transmitted through the water.

The size of the transducer is determined by the frequency that is to be transmitted. The lower the frequency of the signal to be generated, the bigger the diameter of the transducer. This is because the piezo-ceramic tube will be most efficient when operating at its natural resonant frequency and the larger the diameter of the tube, the lower its resonant frequency will be. For example, a 12 KHz transducer may have a typical diameter of 4 in (10 cm) whereas a 50 KHz transducer will have a diameter of less than 1 in (2.5 cm). An analogy would be if you bang on a large drum, you will get a low tone or low frequency note; whereas, if you bang on a small drum you will get a high tone or high frequency note. When a transducer is being used as a receiving antenna, it is generally referred to as a *hydrophone*. It receives or detects an acoustic signal and then converts this signal from mechanical energy into electrical energy which is then processed by electronics.

Most hydrophones receive in an omnidirectional pattern, which means they can detect signals from any direction. These types of hydrophones are called listening hydrophones. They are commonly used to study biological noise as a reference standard for testing, for detection of noise made by vehicles such as submarines, or for underwater acoustic survey work. When it is necessary to determine the direction or the source of an underwater acoustic signal, a specially designed hydrophone is utilized.

3.1.2 The Features of Sound Waves

- A sound wave is created by converting electrical energy into mechanical energy via a transducer.

- Sound in water travels about five times as fast as it travels in air.

- Sound waves travel in direct paths based on the "line-of-sight" concept.

- Sound waves will be reflected or distorted when traveling through a thermocline.

- Sound waves will be affected by biological or man-made noise, air bubbles, and salt.

- Most underwater acoustic equipment operates in the frequency range of 10–100 KHz

- The lower the frequency of a sound wave, the farther it will travel.

- The higher the acoustic power output of a sound wave, the greater the distance it will travel.

3.2 TRANSMISSION OF DATA UNDERWATER USING SOUND

Computers use digital data to transmit and receive information, including e-mail messages and Internet web pages. Submarines do not have telephone or cable connections, and radio signals do not propagate underwater, so a submarine relies on sound to send and receive digital data. Telephone modems allow computers to transmit and receive information over telephone lines; however, they do not work very well with poor connections. Special acoustic modems that can successfully transmit digital data underwater have been developed. These modems convert digital data into special underwater sound signals that can be transmitted between two submerged submarines or between a submerged submarine and a surface ship, as in Figure 3.2. These digital signals can represent words and pictures, just as on land, allowing submarines to send and receive e-mail. Underwater acoustic modems are relatively slow compared to telephone or cable modems on land. Nonetheless, this technology is very important because it provides an accurate and efficient means to send and receive data underwater.

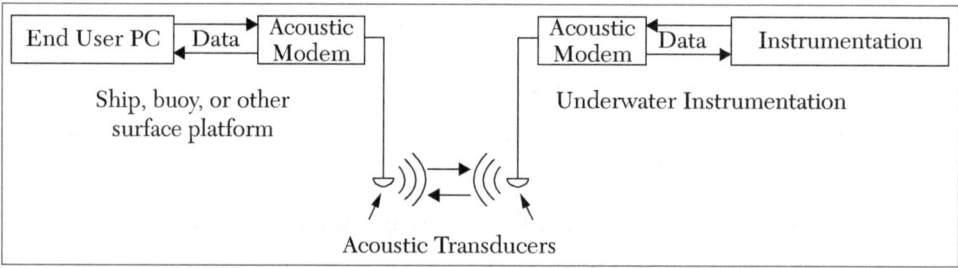

FIGURE 3.2 Basic acoustic communication model.

Oceanographers use acoustics to control underwater instruments and acquire the data that they collect remotely. This technology can also be used to control small, unmanned submarines, called *autonomous undersea vehicles* (AUVs), and get data back from them in real time. These vehicles are currently under intensive development and are beginning to be widely used for oceanographic research and other purposes.

Autonomous vehicles working under the ice can be controlled and their data can be transmitted to a topside station using underwater acoustic links, as in Figure 3.3.

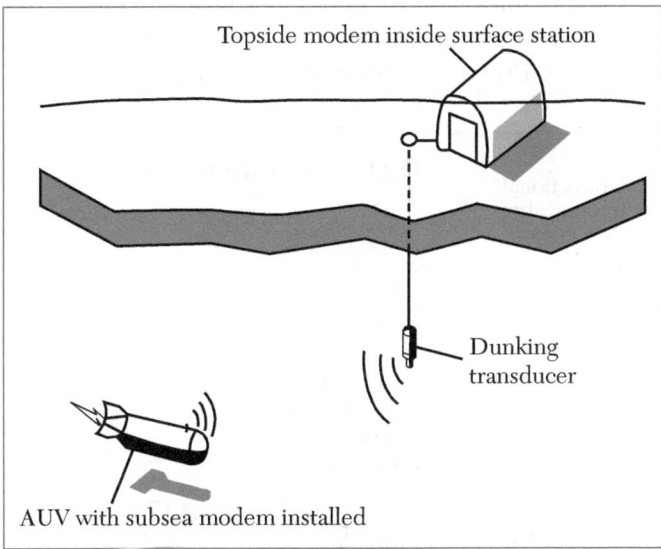

FIGURE 3.3 AUV control using sound.

Acoustic links are used to control underwater instruments and acquire the data remotely, as in Figure 3.4. Underwater data links can also be

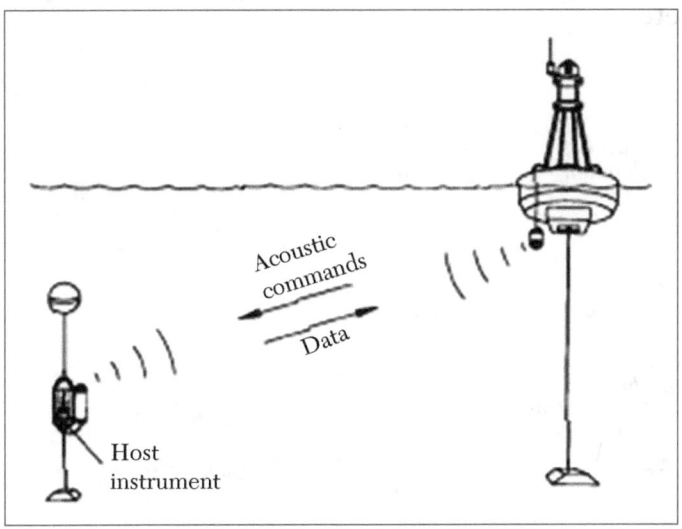

FIGURE 3.4 Underwater instrument control using sound.

combined with satellite data links to provide data in real time from instruments on the seafloor to scientists ashore. One application of this technique is to provide early warnings of tsunamis generated by undersea earthquakes. Tsunami waves are generated when an earthquake causes the seafloor to move. They can cause great damage when the waves build as they come ashore. Pressure sensors that are deployed on the seafloor can

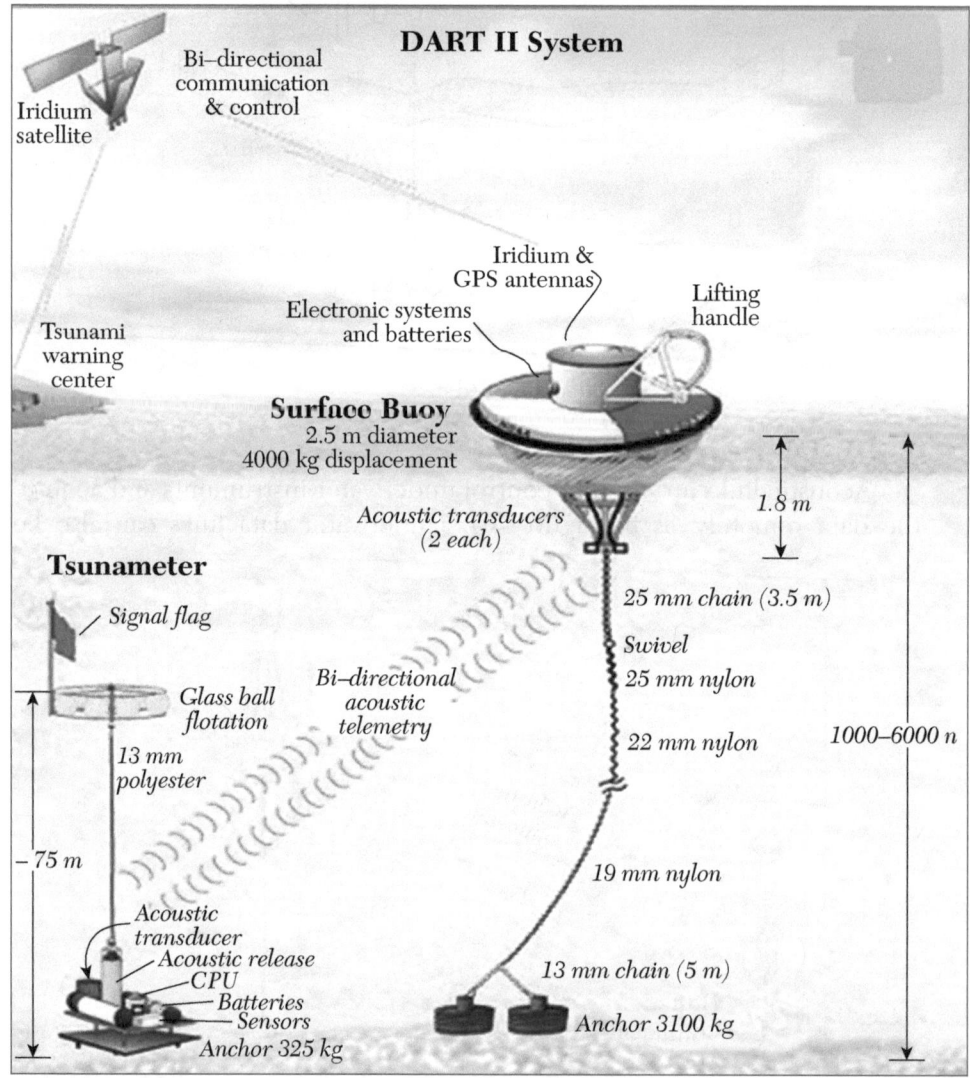

FIGURE 3.5 DART to early warning of tsunamis.

detect tsunamis. Pressure data are transmitted to a nearby surface buoy via an acoustic data link using underwater modems. The data are then relayed to researchers on land in real time via satellite. Researchers can also request real time data independent of the automatic detection system. The data are used to provide early warnings of a tsunami before it comes ashore.

Each *deep-ocean assessment and reporting of tsunamis* (DART; Figure 3.5) station consists of a bottom pressure sensor anchored to the seafloor and a moored surface buoy.

Another practical example for the use of acoustic communication technology is in the search for underwater objects. A robot crawler carries a modem, a camera, and a digital signal-processing unit [Figure 3.6]. The robot, traversing the seafloor, searches for an object. When an object is found, the robot sends an acoustic signal to a ship- or shore-based

FIGURE 3.6 Robot to find underwater objects.

station. The robot can then be commanded to take a still frame photo, compress it, and transfer the image to an acoustic signal that is sent back to the investigator. This technology will allow archaeological expeditions to save money in diving time. Robotic crawlers can carry sensors into very shallow water or even into the surf zone. This robotic crawler, equipped with a camera and modem, can be controlled from substantial distances via acoustic communications

3.3 MEASUREMENT OF THE UPPER OCEAN

Measuring the processes at the sea surface and the region immediately below it is important for both practical and scientific reasons. Large, steep waves can be hazardous to offshore structures, such as oil platforms. Offshore structures and buoys must be designed to survive the largest waves that are likely to occur. Bubbles created by breaking waves play a role in the transfer of gases between the atmosphere and ocean. The amount and types of gases that move between the atmosphere and ocean are important to determining greenhouse gas levels in the atmosphere. Measuring these properties when the winds are strong and the waves high is difficult. *Sonar* [sound navigation and ranging] can measure the upper ocean while being located safely below the ocean surface.

3.3.1 Wave Height Measurement

Wave heights can be measured using sonar that transmits an acoustic pulse up toward the sea surface [Figure 3.7]. The pulse is reflected from the surface and returns to the transmitter, which switches into receiving mode after transmitting. The time that it takes the signal to travel to the sea surface and back depends on the distance to the surface from the transmitter and can provide a measurement of wave height.

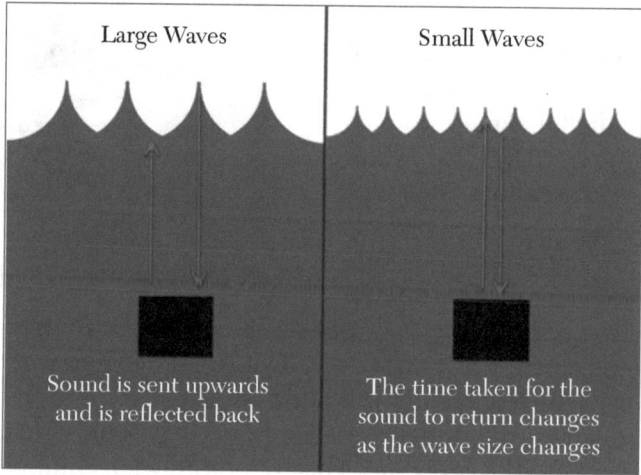

FIGURE 3.7 Wave height measurement.

Upward-looking sonar can be used to determine wave height. The time that it takes a sonar signal to travel to the sea surface and back depends on the distance to the surface from the transmitter and can provide a measurement of wave height.

3.3.2 Wave Velocities Measurement

The velocities in the crests of large, steep waves can also be measured. A sonar placed on the seafloor transmits an acoustic pulse up toward the sea surface. Additional hydrophones are placed on the seafloor a short distance from the sonar. The pulse is not only reflected from the sea surface, but also scattered from bubble clouds caused by breaking waves. The scattered signals have slightly different frequencies than the transmitted signals, because of the motion of the bubbles. The velocities of the bubbles can be determined by measuring the small differences in frequencies. The frequency shifts are due to the *Doppler effect,* and the sonar is called a Doppler sonar [Figure 3.8].

Small bubbles largely move in response to the currents in the waves. Therefore, by measuring the velocities of the bubbles, one can measure

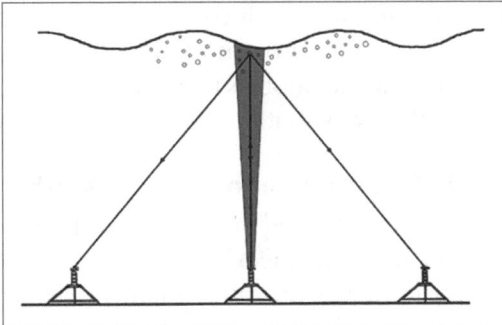

FIGURE 3.8 Wave velocity measurement.

the currents. The currents in the crests of the largest waves are important for determining the greatest forces that an offshore structure must be designed to withstand without collapsing.

3.3.3 Bubbles Study

A simple upward-looking sonar of the type used to measure wave heights can also be used to study the bubble clouds in the upper ocean caused by breaking waves. Bubbles are very effective acoustic *scatterers* at specific frequencies that depend on the size of the bubbles. The frequency of the sonar therefore determines the size of the bubbles to which it will be most sensitive. Bubble clouds have been found to extend 10 meters or more below the surface when winds are strong.

Bubbles in the open ocean are shown in Figure 3.9. At the time the image was taken, wave heights

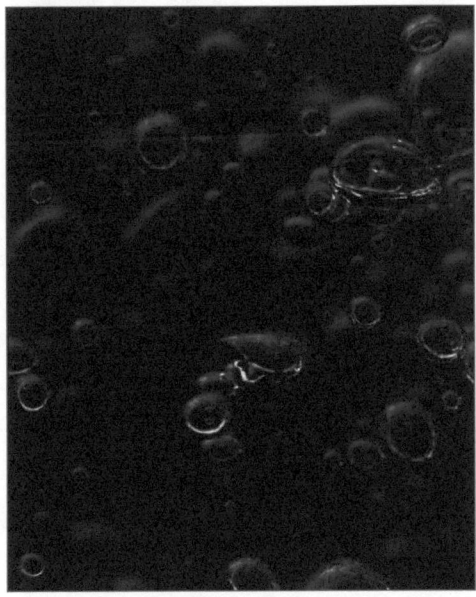

FIGURE 3.9 Bubbles study using sonar.

were approximately 2–3 meters and wind speeds were about 15–20 m/s. To determine the role of bubbles in the transfer of gases between the ocean and atmosphere, one needs to know the sizes of the bubbles, as well as the depths to which they penetrate. Acoustic systems use multiple frequencies to measure the size distribution of the bubbles. The frequency at which bubbles scatter or absorb sound most strongly depends on their size. The transfer of gases due to bubbles is small at low wind speeds, but at higher wind speeds the bubble contribution may dominate in the processes of air-water gas transfer.

3.3.4 Measurement of Water Depth

Sonar is used for seafloor mapping. A combined transmitter and receiver, called a transducer, sends a sound pulse straight down into the water. The pulse moves down through the water and bounces off the ocean bottom. The transducer picks up the reflected sound. Computers precisely measure the time it takes for the sound pulse to reach the bottom and return. In shallow water the sound waves will return very fast and in deeper water it will take more time to receive the echoes. The depth of the ocean is calculated by knowing how fast sound travels in the water (approximately 1,500 meters per second). This method of seafloor mapping is called *echo-sounding*. Echo-sounders can use different frequencies of sound to find out different things about the ocean. Water depth is typically measured by echo-sounders that transmit sound at 12 kHz. Lower frequencies (3.5 kHz) can be used to look at the layers of sediments below the seafloor. Higher frequencies (200 kHz) can be used to identify fish and plankton that are in the water column. Echo sounders calculate water depth by measuring the time it takes for the acoustic signal to reach the bottom and the echo to return to the ship.

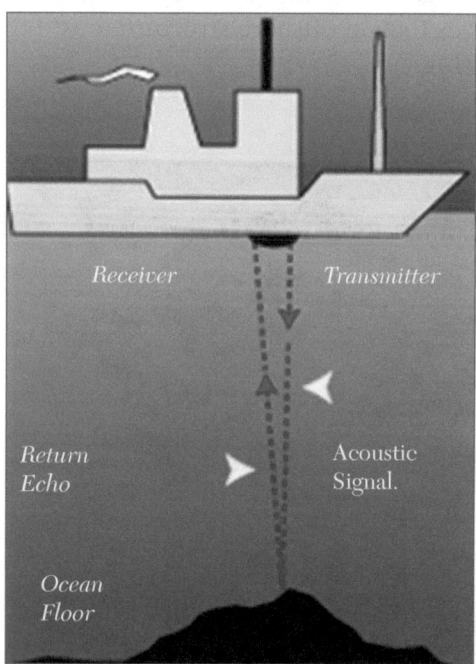

FIGURE 3.10 Echo-sounders to find water depth measurement.

3.4 LOCATING FISH

Sonars send sound waves or signals into the water that rebound when they strike an object. The fish reflects some of the signal back to the boat, the remainder of the signal continues to the seafloor then it bounces back to the boat. Some *sonar* systems are especially designed to locate fish, as shown in Figure 3.11(a). These systems use the same basic principle as other sonar systems—they transmit sound pulses, measure the time it takes for ech-

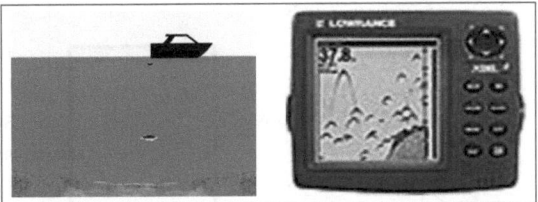

FIGURE 3.11 (a) Sonar to locate fish and (b) display of the deep sea.

oes to return, and calculate the distance to the objects. Fish finding sonar units send and receive signals many times per second. They concentrate sound into a beam that is transmitted from a *transducer*. These units include visual displays that print the echoes. The bottom appears as a continuous line drawn across the display. In addition, any objects that are in the water between the surface and the bottom may also be displayed, as in Figure 3.11(b).

Fish finders detect the presence of fish primarily by detecting the air in their *swim bladders*. The air conserved in the swim bladder changes the sound path and reflects energy back. The fish finder detects this reflected energy and converts it into fish images on the screen. Fish finders operate at high *frequencies* of sound, approximately 20–200 kHz (20–200,000 cycles per second). This helps define targets and can even display two fish as two separate echoes or arches. Lower frequencies (i.e., 50 kHz) can penetrate deeper

waters but may not be able to define individual targets. Putting more energy into the pulse sent out by the transducer increases the probability of getting a signal to return in deeper water. Images are formed on the visual display as arches due to the movement of the boat or the fish. When sound is transmitted from the transducer it is concentrated into a beam. As the sound passes into deeper water, the beam spreads out and covers a wider area. If the transmitted sound were plotted, it would look like a traffic cone with a pointed top and a broad base, as shown in Figure 3.12.

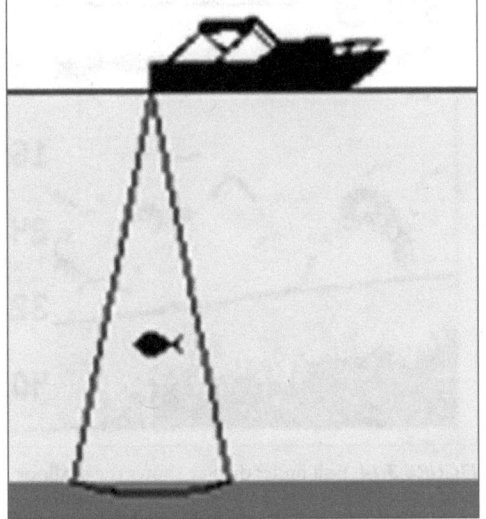

FIGURE 3.12 Sound transmitted from the boat's transducer spreads out in a conical shape.

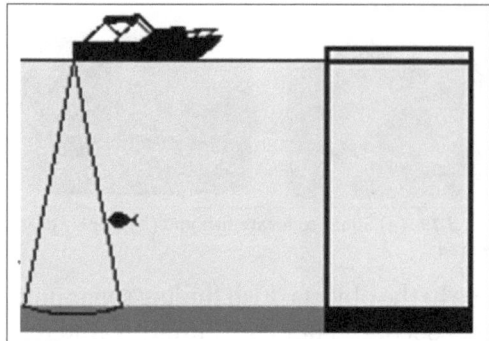

FIGURE 3.13 Example of how an arch is formed as a fish passes through the sonar beam.

Fish that swim within this cone may reflect some of the sound back to the transducer. The reflected sound, or echo, appears on the sonar's chart display. A school of fish will appear as many different shapes or formations, depending on how much of the school are within the transducer's cone. Individual fish, especially those in deeper water, may appear as arches on the display. The following illustrations [Figure 3.13] demonstrate how a fish arch forms as the fish moves through the sonar beam.

A fish arch forms as the fish moves through the sonar beam. A mark appears on the chart display (on right) when the fish enters the outer edge of the cone. As the fish swims through the cone, the distance between the transducer and the fish decreases, and the mark begins to curve up. When the fish is at the center of the beam it is directly beneath the transducer. The mark begins to flatten out as the fish reaches its closest point to the transducer. As the fish continues to move through the beam to the opposite end of the cone the distance increases. The mark begins to curve downward because the fish is moving further and further away from the transducer. An arched mark appears as the chart display graphs this distance change.

3.4.1 Identifying Fish

Experienced fishermen can use fish finders to identify species of fishes [Figure 3.14]. As a fish moves through the sonar beam of a fish finder, a mark specific to that fish species appears on the chart display. To determine what species of

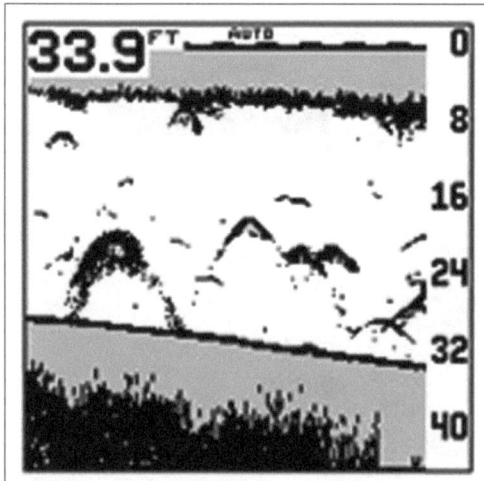

FIGURE 3.14 Fish finder display shows the seafloor gently sloping down to 33.9 feet (black horizontal line with gray beneath) while individual fish appear as arches on the display.

fish produces what kinds of marks, fishermen must be familiar with the area they are fishing, the fish that swim there, and the swimming patterns of different schools of fish. For instance, slow swimming carp produce short fat marks. *Stripers* swim much faster and move around a lot more, so they produce dotted lines. Baitfish may swim in circles as other fish herd them or as they swim to the surface, producing a third type of mark. These are just examples, and each fish finder will have different markings for a particular fish. Therefore, experience and close watching of patterns will help to perfect a fisherman's skill in using fish finding sonars to identify fish.

Scientists are developing new and improved methods to differentiate between the marks, also called *echo signatures*. Each species of fish has a unique size and shape of its *swim bladder*. The differences in swim bladders cause differences in the return echo of a sonar signal. Echo signatures for specific species can then be determined and used to identify fish.

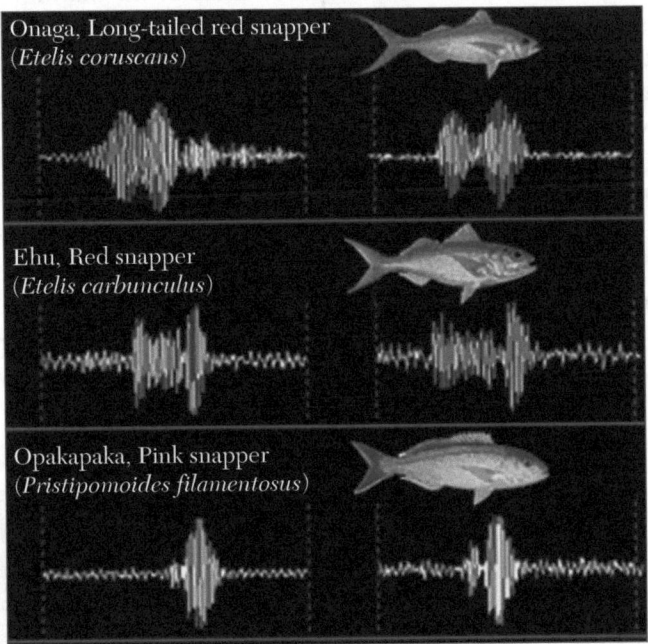

Onaga, Long-tailed red snapper
(*Etelis coruscans*)

Ehu, Red snapper
(*Etelis carbunculus*)

Opakapaka, Pink snapper
(*Pristipomoides filamentosus*)

FIGURE 3.15 Echo signatures.

The unique echo signatures of three different species of Hawaiian snapper are shown in Figure 3.15. Echo signatures on the left were taken from an anesthetized fish under controlled conditions at the surface. Echoes on the right were taken from free-swimming fish at 250 meters deep. Differences in echo signature structure are observed between species, but differences between control and free-swimming measurements are minimal for each species. Echoes measured at the surface, under controlled conditions, can therefore be applied to identify different fish species at depth. This type of acoustic data is important for fisheries surveys. This method is especially useful when studying deep-sea bottom fish. Many fish are found below diving depth, and therefore can only be studied using submersible vehicles or fishing gear. Acoustics provide an additional means for scientists to identify bottom fish and monitor them in their natural environment.

It is important to determine the factors that affect echo signatures. As water gets deeper, the *pressure* increases. The increased pressure can compress the air in the swim bladder as the fish dives. Using acoustics in conjunction with video and low light cameras, scientists have found that many fish can regulate the size and shape of their swim bladders, even under high pressure. The goal of the research is to identify the echo signatures of different species and make sure that these signatures do not change with movement. This will allow new fish finders to differentiate one species from another very accurately and quantitatively.

3.4.2 Communicating Underwater

Special underwater communication systems have been developed to allow divers to talk to each other underwater. A *transducer* is attached to the diver's face mask, which converts his or her voice into an ultrasound signal. A fellow diver has an ultrasound receiver, which accepts the signal and converts it *back* to a sound that the diver can hear, allowing for communication. The same system can be used for communication between the diver and a surface ship.

FIGURE 3.16 Acoustic communication systems allow divers to talk to each other underwater.

3.5 MEASUREMENT OF OCEAN TEMPERATURE

The speed of sound in water depends on the water properties of temperature, *salinity*, and pressure (directly related to the depth). A typical speed of sound in water near the ocean surface is about 1,520 meters per second. That is more than four times faster than the speed of sound in air. The speed of sound in water increases with increasing water temperature, increasing salinity, and increasing depth. Most of the change in sound speed in the surface ocean is due to changes in temperature. This is because the effect of salinity on sound speed is small and salinity changes in the open ocean are small. Near shore and in estuaries, where the salinity varies greatly, salinity can have a more significant effect on the speed of sound in water. As the depth increases, the pressure of the water has the largest effect on the speed of sound.

The *approximate* change in the speed of sound with a change in each property:
Temperature 1°C = 4.0 m/s
Salinity 1PSU = 1.4 m/s
Depth (pressure) 1 km = 17 m/s
Note: Changes in the speed of sound for a given property are not linear.

Under most conditions the speed of sound in water is simple to understand. Sound will travel faster in warmer water and slower in colder water. To measure the temperature of the water, a sound pulse is sent out from an underwater sound source and heard by a *hydrophone* in the water some distance away (up to thousands of kilometers). The time the sound takes to go from the source of the sound to the listening device (a hydrophone) is measured. From the travel time, the speed of sound between the source and the hydrophone can be calculated. If the salinity and depth where the sound traveled are known, the temperature of the water can be calculated. Two specific methods of measuring the temperature of the ocean with sound are explained below.

3.5.1 Acoustic Tomography

Acoustic tomography uses precise measurements of acoustic travel times to draw ocean temperature maps, showing ocean temperatures just as weather maps show temperatures in the atmosphere. Data from many crossing acoustic paths are used to generate these maps of ocean temperatures. In Figure 3.17, for example, four acoustic sources (S) are shown transmitting to five acoustic receivers (R), giving 20 acoustic paths through a region

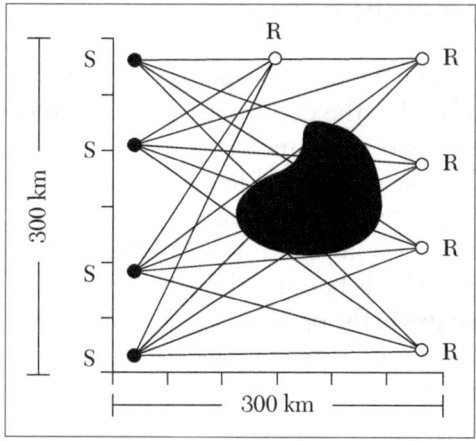

FIGURE 3.17 Acoustic tomography.

roughly 300 kilometers on a side. Suppose that the shaded region is warmer than its surroundings. Sound that travels through the warm region will travel slightly faster than sound that does not, because sound speed increases with increasing temperature, as described above. The travel times of sound pulses traveling through the warm region will therefore be slightly shorter than they would have been if the warm region was not there. By combining all of the different travel times it is possible to draw a map showing the warm and cold regions through which the sound has traveled. This is important because the ocean has "weather" just as the atmosphere does. Warm and cold *eddies* that are the oceanic equivalent of atmospheric storms move around, grow, and weaken. There are subsurface oceanic cold and warm fronts just as there are cold and warm fronts in the atmosphere. These eddy and subsurface fronts have important effects on marine life, with marine animals that prefer warm water tending to remain in warm eddies, for example.

The basic principles used in acoustic tomography are closely related to those used in CAT (*computed axial tomography*) scans in medicine. In a CAT scan, the absorption of X-rays is used to map a "slice" through the human body. ("Tomo" is derived from the Greek word for "cut or slice.") In acoustic tomography, the travel time of sound waves are used to map temperatures in a "slice" of the ocean. Even when it is not feasible to have enough sources and receivers to make detailed temperature maps, acoustic travel times can be used to obtain the average temperatures along the paths which the sound traveled. This is sometimes called *acoustic thermometry*.

3.5.2 Inverted Echo-Sounders

Inverted echo sounders (IES) measure the temperature of the water column at a single point. The IES is attached to the ocean bottom. It emits a sound pulse aimed toward the surface of the ocean. The sound pulse will reflect off the surface of the ocean and return to the bottom. The IES listens for the return of the sound pulse from the ocean surface. The travel time of the sound is used to calculate the speed of sound through the water. The temperature profile is calculated from the speed of sound through the water. The IES must be calibrated with a measurement of the water column properties. Sometimes a pressure sensor is used with the IES to make the calibration. Inverted echo-sounders are often used to monitor a specific region of the ocean. They are often placed in groups (or arrays) to cover a wider area. The IES block diagram is shown in Figure 3.18.

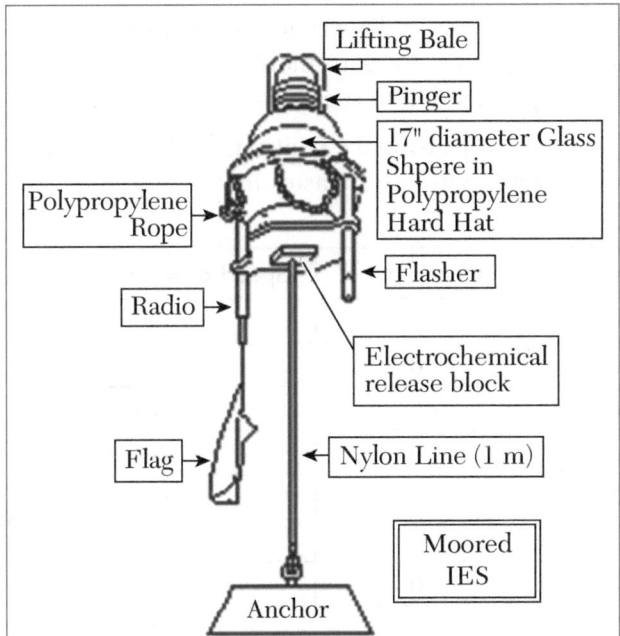 

FIGURE 3.18 Inverted echo-sounders.

3.5.3 Measuring Global Climate Change

One way the ocean will respond to global climate change is with a change in temperature. The average temperature of the ocean will rise as global climate warms. Where the warming occurs and the rate at which it occurs are of great interest to *climatologists*. There are several difficulties in measuring the kinds of temperature changes that interest climatologists.

1. The ocean is very large. To get a good picture of temperature change in the ocean, measurements of the whole ocean are needed.

2. The ocean is filled with warm and cold eddies, like storms in the atmosphere. The temperature changes associated with these eddies are large compared to the small changes in ocean temperature expected from

climate change. Large-scale average temperatures are needed to see climate change.

3. The temperature measurements must be continuous in time.

All these requirements mean that lowering temperature sensors from ships will not provide the kind of information on climate change that is needed. Although measurements made by lowering a temperature sensor from a ship are very accurate, many such measurements must be combined to provide the large-scale averages needed to see climate change.

There are simply not enough ships to make continuing measurements of temperature all over the ocean. Satellite measurements will help, but

Temperature (°C)	Salinity (S)	Depth (km)	Speed of Sound (m/s)
0	0	0	1402
0	35	0	1449
5	35	0	1470
5	35	0	1470
10	35	0	1490
5	35	0	1470
20	35	0	1521
30	35	0	1545
5	35	0	1470
20	5	0	1488
5	35	0	1470
20	10	0	1493
20	20	0	1505
20	35	0	1521
5	35	1	1487
5	35	2	1503
5	35	3	1521
5	35	4	1539

TABLE 3.1 The Speed of Sound Calculated under Different Ocean Conditions

Note: Changes in the speed of sound for a given property are not linear.

they only provide information at the ocean surface. Measurements made by drifting buoys that send data back by satellite will also help, but data from many buoys have to be averaged. Using sound to measure ocean temperatures directly provides the large-scale average temperatures needed to study climate change. In warmer water, sound travels faster. By measuring the travel time of sound between two points, the average temperature of the water between those points can be determined. Very precise measurements of the average temperature can be made with sound.

Acoustic tomography can be used to measure the temperature of the ocean over large areas. A sound source sends out a sound signal at known times. The sound is sent out in the *SOFAR channel* so that it will travel as far as possible. Low-frequency sound can travel thousands of kilometers in the ocean. *Hydrophones* all over the ocean basin listen for the sound. The time taken to travel to each hydrophone is then calculated and the average temperature determined. The results apply to large areas of the ocean, necessary for understanding climate change.

The time taken to travel to the hydrophone can be measured to within 20–30 milliseconds (20–30 thousandths of a second) at ranges of up to 5,000 kilometers. The travel time over 5,000 kilometers (3,100 miles) is about an hour. This travel time accuracy lets us determine the average temperature to within a few millidegress (thousands of a degree) Celsius. Changes in travel time are due to changes in ocean temperature. The change in temperature can be measured to within 0.001 degrees Celsius per day, as shown in Figure 3.19.

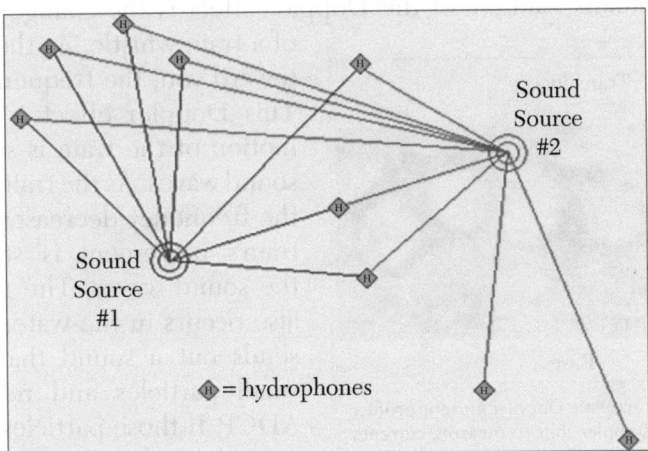

FIGURE 3.19 Climate change study using acoustic tomography.

3.6 MEASUREMENT OF OCEAN CURRENTS

Currents are commonly measured with sound. There are several different ways to measure currents with sound. An instrument called an *acoustic Doppler current profiler* or ADCP is often used to measure the current in specific places like shipping channels, rivers and streams, and at buoys. They are also called *acoustic Doppler profilers* (ADP). ADCPs can be placed on the bottom of the ocean, attached to a buoy or mounted on the bottom of ships. *RAFOS floats* (SOFAR spelled backwards) also use sound to measure currents. RAFOS floats are typically used in the open ocean to measure a current like the Gulf Stream. A technique called *reciprocal transmission* can also be used to measure currents with sound.

3.6.1 Acoustic Doppler Current Profiler

An ADCP sends out a sound pulse. The sound pulse is at a very high *frequency*, from 40kHz to 3,000 kHz. The human ear can hear frequencies up to 20kHz and even dolphins only hear frequencies up to 120kHz. At such high frequencies the *wavelength* is very small, about 6 mm to 0.5 mm.

The sound pulse from the ADCP will reflect off small particles in the water. These small particles may be fine silt or small living creatures like plankton. Even very clear water has many small particles in it. The ADCP listens with a *hydrophone* for the sound that is bounced off the small particles. [Figure 3.20]. The measurement of currents with sound depends on the *Doppler effect*. The Doppler effect is a change in frequency of a sound due to the motion of the source of the sound relative to the listener. The most common example of the Doppler effect is the change in frequency of a train whistle. As the train comes toward you, the frequency increases. This Doppler effect is because the motion of the train is squeezing the sound waves. As the train moves away, the frequency decreases because the train's movement is stretching out the sound waves. The Doppler shift also occurs in the water. The ADCP sends out a sound that reflects off small particles and returns to the ADCP. If those particles are in a current, then those particles are moving

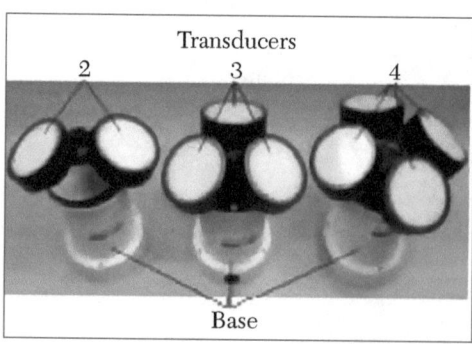

FIGURE 3.20 An acoustic Doppler current profiler (ADCP) uses the Doppler shift to measure currents in the ocean.

with the current. There will be a Doppler shift in the frequency of the sound that reflects off the small particles and returns to the ADCP. That Doppler shift can be used to calculate the current speed. Most ADCPs have three or four sound sources that work together. By using several sources, the ADCP can tell the direction of the current as well as its speed [Figure 3.21]. The ADCP can also tell at what depths in the water column the current is moving by how long it takes the sound to return to the ADCP. The detailed working of ADCP is explained in Chapter 9.

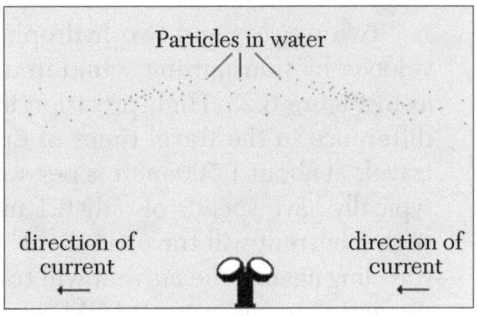

FIGURE 3.21 ADCP operation.

3.6.2 RAFOS Floats

RAFOS floats (SOFAR spelled backward) are floating instruments designed to move with a current and track the current's movements. The RAFOS float keeps track of its own position by listening for the signal from sound sources in the water near the study area and uses the time of travel and the phase of the sound to determine its position. Because the RAFOS float moves with the current, the float's position tracks the path of the current. The RAFOS float

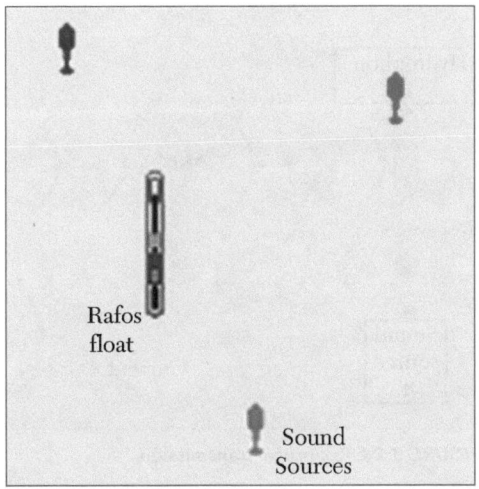

FIGURE 3.22 RAFOS float.

can be designed to float at different depths, allowing the full structure of the current to be studied, as in Figure 3.22

3.6.3 Reciprocal Transmission

A boat going downstream with the current in a river travels faster than a boat going upstream against the current. In same way, a sound pulse moving in the same direction as a current travels faster than one is moving against the current. Sound pulses transmitted in opposite directions at the same time (called *reciprocal transmissions*) will therefore have different travel times. The pulse traveling with the current will have a shorter travel time than the pulse traveling against the current. The difference between the two travel times can be used to compute the current.

Two sources and two hydrophones are necessary to measure current velocity by transmitting sound in opposite directions through the current, as in Figure 3.23. High-precision measurements are required because the difference in the travel times of oppositely traveling pulses is tiny. Sound travels at about 1,500 meters per second in the ocean, while ocean currents typically have speeds of only 0.1 meters per second. Sound traveling with such a current will travel at about 1,500.1 meters per second, while sound traveling against the current will travel at about 1,499.9 meters per second. *Acoustic current meters* (ACM) apply this basic principle to measure ocean currents without using propellers or any other moving parts.

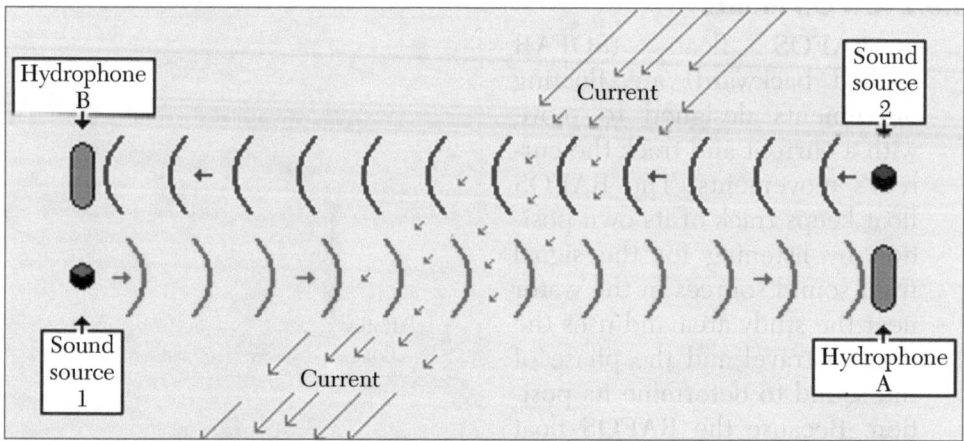

FIGURE 3.23 Reciprocal transmission.

3.6.4 Measuring Waves in the Surf Zone

Waves generated in distant storms roll across the ocean, eventually reaching shore. As the water becomes shallow, the waves break, spraying foam and running up on the beach. Understanding beaches and the *surf zone* is vital to coastal development and living wisely along the coast. The surf zone is a complex place. Breaking waves create currents that run off-shore (*rip currents*) and other currents that flow along the beach (along-shore flows). The direction that the incoming waves travel changes as the water depth changes. The waves are *reflected* and *refracted* because the speed at which the waves move depends on water depth. In particular, *submarine canyons* located near shore can have strong effects on the incoming waves and the surf zone. Submarine canyons are narrow, steep-sided valleys on the sea floor that resemble river canyons on land. Acoustic Doppler current profilers (ADCPs) and acoustic Doppler velocimeters (ADVs) both measure the speed and direction of ocean currents by measuring how the frequency of a sound changes as it reflect from a moving object. ADCPs are used to measure ocean currents and how they vary with depth over distances extending up to several hundred meters away from the instrument.

3.6.5 Sound Use to Make Long-Term Measurements of the Ocean

For the past century, most ocean measurements have been made from research vessels. Small groups of scientists and technicians made measurements during research expeditions lasting for a month or so, and then returned home to analyze the data that had been collected. Although much has been learned about the ocean using this approach, it is not very satisfactory for studying long-term changes, such as those associated with climate change, or short-lived events that occur intermittently, such as undersea volcanic eruptions or severe storms that mix the upper ocean layers. Scientists are now constructing ocean observatories that use undersea cables to connect instruments to shore. These new observatories are designed to make long-term, continuous measurements for decades. They will be used to study biological, chemical, physical, and geological processes in the ocean, at the seafloor, and at the boundary between the ocean and the atmosphere. These measurements of ocean properties are like the long-term, continuous measurements of the atmosphere that have long been provided by weather stations around the globe. The undersea cables are like those used to carry telephone calls under the ocean. They provide power to the instruments, transfer data from the instruments to shore, and allow scientists to send commands to control the measurements that the instruments make. This approach is like using underground cables and a cable modem to connect home computers to the Internet. The cabled observatories essentially extend the Internet far offshore, into the deep ocean. Sound is used for many purposes in these ocean observatories, including making measurements of ocean currents and other ocean properties, and performing various engineering tasks, such as conducting surveys of potential undersea cable routes.

3.7 SOUND USE TO STUDY THE EARTH'S HISTORY

At the bottom of the ocean, there are layers of sediments that contain information about the Earth's history. Sound is used to map and characterize these sediment layers, selecting the precise locations for coring or drilling sediment samples to study the history of Earth's climate and ocean. Scientists also use sound to answer questions such as how and where earthquakes occur, how submarine volcanoes form, and what effects their eruptions have. The layers of the seafloor are examined with *seismic reflection* and *seismic refraction* (also called *wide angle seismics*). *Echo-sounding* is a basic type of seismic reflection. Echo sounding is used to measure the depth of the water. High-frequency echo-sounders (12,000 Hz) are used to measure the depth to the seafloor. A sound pulse is sent from a ship and that sound reflects off the seafloor and returns to the ship. The time the sound takes to travel to the bottom and back is used to calculate the distance to the seafloor. Low-frequency echo sounders (1,000 to 6,000 Hz) can penetrate a short distance into the seafloor, up to approximately 100 meters, to study the upper sediment layers.

Seismic reflection uses a stronger sound signal and lower sound frequencies (10–50 Hz) than echo-sounding in order to look deeper below the seafloor. The sound pulse is often sent from an *airgun* array towed behind a slowly moving ship. An airgun uses the sudden release of compressed air to form bubbles. The bubble formation produces a loud sound. The sound from the airgun travels down to the seafloor. Some of the sound reflects off the seafloor but some of the sound penetrates the seafloor, sometimes as much as 20–30 km below it, depending on how the array is designed. The sound that penetrates the seafloor may also reflect off layers of sediment or rock within the seafloor. The reflected sounds travel back up to the surface.

The ship also tows hydrophones (called a *towed array* or *streamer*) which detects the reflected sound signal when it reaches the surface. They use many hydrophones to hear weaker reflections from deeper in the Earth. The time it takes the sound to return to the ship can be used to find the thickness of the layers in the seafloor and their position (sloped, level, etc.). It also gives information about the composition of the layers. By towing multiple hydrophone streamers separated by 50–200 meters, scientists can create 3-dimensional images of the Earth's sediment layers. This technique is called *multi-channel seismics*.

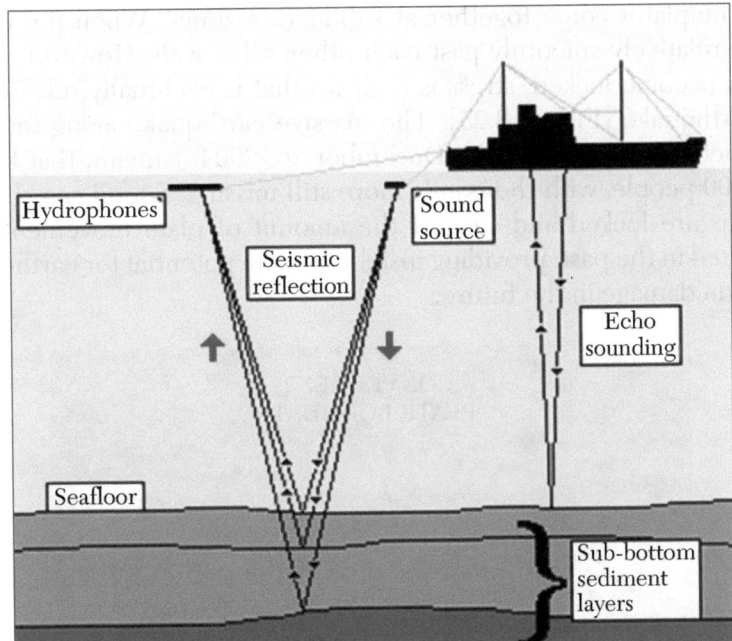

FIGURE 3.24 Seismic reflection to study Earth layers.

Seismic reflection [Figure 3.24] gives more information about the layers. Sound pulses that enter the seafloor are both reflected and refracted (or bent) as they pass into different layers. The refracted sound pulse follows a complex path. With seismic refraction, the density of the layers can be determined. Seismic reflection and refraction can also be done with an instrument on the seafloor called an *ocean bottom seismometer* (OBS). This instrument is placed on the seafloor and uses sound from artificial and natural sources. A seismic survey may make use of both ship board measurements and measurements from an array of ocean bottom seismometers.

Scientists use the seismic reflection and refraction data for many studies, including reconstructing past sea-level changes, predicting the location of future undersea earthquakes, and understanding how oceanic crust is formed at mid-ocean ridges. As sea level rises, the water depth increases, submerging land where new ocean sediments are deposited. However, if sea level falls, sediments on the seafloor are exposed and may be eroded away. By using sound to map the layers and sampling sediments to date the layers, the history of sea-level change can be estimated. The world's largest earthquakes occur tens of kilometers below the seafloor where two tectonic plates come together at subduction zones. When the two plates move relatively smoothly past each other, all is well. However, if the two plates become locked, stress is built up that is eventually released during an earthquake [Figure 3.25]. The massive earthquake along the Sumatra subduction zone caused the December 26, 2004, tsunami that killed over 185,000 people, with thousands more still missing. Sound can show which regions are locked and suggest the amount of plate movement that has occurred in the past, providing insight into the potential for earthquake and tsunami damage in the future.

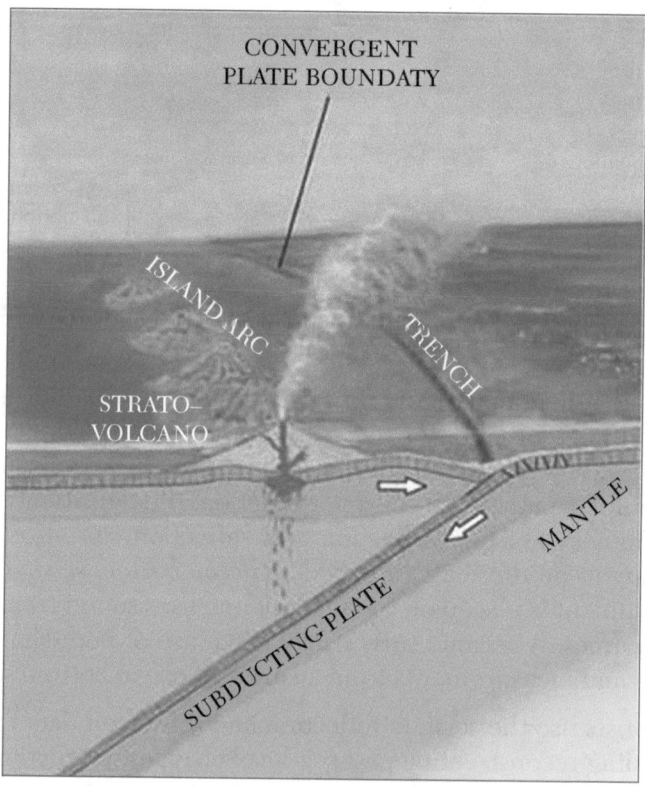

FIGURE 3.25 Ocean subduction zone.

EXERCISES: PART A (ANSWER IN A WORD OR A SENTENCE)

1. Define underwater acoustics.

2. Define acoustic transducer.

3. What is the use of hydrophone?

4. What is the expansion of SONAR?

5. What is Doppler effect?

6. _____ sonar is used to measure wave velocities.

7. The sound wave travels in the water _____ per second approximately.

8. The method of sea floor mapping is called _____

9. Sound will travel _____ in warm water and _____ in cold water.

10. Define acoustic tomography.

11. _____ in the frequency of sound is used to measure ocean current.

12. _____ profilers measure ocean currents.

13. _____ is used to measure the depth of water.

EXERCISES: PART B (ANSWER IN A PAGE)

1. Write about the interaction of sound with sea floor.

2. List the sound wave features.

3. How the data transmission is done using underwater sound.

4. What is the role of SONAR in wave height measurement?

5. Write in detail about the underwater objects detection.

6. How will you measure the wave velocities?

7. Write about the bubble study in short.

8. Explain about the water depth measurement technique.

9. How the divers communicate each other?

10. What is the role of sound in measuring temperature?

11. Write about invert echo sounders to measure temperature.

12. Write about the method to measure global climate change.

13. Write about how to measure waves in the surf zone?

14. Write the role of sound to study the earth's history.

EXERCISES: PART C (ANSWER WITHIN 2 OR 3 PAGES)

1. Write about underwater acoustic communication in controlling tsunamis.

2. Write about locating and identifying fish using sound.

3. Write the working principle of acoustic tomography.

4. Explain the different methods to measure temperature using sound.

5. Write about the ADCP and RAFOS working to measure the ocean current.

6. Write the 3 different methods to measure the ocean current.

7. Write in detail the use of sound to measure earth layers and subduction zone.

WEB LINKS

http://nctr.pmel.noaa.gov/Dart2

http://www.dosits.org

http://www.whoi.edu/instruments

CHAPTER 4

UNDERWATER COMMUNICATION

This chapter discusses about underwater wireless communication. The acoustic waves and acoustic communication, optical waves and optical communication, underwater acoustic communication, wireless underwater optical communications, underwater mobile communication, types of modulation, Internet, and GPS are also discussed.

4.1 UNDERWATER WIRELESS COMMUNICATION

Underwater wireless communication networks (UWCNs) consist of *sensors* and *autonomous underwater vehicles* (AUVs) that interact, coordinate and share information with each other to carry out sensing and monitoring functions. Its range of applications include coastal surveillance systems, environmental research, autonomous underwater vehicle (AUV) operation, oil-rig maintenance, collection of data for water monitoring, and linking submarines to land. In the underwater world, there are three types of carrier wave that are most commonly used in wireless communication.

1. *Electromagnetic wave:* Using electromagnetic waves, communication can be established at a higher frequency and bandwidth. Its limitation is due to high absorption/attenuation, which has significant effect on the transmitted signal. Big antennas are also needed for this type of communication, which affects the design complexity and cost.

2. *Optical wave:* Optical waves also offer high data rate transmission. Nevertheless, the signal is rapidly absorbed in water and suffers from the scattering effect. This will affect the data transmission accuracy.

3. *Acoustic wave:* Acoustic wave is the most preferred signal by many applications, owing to its low absorption characteristic for underwater communication. Even though the data transmission is slower compared to other carrier signals, the low absorption enables the carrier to travel at longer range.

4.1.1 Environment/Propagation Medium

In comparison to communication in terrestrial applications, for underwater wave propagation, the challenges are quite different. Water itself becomes the main source for signal interference. The type of water (freshwater/sea water), depth pressure, dissolved impurities, water composition, and temperature affect sound propagation. Common terrestrial phenomena like scattering, reflection, and refraction also occur in underwater communication.

4.2 FUNDAMENTALS OF WAVES

Understanding the first principles of each physical wave used in UWSN wireless communication is critically important.

4.2.1 Acoustic Waves and Physical Properties

Among the types of waves, acoustic waves are used as the primary carrier for underwater wireless communication systems due to their relatively low absorption in underwater environments. An acoustic wave has several propagation characteristics that are unique from other waves, two of which are highlighted below:

Propagation velocity: The extremely slow propagation speed of sound through water is an important factor that differentiates it from electromagnetic propagation. The speed of sound in water depends on the water properties of temperature, salinity, and pressure (directly related to the depth). A typical speed of sound in water near the ocean surface is about 1,520 m/s, which is more than four times faster than the speed of sound in air, but five orders of magnitude smaller than the speed of light. The speed of sound in water increases with increasing water temperature, increasing salinity, and increasing depth. Most of the changes in sound speed in the surface ocean are due to changes in temperature. This is because the effect of salinity on sound speed is small, as are salinity changes in the open ocean. Near shore and in estuaries, where the salinity varies greatly, salinity can have a more significant effect on the speed of sound in water. As depth increases, the pressure of water has the largest effect on the speed of sound. Under most conditions, the speed of sound in water is simple to understand. Sound will travel faster in warmer water and slower in colder water. Approximately, the sound speed increases 4.0 m/s for water temperature. As the depth of water (therefore also the pressure) increases 1 km, the sound speed increases roughly 17 m/s. It is noteworthy that the above assessments are only for rough quantitative or qualitative discussions, and the variations in sound speed for a given property are not linear in general.

Absorption: During propagation, wave energy may be converted to other forms and absorbed by the medium. The absorptive energy loss is directly controlled by the material imperfection for the type of physical wave propagating through it. For acoustic waves, this material imperfection is inelasticity, which converts the wave energy into heat.

4.2.2 Acoustic Communication

Acoustic communication is defined as communication from one point to another using acoustic signals. Acoustic signal is the only physically feasible tool that works in underwater environment. Compared with it, electromagnetic waves can only travel in water a short distance due to the high attenuation and absorption effect in an underwater environment. It is found that the absorption of electromagnetic energy in sea water is about $45 \times f$ dB per kilometer, where f is frequency in Hertz. In contrast, the absorption of acoustic signal over most frequencies of interest is about three orders of magnitude lower. There are some investigations into utilizing optical signals for underwater applications. However, the optical signal can only pass through a limited range in a very clean water environment (deep water, for example). Thus, it is not a proper tool for long-distance transmission underwater, or in a not-so-clean, e.g., shallow water environment.

Underwater acoustic networks, underwater acoustic sensor networks (UASNs), and *autonomous underwater vehicle networks* (AUVNs), are defined as networks composed of more than two nodes, using acoustic signals to communicate for underwater applications. UASNs and AUVNs are two important kinds of UANs. The former is composed of many sensor nodes, mostly for monitoring. The nodes are usually immobile or with limited capacity to move. AUVNs are composed of autonomous or unmanned vehicles with high mobility, deployed for applications that need mobility, e.g., exploration. A UAN can be an UASN, an AUVN, or a combination of both. UAN is explained in detail in Chapter 5.

4.2.3 Optical Waves

Optical wave communication obviously has a big advantage in data rate. However, there are a couple of disadvantages for optical communication in water. Firstly, optical signals are rapidly absorbed in this environment. Secondly, optical scattering caused by suspended particles and planktons is significant. *Multi-scattering* causes the optical pulse to widen in the spatial, temporal, angular, and polarization domains. Thirdly, the high level of ambient light in the upper part of the water column is another adverse effect for using optical communication. Optical wave transmission requires high precision in pointing narrow laser beams. In very clean water, e.g., the deep sea, blue-green wavelengths may be used for short-range connection. The advantage of optical signaling lies in its high data rate at distances up to 100 meters. As of now, the only practical solution for underwater communication with acceptable range utilizes acoustic signals, which travel underwater with longer distance, less attenuation, and higher reliability. However, the available bandwidth is extremely limited for acoustic signals. For a very long distance, at the order of 1,000 km, the available bandwidth falls below 1 kHz; only at very short ranges, below about 100 meters, may more than 100 kHz of bandwidth be available. A high bit error rate is common in underwater channels, due to the multi-path interference and time-varying nature of underwater acoustic channels.

4.2.4 Optical Communication

The present acoustic underwater communication is a legacy technology that provides low data-rate transmissions for medium-range communication. Data rates of acoustic communication are restricted to around tens of thousands of kilobits per second for ranges of 1 km, and less than a thousand kilobits per second for ranges up to 100 km, due to severe frequency-dependent attenuation and surface-induced pulse spread. In addition, the speed of acoustic waves in the ocean is approximately 1,500 m/s, so that long-range communication involves high latency, which poses a problem for real-time response, synchronization, and multiple-access protocols. In addition, acoustic waves could distress marine mammals such as dolphins and whales. Thus, acoustic technology cannot satisfy emerging applications that require around-the-clock, high-data-rate communication networks in real time. Examples of such applications are networks of sensors for the investigation of climate change; monitoring biological, biogeochemical, evolutionary, and ecological processes in sea, ocean, and lake environments; and unmanned underwater vehicles used to control and maintain oil production facilities and harbors. As already pointed out, water quality plays a key role in deciding whether optical waves can be used for underwater communication. Therefore, the applicability of optical communication heavily depends on environments. Using the same analogy for acoustic and electromagnetic waves, optical communication works in an environment-limited region. So far, there are not many commercial activities using underwater optical communication, and no commercial optical modems are available specifically for underwater use. Recent interest in underwater sensor networks and sea floor observatories have greatly stimulated the interest in short-range, high-rate optical communication in water. Improvements in the availability of the network could be achieved by a hybrid communication system that would include an optical transceiver and an acoustical transceiver. A hybrid communication system can provide high data-rate transmission by using the optical transceiver. When the water turbidity is high or the distance between the terminals is large, the system can switch to a low data rate using the acoustic transceiver, thereby increasing the average data rate and availability. However, the complexity and cost of the system are increased. In this kind of system, smart buffering and prioritization could help to mitigate short-term data rate reduction. Table 4.1 gives the performance comparison between acoustic and optical communication.

Telemetry method	Range	Data rate	Efficiency
Acoustic	Several Km	1 kbps	100 bits/Joule
Optical	100 meters	1 Mbps	30,000 bits/Joule

TABLE 4.1 Comparison of Acoustic and Optical Communication

Parameters	Acoustic	Electromagnetic	Optical
Nominal speed (m/s)	~ 1,500	~ 33,333,333	~ 33,333,333
Power Loss	> 0.1 dB/m/Hz	~ 28 dB/1km/100MHz	α turbidity
Bandwidth	~ kHz	~ MHz	~ 10–150 MHz
Frequency band	~ kHz	~ MHz	~ 10^{14}–10^{15} Hz
Antenna size	~ 0.1 m	~ 0.5 m	~ 0.1 m
Effective range	~ km	~ 10 m	~ 10–100 m

TABLE 4.2 Comparison of Acoustic, EM, and Optical Waves in Seawater Environments

4.3 UNDERWATER ACOUSTIC COMMUNICATION

Underwater acoustic communication is a technique of sending and receiving messages below water. There are several ways of employing such communication but the most common is using *hydrophones*. Underwater communication is difficult due to factors like multi-path propagation [Figure 4.1], time variations of the channel, small available bandwidth, and strong signal attenuation, especially over long ranges. In underwater communication there are low data rates compared to terrestrial communication, since underwater communication uses acoustic waves instead of electromagnetic waves.

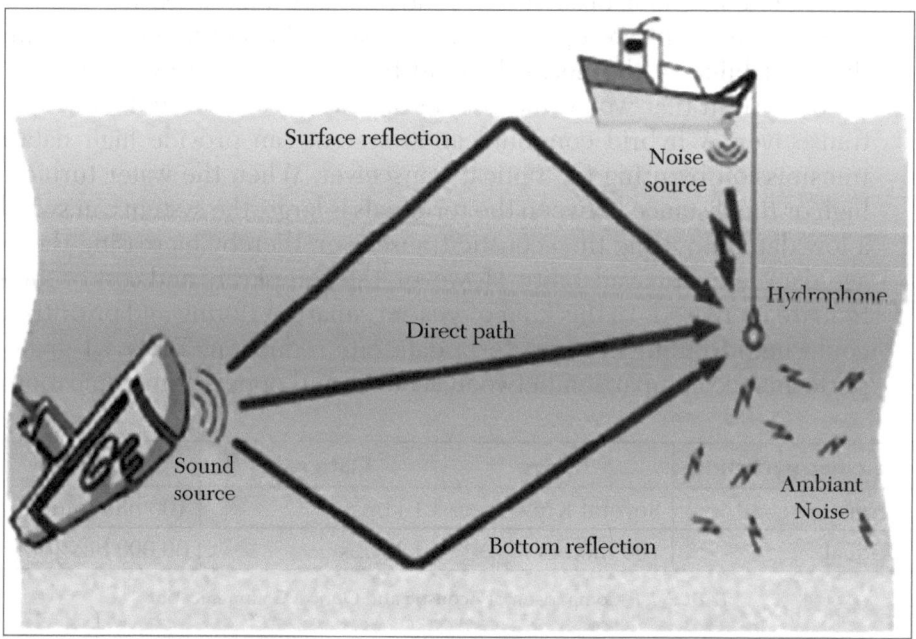

FIGURE 4.1 Example of multipath propagation.

4.3.1 Limitations of Underwater Acoustic Communication

High delay: The signal propagation speed in the underwater acoustic channel is about 1.5×10^3 m/sec, which is five orders of magnitude lower than the radio propagation speed (3×10^8 m/sec). The large propagation delay seriously reduces the throughput of the system considerably and determines the instability of the underwater control network system.

Limited bandwidth: The acoustic band underwater is very small due to absorption, so most acoustic communication system operates below 30 kHz. As a result, the bandwidth of underwater acoustic channels operating over several kilometers is about several tens of kbps, while a short-range system over several tens of meters can reach hundreds of kbps.

High bit error rate: Because of path loss, multi-path fading, Doppler spread, and noise (from man and environment) in the underwater acoustic channel, there is a large bit error rate in the underwater acoustic channel, which is on the order of 10^{-2}–10^{-5}. To prevent serious errors in communication, special ARQ (*automatic repeat request*) techniques and FEC (*forward error correction*) techniques must be adopted, which improve complexity in the underwater acoustic sensor network.

High energy consumption: The power consumed in underwater acoustic communication is greater than in terrestrial radio communication, because more power is consumed in the complex signal processing at receivers to compensate for the impairments of the channel.

Affected by the above factors, the current underwater acoustic sensor network only provides limited communication for different applications, which can communicate information among different sensor nodes without any quality of services. At the same time, the above factors cause the efficiency of the underwater acoustic sensor network to be very low and the complexity of protocol stack to be high.

4.4 UNDERWATER OPTICAL COMMUNICATIONS

This communication technology is expected to play an important role in investigating climate change, in monitoring biological, biogeochemical, evolutionary, and ecological changes in sea, ocean, and lake environments, and in helping to control and maintain oil production facilities and harbors using unmanned underwater vehicles (UUVs), submarines, ships, buoys, and divers. However, the present technology of underwater acoustic communication cannot provide the high data rate required to investigate and monitor these environments and facilities. Optical wireless communication has been proposed as the best alternative to meet this challenge. *Laser* and *LED* are used as optical light sources.

Limitations of acoustic communications are:

1. The speed of acoustic waves in the ocean is approximately 1,500 m/s.

2. Long-range communication involves high latency, which poses a problem for real-time response, synchronization, and multiple access protocols.

3. Acoustic waves could distress marine mammals such as dolphins and whales.

Acoustic technology cannot satisfy emerging applications that require around-the-clock high data-rate communication networks in real time. Figure 4.2 is an alternative means of underwater communication based on optics, wherein high data rates are possible. However, the distance between the transmitter and receiver must be short, due to the extremely challenging underwater environment, which is characterized by high multi-scattering and absorption. Multi-scattering causes the optical pulse to widen in the spatial, temporal, angular, and polarization domains. Although high data rates are threatened by extremely high absorption and scattering, there is evidence that broad-band links can be achieved over moderate ranges.

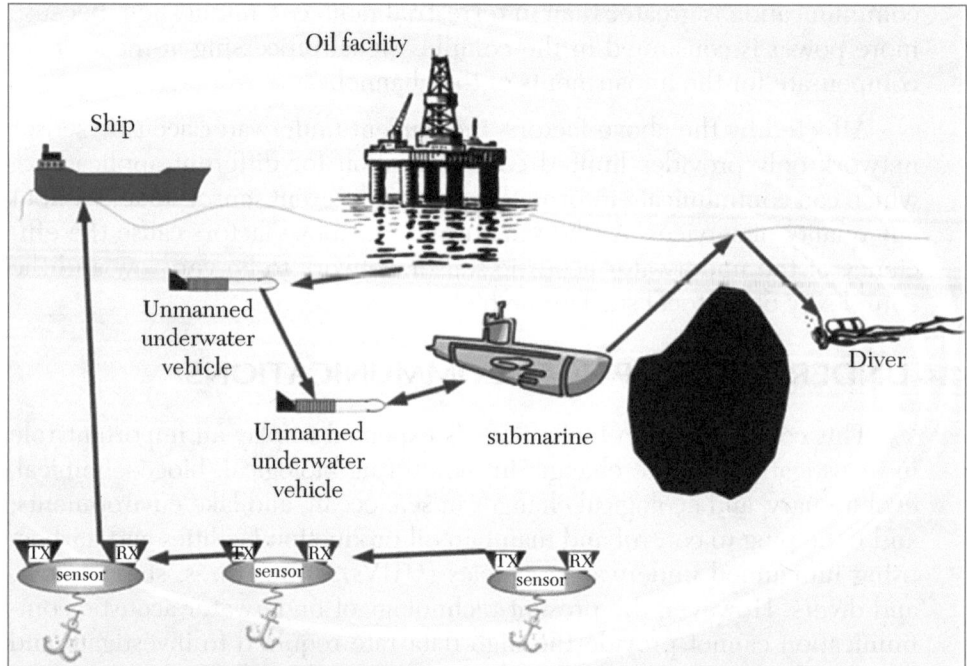

FIGURE 4.2 The line-of-sight communication scenario.

4.4.1 Using Laser as Optical Communication Above Water and Underwater

There is a growing need in maritime applications to quickly transfer large volumes of information between different units or in a sensor network. *Radio frequency* (RF) channels and satellite links on ships are often limited to data rates of some hundreds of kilobits per second to some megabits per second. Laser links offer the opportunity to overcome these restrictions for optical line-of-sight communications due to the very high frequencies used and, when compared to conventional RF links, the angle of the transmitting laser beam is small. Restrictions arise due to precipitation, like rain or fog. Speed limitations are more pronounced underwater. Underwater acoustic modems can transfer data up to a few kilobits per second over distances of some kilometers, depending on the channel characteristics and the frequencies used. Lasers operating in the blue-green spectral range of the electromagnetic spectrum offer an alternative to realizing much higher data rates. Depending on the turbidity of the water, communication ranges are limited to 10 to 100 meters. For testing underwater laser communication [Figure 4.3], a continuous wave laser operating at 532 nanometers wavelength was chosen. In coastal waters with high turbidity and high content of organic matter, the green light attenuation is less than that for blue laser light, making blue preferable for operations in clear water masses of the open ocean.

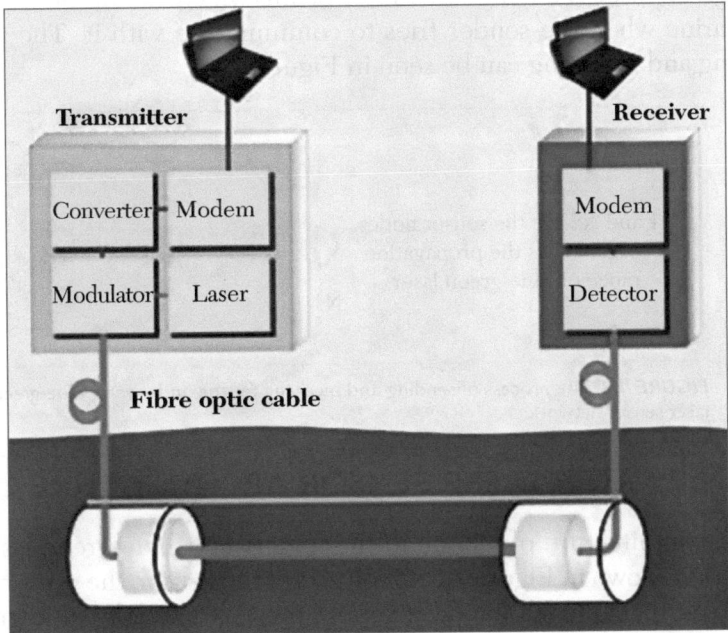

FIGURE 4.3 Underwater laser optical communication (experimental setup).

Digital data from a notebook computer can be transferred via Ethernet to a modem of the transmitting unit of the underwater communication system. A converter drives an *acoustic optical modulator* (AOM), which is responsible for the intensity modulation of the laser beam. The system was designed for a data rate of 10 megabits per second. The modulated laser light was transferred via a fiber optic cable to an underwater housing that contains a collimator. The transmitted laser beam travels 3 meters in the water column before it is coupled back into another fiber optic cable. With the help of a detector and a second modem, the optical signals are converted back to a digital data stream. Free-space optical links have proven to be an effective addition to established communication technologies, both above water and underwater. They allow much higher data rates in line-of-sight communication.

Blue-green laser in underwater communication: Most laser light cannot penetrate through the sea due to being absorbed, but the blue-green laser (the length of wave is about 470~570 nm) has minimum energy fading in the sea, about 0.155~0.5 db/m. Hence, the blue-green laser can propagate from several hundreds of meters to kilometers in the sea, and this feature is called the *window effect* and some submarine communication systems have been developed based on it. In these communication systems, the blue-green laser is a collimated laser beam, which should be aimed at the submarine when the sender tries to communicate with it. The process of sending and receiving can be seen in Figure 4.4.

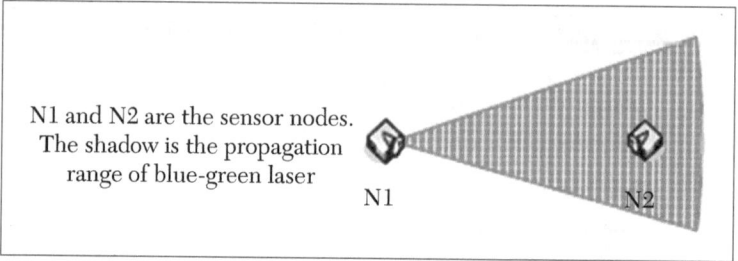

FIGURE 4.4 The process of sending and receiving for the underwater blue-green laser sensor network.

4.5 UNDERWATER LASER SENSOR ARCHITECTURE

The architecture of a node in the underwater blue-green laser sensor network is shown in Figure 4.5. It consists of the sensor, the sensor interface circuitry, the memory, the power supply, the CPU, and the blue-green laser modem. The CPU gets the data in the sensor through the sensor interface circuitry, and then the CPU can store the data in the memory, process the data, and send/receive the data by controlling the blue-green laser modem.

4.5.1 Protocol Stack for the Underwater Laser Sensor Network

The protocol stack for the underwater laser sensor network should consist of physical layer, data link layer, network layer, transport layer, and

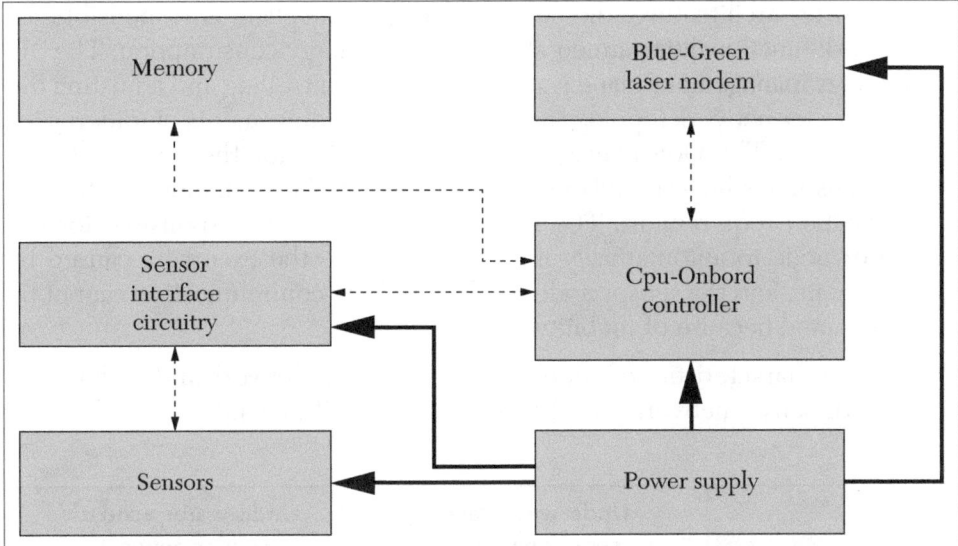

FIGURE 4.5 Internal architecture of a node in the underwater blue-green laser sensor network.

application layer functionalities. Considering the critical underwater environments, the underwater laser sensor network is different from the terrestrial sensor network. The protocol stack should also include the energy management plane, 3D topology management plane, QoS management plane, and the mobile management. The architecture of the protocol stack for the underwater laser sensor network is pictured in the Figure 4.6.

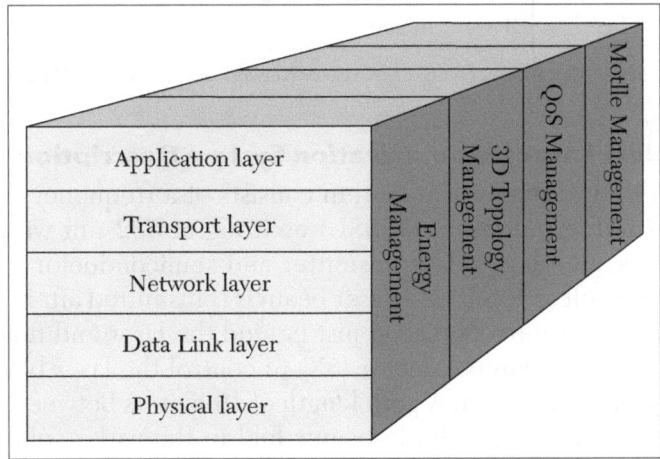

FIGURE 4.6 The architecture of protocol stack for the underwater laser sensor network.

In this architecture, the energy management plane is responsible for network functionalities aimed at minimizing energy consumption. The 3D topology management plane is responsible for controlling and adjusting the underwater network topology according to the requirements of underwater exploration. The QoS management is responsible for the quality of data transmission, which should ensure the transmitted information satisfies the application requirement. The mobile management is responsible for the sensor node to automatically move to overcome the excursion caused by the stream, and the sensor node will ensure laser communication cannot be interrupted because of mobility.

The characteristics of underwater laser sensor network and underwater acoustic sensor network have been compared in Table 4.3.

	Underwater laser sensor network	**Underwater acoustic sensor network**
Delay	Little	High
Bandwidth	Several hundreds of kbps per kilometer	10.75 pt
Bit error rate	Low	High
Energy consumption in communication	Low	High
Propagation distance	1–2 kilometers	Several tens of kilometers
Propagation speed	3×10^8 m/sec	1.5×10^3 m/sec

TABLE 4.3 Characteristics of Underwater Laser Sensor Network and Underwater Acoustic Sensor Network

4.5.2 Wireless Laser Communication System Description

Our optical transceiver system consists of a frequency-doubled *diode-pumped solid-state laser* (DPSSL) emitting at 532 nm with a processing electronics unit as optical transmitter and semiconductor detector with a processing unit as receiver. A laser beam is transmitted after reflection from two adjustment mirrors placed just behind the laser and finally is made to pass through a beam collimator (5X) to control the laser beam divergence and spot size at receiver. A path length of 10 meters between two transceivers is achieved after multiple beams fold in the water cell. Turbulence is created using a water churning motor.

Electronics processing unit: Transmitter electronics modules comprise a level converter followed by a buffer to make an RS 232 signal level compatible with laser driving TTL signal. At the receiver, detector output is passed through a comparator to compensate the laser beam attenuation and followed by level converter to make the inverted TTL signal compliant to RS 232 level.

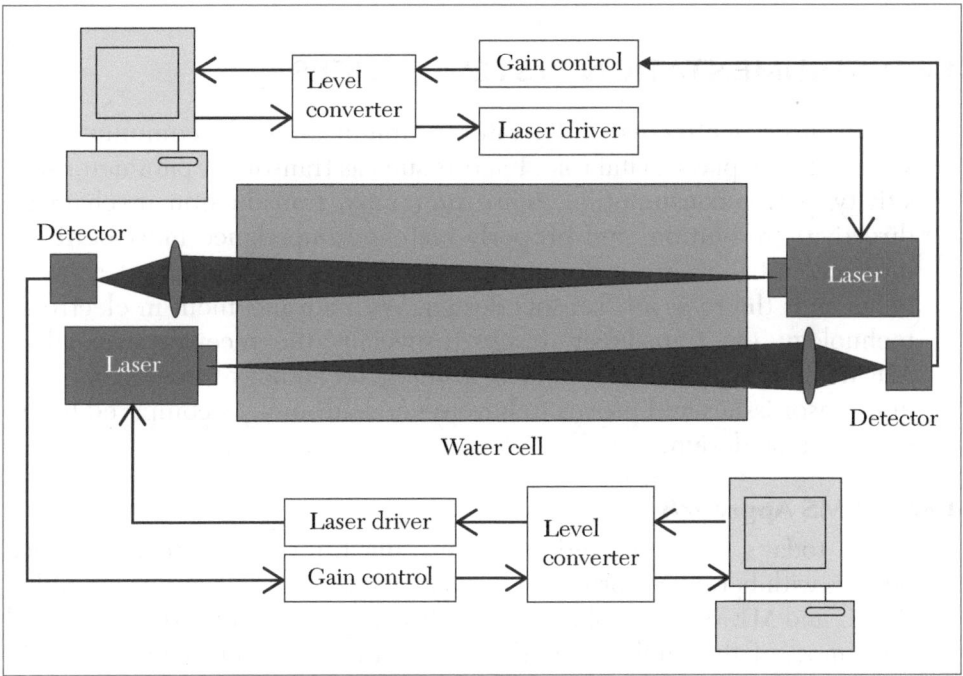

FIGURE 4.7 Block diagram of overall wireless laser communication system.

Operation: As the designed system operates in full duplex hand shaking mode, both transceiver human interfaces (PCs) are set at same baud rate and common packet formats. Selection of baud rate solely depends upon the *laser on-off stabilized rate*. Laser on-off stabilized rate signifies the number of missing/distorted pulses per second at a particular specified rate and adds to *bit error rate* (BER). Generally, laser supports a lower stabilized on-off rate in comparison to a specified rate. The designed system can support data rates up to 100 kbps but the limiting factor is the stabilized on-off rate of the laser and should be checked before integrating the laser in the system. After matching the baud rates and data formats of transmitter and receiver, alignment of both transceiver systems is checked in the chat mode. A character sequence is transmitted from one end, and from the other end the received character sequence is retransmitted to the first end. If the sequence is the same, the system is aligned and ready for the file transmission. Raw file formats like Notepad or Word Pad can be transmitted at higher data rates in comparison to other formats. The system is also equipped with independent transceiver testing without turning the laser on.

4.6 INSTRUMENTATION SYSTEM DEVICES

In ensuring effective underwater communication, the communication system design plays a vital role. Factors such as transducer parameter sensitivity, power consumption, noise immunity, transduction mechanism, directivity, resolution, and properly matched impedance must be taken into account during the design process. One of the important areas to focus on is the receiver (sensor) design. With advancement in electronic technology, the transducer design (especially the receiver) can adopt MEMS (*micro electro mechanical systems*) technology to overcome several sensor issues and proves to have several advantages compared to the conventional design.

4.6.1 MEMS Approach

In today's electronic industries, manufacturers compete to produce devices with better performance in smaller size. This scenario has enabled the IC and MEMS technology to grow faster in electronic industries. Until now, most of the applications that utilize this approach mainly focus in imaging industries, owing to the fact that this type of sensor can offer a high bandwidth and sensitivity.

However, the realization of MEMS in underwater communication—especially in sensor design—could be a worthy effort to bridge the gap between terrestrial and underwater communication systems. Reduction in size has offered a lot of advantages in terms of power consumption, portability, production, and cost. Even though the exploration of this approach in underwater communication is still new, it is technically possible. Its main contribution is the ability to overcome the problems caused by size and power consumption. In autonomous underwater vehicles (AUVs), for example, the utilization of a MEMS device with a smaller battery will reduce the overall weight of the AUV, thus reducing the power needed to drive the vehicle. Many researchers in underwater communication systems are concentrating on other aspects, such as overall system development, communication protocol, signal processing, and conventional transducers. A MEMS-based sensor for underwater communication can provide a new platform for researchers to explore more of what this technology can offer and it can become a new research area that requires extensive study and could contribute to many novel outcomes.

There has been a growing interest in monitoring underwater mediums for scientific exploration, commercial exploitation, and attack protection to contribute to human wellbeing. Industries are increasingly interested in technologies like wireless sensor networks. Underwater sensor networks consist of a variable number of sensors and vehicles that are deployed to perform collaborative monitoring tasks over a given area.

4.7 UNDERWATER MOBILE COMMUNICATION

Underwater communication, especially on mobile platforms, forms point-to-point links and requires definite pointing and tracking. Systems that use collimated laser links generally employ such links. There are systems that use very large aperture (approximately 20 in) *photomultiplier tubes* (PMTs) that enlarge the receiver *field of view* (FOV). Large area PMTs have the disadvantage of being expensive and bulky. Hence, compact systems are desired, which do not have much volume budget or energy budget for sophisticated pointing and tracking. Smart antennas are used in traditional RF wireless systems, which make them capable of signal processing to provide angle of arrival information and broadcast beam-forming. In indoor optical wireless communication, several antennas with spatial diversity and angular diversity are employed for non-line-of-sight communications, ambient light rejection, electronic tracking and pointing, corresponding localization, and multi-hop networking. Modern networks are also very much required to be energy efficient. It is obvious to consider the benefits of extending such techniques to the underwater environment.

4.7.1 Benefits of Smart Optical Systems for Underwater Vehicles

Smart optical transmitters and receivers can evaluate and estimate the obvious optical effects of water, transmit a beam of light in a fixed direction, and find out the direction of the light beam and the peculiarity of the light beam that is received. A receiver transmission's gain and power during detection and acquisition of another platform can be changed by evaluating water quality. A device's orientation, identity, and relative angle can be utilized to localize and evaluate its relative position. Possible benefits include:

Non-mechanical pointing and tracking on a moving underwater device: An optical transmitter or receiver mounted on a device can go in and out of sighting with another stationary or fixed platform. This process depends upon the state of the sea and commands of the underwater device. An optical front end capable of varying its effective field of view (FOV), detecting the angle of arrival at its receiver and electronically directing its output beam, can possibly maintain a communications link in such an environment. Furthermore, one can use signal diversity expertise to improve and enhance signal reliability.

Maintaining a link with a stationary node as an underwater device drives by: It is quite difficult for underwater devices to maintain a precise relative position. The ability to interrogate and obtain information from a stationary sensor node as a device drives by can add significant operational capability. Thus, a quasi-omni-directional receiver is valued which can continually adapt its FOV and optical power.

Providing sensory information to underwater devices: In a swarm environment, localization information can be collected from angle of arrival information as different nodes communicate with one other. This information can be transmitted to the device to augment its other sensory data for navigation and avoiding collisions. A smart optical front-end can also contribute to other sensory information such as water quality measurements obtained from the communications link.

Duplex multi-user system: Each transceiver is composed of a smart receiver and a smart transmitter which allow synchronous reception from two non-colocated transmitters. Since each transmitter is *code division multiple access* (CDMA) coded, the receiver at one location is also capable of associating data streams of another smart receiver with a different location by its corresponding directions. Whenever two smart receivers lie on the same line, the CDMA code still permits for dividing the two transmit streams at the receiver on the first smart receiver. In a mesh network scenario, as illustrated in Figure 4.8, node A and node C are not in the range of each other.

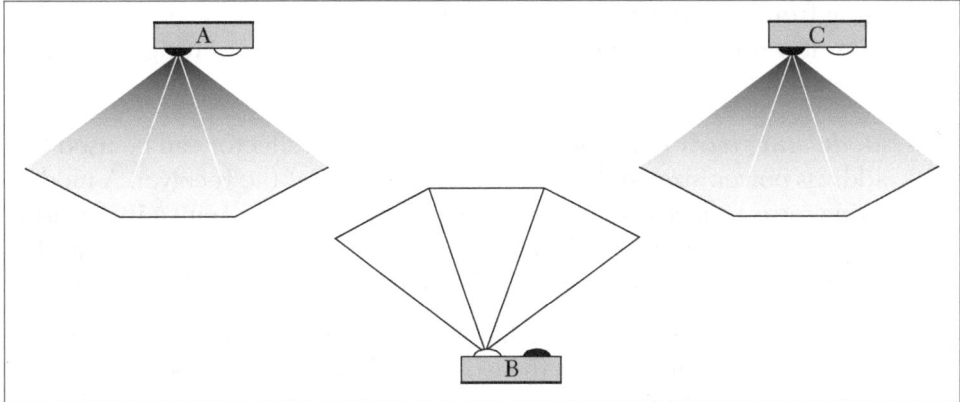

FIGURE 4.8 Multi-user reception system using three nodes: A, B, and C; A and C are transmitting nodes while B is receiving node.

Supposing localization data from angle of arrival is kept at each node, node B can broadcast messages between the node A and the node C through a hop network. If B is a mobile node, it can be placed to adequately expand the optical communication range between A and C when needed.

Optical backscatter estimation and evaluation to assess water quality: The bidirectional system delivers a way for a receiver to observe optical backscattering while its colocated transmitter is active. Background noise and unmodulated light are isolated based on the modulated schemes used. Using volume scattering information, an estimation of the attenuation coefficient can be made found on the measured amount of backscatter. Also, SNR (*signal-to-noise ratio*) measurements can be obtained from the transmitted and received signals.

Electronic switched pointing and tracking: The transmitter receives the information about angle of arrival from its colocated receiver. The transmitter can hence switch to a light beam which points its output in the direction of the beam to be received to optimize the link.

4.7.2 Systems and Methods in Underwater Optical Communication

Photomultipliers tubes (PMTs) are used to achieve wide FOV since they have very large apertures. They have an advantage of short rise time and wide spectral response, not to mention the blue-green window used in optical communication. PMTs also have a wide extent of aperture sizes ranging from 10 mm to 500 mm (20 in) in diameter. These are utilized in underwater optical communication systems to elude pointing and tracking needs.

1. *Modulating retro-reflector*: A modulating retro reflector can be used to address power, size, and pointing requirements at the receiver. A modulating retro-reflector strikes out the requirement for a transmitting laser on a platform containing data and reduces the pointing specifications by retro-reflecting the modulated light again to the communicating source.

2. *Indoor optical wireless:* There has been some exploration in the field of indoor optical wireless in the work of spherical photodiode arrays for enlarging FOV. An improvement in range by a diminution in path loss, multi-path distortion, and background noise can be made possible by optimally combining the photodiode outputs.

3. *RF communication:* Terrestrial RF communications have gained from recent growth in spatial diversity and smart antennas. Mobile communications also give an idea about some of the implementations workable with an antenna. However, in optical systems, we do not have the RF implementation of being able to use cogent beam-forming or phased arrays.

4. *Smart transmitter:* The smart transmitter has the following both electronic switched beam-steering and increased directionality. The LED (*Light Emitting Diode*) is a semiconductor device that produces a relatively narrow spectrum light, dependent on the material used with a specific brightness dependent on the forward bias current that is applied. The speed at which an LED can be modulated is usually limited by the die size for high-brightness LEDs. This implies a tradeoff between power and speed, since larger die size provides higher brightness. The smart transmitter is composed of a shortened hexagonal pyramid with a large number of LEDs. Each LED in the transmitter is coupled with its own lens that converges the extensive FOV of the LED to a limited beam in a particular direction. Each LED is uniquely addressed and driven, which allows the modulator to select an output direction. This constructs the procedure for a basic switched beam-steering at the transmitter side. For a multi-user environment, it is mandatory to provide multiple access to the medium. LEDs at different wavelengths can be used, but receivers would require multiple filters. Time Division Multiple Access would thus need synchronous clocks.

5. *Smart receiver:* Like smart transmitter, the goal of the smart receiver is to develop a quasi-omni-directional system to reduce the pointing and tracking requirements generally associated with free-space optical systems. Further, to potentially reduce pointing and tracking requirements, this design also potentially allows one to estimate and evaluate angle of arrival. This can be used in combination with a CDMA-type multiple access system. Thus, the signals from distinct platforms can be differentiated from their coded signals and have a demonstration of their location. This increases the number of applications and includes applications such as localization, navigation assistance, and mesh networking. Using *multi-input multi-output* (MIMO) techniques, this optical approach possibly also imparts angle and spatial diversity for enhancing the representation of point-to-point links. The smart receiver has increased field of view and angle of arrival estimation.

There are many design considerations that must be kept in mind due to their significance to underwater free-space optical communication. First, unlike optical front-end arrays in terrestrial free-space optics and indoor optical wireless that use photodiode arrays with no lenses, the smart receivers that are used in the underwater communication need to be mounted with an array of lenses. This is done to estimate the angle of arrival of signals focused on the receiver. Free space optical communication underwater has always required an improved FOV and this is one of the primary issues to work upon. A significant improvement in the FOV can be made by using quasi-omni-directional lenses at the receiver side.

A smart transmitter can evaluate water quality by utilizing its backscattered return light and a colocated receiver to estimate the *attenuation coefficient* (channel state) of the channel at the transmitter. This expertise has the benefit of knowing the water quality without counting on a backchannel for back-telemetry or even a different instrumentation sensor. Knowing this information allows the transmitter to, for example, adaptively change its transmitting power, data rate, code rate, or other parameters. The question for this expertise is that the return beam from backscatter, depending on the attenuation coefficient of the channel, can be as low as roughly six orders of magnitude below the output power of the transmitter. To some degree, this can be elucidated by a few methods, including sending a higher power training sequence to enlarge the amount of backscattered light used for estimation and evaluation, the receiver associating the captured light with the genuine information being transmitted, or even temporarily increasing the receiver gain. Expertise such as the use of a lock-in amplifier can be used and is aided by the fact that the transmitter and the backscatter-receiver are colocated.

4.8 TYPES OF MODULATION

In general, the modulation methods developed for radio communications can be adapted for underwater acoustic communications (UAC). However, some of the modulation schemes are more suited to the unique underwater acoustic communication channel than others. Some of the modulation methods used for UAC are as follows:

- *Frequency shift keying* (FSK)

- *Phase shift keying* (PSK)

- *Frequency hopped spread spectrum* (FHSS)

- *Direct sequence spread spectrum* (DSSS)

- *Frequency and pulse position modulation* (FPPM and PPM)

- *Multiple frequency shift keying* (MFSK)

4.8.1 Frequency Shift Keying as Applied to UAC

FSK is the earliest form of modulation used for more advanced forms of UAC by acoustic modems and this method has been used to measure the speed of sound in water. FSK usually employs two distinct frequencies to modulate data, for example, frequency F1 to indicate bit 0 and frequency F2 to indicate bit 1. Hence a binary string can be transmitted by alternating these two frequencies depending on whether it is a 0 or 1. The receiver can be as simple as having analogue matched filters to the two frequencies and a level detector to decide if a 1 or 0 was received. This is a relatively easy form of modulation and was used in the earliest acoustic modems. However, more sophisticated demodulators using *digital signal processors* (DSP) can be used in the present day. The biggest challenge FSK faces in the UAC is multi-path reflections. With multi-path (particularly in UAC), several strong reflections can be present at the receiving hydrophone and the threshold detectors become confused, thus limiting the use of this type of UAC to vertical channels. Adaptive equalization methods have been tried with limited success. Adaptive equalization tries to model the highly reflective UAC channel and subtract the effects from the received signal.

The success has been limited due to the rapidly varying conditions and the difficulty to adapt in time.

In FSK modulation, information bits are used to select the carrier frequencies of the transmitted signal. The receiver compares the measured power at different frequencies to infer what has been sent. Using only the energy detector at the receiver, this scheme bypasses the need for channel estimation and is thus robust to channel variations. However, guard bands are needed to avoid the interference caused by frequency-spreading, and a guard interval is inserted between successive symbol transmissions for channel clearing to avoid the interference caused by time-spreading. As a result, the data rate of FSK is very low. Frequency hopped (FH) FSK improves the data rate as it does not need to wait the channel clearing corresponding to the previous symbol transmission on a different frequency. However, due to the bandwidth expansion via frequency hopping, the overall bandwidth efficiency remains low, typically much below 0.5 bits/sec/Hz.

4.8.2 Direct Sequence Spread Spectrum (DSSS)

In *DSSS modulation*, a narrow band waveform of bandwidth W is spread to a large bandwidth B before transmission. This is achieved by multiplying each symbol with a spreading code of length $B = W$, and transmitting the resulting sequence at a high rate as allowed by bandwidth B. Multiple arrivals at the receiver side can be separated via the despreading operation, which suppresses the time-spreading induced interference and auto-correlation properties of the spreading sequence. Channel estimation and tracking are needed if phase-coherent modulation such as phase shift keying (PSK) is used to map information bits to symbols before spreading. For noncoherent DSSS, information bits can be used to select different spreading codes to be used, and the receiver compares the amplitudes of the outputs from different matched filters, with each one matched to one choice of spreading code. This avoids the need for channel estimation and tracking. Due to the spreading operation, the data rates are often in the order of hundreds of bps while using bandwidth of several kHz, resulting in bandwidth efficiency well below 0:5 bits/sec/Hz.

Single carrier phase-coherent modulation with adaptive channel equalization: One major step towards high rate communication is the direct transmission of phase-coherent modulations, including phase shift keying (PSK) and *quadrature amplitude modulation* (QAM). The channel introduces a great deal of *inter-symbol interference* (ISI) due to multipath propagation. Advanced signal processing at the receiver side is used to suppress the interference; this process is termed as channel equalization. Although widely used for slowly varying multi-path channels in radio applications, channel equalization for fast-varying underwater channel is a big challenge. The canonical receiver is successfully combined a second-order phase-locked-loop to track channel phase variations with an adaptive decision feedback equalizer to suppress the ISI. Without the guard interval insertion and the spreading operation, much higher data rates can be achieved with single carrier phase-coherent modulation than those of FSK and DSSS. One concern about single carrier transmission is that the receiver may be less robust as the parameters in the adaptive receiver need to be fine-tuned depending on channel conditions. When data symbols are transmitted at a higher rate, the same physical channel leads to more channel taps in the discrete-time equivalent model. The complexity of time-domain equalization grows quickly as the number of channel taps increases, which will eventually limit the rate increase for single carrier phase-coherent transmission.

4.8.3 Multicarrier Modulation

The idea of multi-carrier modulation is to divide the available bandwidth into a large number of overlapping sub bands, so that the waveform duration for the symbol at each sub band is long compared to the multipath spread of the channel. Consequently, inter-symbol interference may be neglected in each sub band, greatly simplifying the receiver complexity of channel equalization. Precisely due to this advantage, multicarrier modulation in the form of *orthogonal frequency division multiplexing* (OFDM) has prevailed in recent broadband wireless radio applications. However, underwater channels entail large Doppler spread, which introduces significant interference among OFDM subcarriers. Lacking effective techniques to suppress the intercarrier interference (ICI), early attempts at applying OFDM to underwater environments had a very limited success.

Recently, there have been extensive investigations of underwater OFDM communication, including noncoherent OFDM based on on-off keying, on a low-complexity adaptive OFDM receiver, and on a pilot-tone-based block-by-block receiver. The block-by-block receiver does not rely on channel dependence across OFDM blocks, and thus it is robust to fast channel variations across OFDM blocks. In contrast to single carrier phase-coherent transmission, OFDM has the desirable property that one signal design can be easily scaled to fit into different transmission bandwidths with negligible changes on the receiver. With bandwidth varying from 3 kHz to 50 kHz, data rates from 1.5 kbps to 25 kbps after rate 1/2 coding and QPSK modulation are reported. Further, with different bandwidths of 12 kHz, 25 kHz, and 50 kHz, data rates of 12 kbps, 25 kbps, and 50 kbps after rate 1/2 coding and 16-QAM (Quadrature Amplitude Modulation) are also achieved. These demonstrate the feasibility and flexibility of OFDM for underwater acoustic communication.

4.8.4. Multi-Input Multi-Output Techniques

A wireless system that employs multiple transmitters and multiple receivers is referred to as a multiple-input multiple-output (MIMO) system. It has been shown that the channel capacity in a scattering-rich environment increases linearly with $\min(N_t, N_r)$, where N_t and N_r are the numbers of transmitters and receivers, respectively. Such a drastic capacity increase does not incur penalty on precious power and bandwidth resources, but rather comes from the utilization of the spatial dimension virtually creating parallel data pipes. Hence, MIMO modulation is a promising technology to offer yet another fundamental advance on high data rate underwater acoustic communication. MIMO has been applied in both single carrier transmission and multi-carrier transmission. For single carrier transmission, existing adaptive channel equalization algorithms are leveraged to deal with MIMO channels. The data rate increases substantially. For example, a 12 kbps rate is achieved with 3 kHz bandwidth at the range of 2 km, leading to a bandwidth efficiency of 4 bits/sec/Hz, using six transmitters and QPSK modulation. Due to OFDM's unique strength in handling long dispersive channels with low equalization complexity, the combination of MIMO and OFDM is another appealing solution for high data rate transmission but with low receiver complexity.

MIMO introduces additional interference among parallel data streams from different transmitters. Also, each receiver has more channels to estimate, which requires more overhead spent on training symbols. For fast varying underwater channels, the number of transmitters might not be large for best rate-and-performance tradeoff. In addition to colocated antennas, distributed MIMO is also possible if clustered single-transmitter nodes could cooperate. Certainly, implementation of distributed MIMO needs to address challenging practical issues such as node synchronization and cooperation.

4.9 INTERNET TO SHIPS USING OCEANOGRAPHIC TOOL (SEANET)

Out in the middle of the ocean, thousands of miles from shore, oceanographers aboard research ships have been cut off from the fiber-optic-based transmission lines that link the rest of the land-based world to the Internet. But in 1995, a new communications system called SeaNet (Figure 4.9) was developed to extend the Internet to ships at sea. SeaNet greatly expands the number of scientists and students that can participate on research cruises. It gives them access to shipboard data, images, and information as they sit at their computer terminals in shore-based laboratories or schools. The SeaNet system takes advantage of high-speed satellite, cellular, and other communications technology, as well as specialized software and hardware tools developed by the SeaNet engineering group, to provide affordable, high-speed data transmission to and from shore.

The SeaNet uses *data pipes* to reduce cost and time. It allows data files to be transferred easily and without errors between the ship and any computer on shore connected to the Internet. Data files containing information or images are first combined into one *batch*. Then they are *compressed*, so that they don't take as much time (and money) to be transmitted via satellite. Then they are put into a computer holding bin. Several times a day, a technician on board ship establishes a connection to a satellite and

FIGURE 4.9 SeaNet is a new communications system developed in 1995 to extend the Internet to ships at sea.

the batch of compressed files is transferred. To maximize the amount of data transferred during each active satellite link, data files are transmitted to and from the ship simultaneously. The files are also transferred at high speed—about 64,000 bytes per second. This fast, two-way transfer makes the system very cost-efficient.

4.9.1 Global Positioning System (GPS) in Ship

Ships use satellite navigation to find their positions in the middle of the ocean. Each satellite is in a constant, or fixed, orbit around the Earth and its position is determined very precisely by the military, using several ground stations and antennas around the world. More than 60 GPS satellites have been launched over the past 30 years. Figure 4.10 shows launching of a GPS satellite. The *global positioning system* (GPS) is

FIGURE 4.10 A GPS satellite being launched on a Delta II Booster rocket.

a satellite-based radio navigation system that permits people on land, at sea, and or in airplanes to determine their three-dimensional position and velocity. GPS also provides users with very accurate time. All this information is available 24 hours a day in all weather, anywhere on Earth. The GPS was developed primarily for military applications during the Cold War, but for the last 25 years oceanographers have also been able to use satellite-based navigation to determine their position. Knowing your precise location is important because all of the data and observations you are collecting have to be referenced to a position on the Earth's surface, or in our case, on the seafloor. In addition, we often want to return to a scientifically interesting seafloor site, which would be almost impossible to relocate without satellite navigation.

Normally four or more satellites are used to fix a ship's position. The satellites and the receivers on the ship all have very precise, synchronized clocks. Satellites send radio waves to the ship and clocks measure how long it takes for these signals to make their trip. The longer it takes, the longer the distance between satellite and ship. In this way, time is converted to distance. Using measurements from at least three satellites (a method called

FIGURE 4.11 GPS satellite nosecone.

triangulation), computers can calculate the unique point on the face of the Earth where all three (or more) of the distances measured from all the satellites intersect. This gives the location of the ship. Figure 4.11 shows a GPS satellite loaded into the nosecone of the delivery rocket. The folded-up gold foil is designed to spread out into the solar panels, which provide power to run the satellite when it is orbiting in space.

Figure 4.12(a) shows the GPS Nominal Constellation, which has 24 satellites in six orbital planes, four satellites in each plane 20,200 km altitude, 55° inclination. Figure 4.12(b) shows GPS contacts to locate the objects.

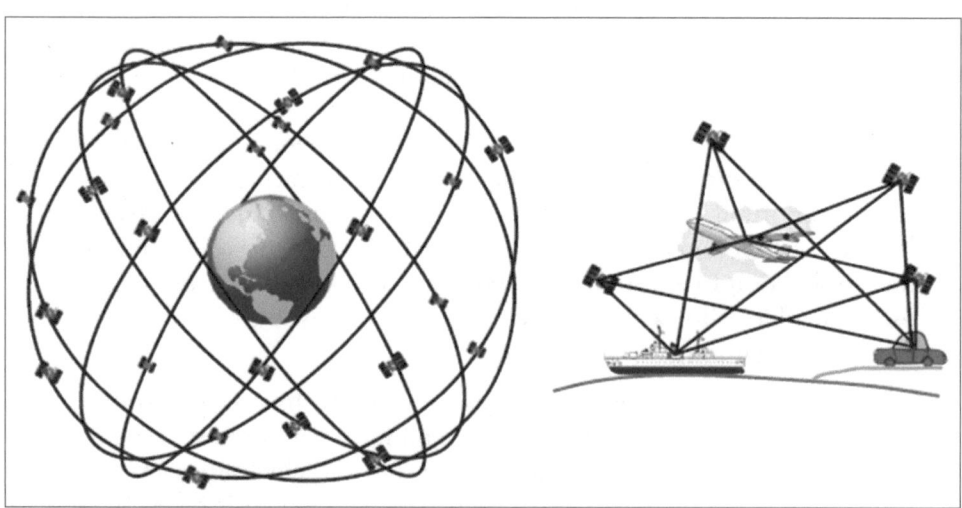

FIGURE 4.12. **a.** GPS satellite orbits and **b.** GPS contacts with receivers.

4.9.2 Internet Access for Voyagers

For better or worse, we live in a connected world. When a person sets off voyaging, access to Internet connectivity changes with place. It depends on factors like the location where we are going, what kind of access is needed, and what is the cost and energy consumption. There are three main considerations here: purchase and installation costs, access fees, and energy consumption requirements to operate.

1. *Wi-Fi:* One thing that every boat can benefit from for starters is improved Wi-Fi access. Even laptops with built-in wireless could add an external Wi-Fi antenna or access point. The biggest impediment to getting a good signal for Wi-Fi on board is the clutter of equipment and metal structure between the computer and the land-based Wi-Fi antenna. For occasional access with little concern about performance, simple USB wireless network adapters are available. They can be suction-cupped to a port for use and stowed when not needed. While very convenient, they will be limited by similar line-of-sight issues to the computer itself. Still, they offer a significant improvement over the computer's built-in antenna. For more reliable Wi-Fi access while in marinas or hot spot areas, consider an Ethernet-based wireless access point with an external antenna.

2. *Smartphones:* One can even use a smartphone as a Wi-Fi hot spot on the boat. Then a laptop will have instant access to the Web and all of its features. Smartphones are very energy efficient and will use little additional power on board.

3. *E-mail offshore:* Wi-Fi and smartphones may be good solutions for some coastal or other near-land voyaging, but what about offshore and remote locations? For any sort of access more than a few miles from home shores, you need to look at satellite or long-range radio solutions. One of the most common forms of offshore communications is e-mail via *single-sideband* (SSB) radio using a global service like SailMail or Winlink. Any of these services assumes as a prerequisite a good quality SSB radio [Figure 4.13] and installation. It also requires purchasing a custom radio modem.

FIGURE 4.13 SSB radio is a good choice for sending e-mail offshore.

4. *Global Internet access:* Radio e-mail services, although global in reach, do not allow for general Internet access—for that, a satellite modem is needed. There are numerous satellite services available, which come in a range of capabilities and costs. For global operation, there are only a few satellite services that give Internet access: Iridium and Inmarsat are probably the two best known. For coastal cruising (or maybe just trans-Atlantic), Globalstar offers an alternative. These providers—Iridium, Inmarsat, Globalstar and value-added providers like KVH—provide different technologies and fundamental capabilities. For instance, Iridium primarily offers satphones for voice communication, but those phones can be used to send data as well. Or, for higher bandwidth applications, Iridium offers a standalone dome antenna for use with the company's OpenPort service. These ranges of capabilities apply to the other satellite networks as well.

The Iridium Pilot unit is a high-end solution for Internet access for large yachts and commercial vessels. An alternative is a single Iridium satphone with an external antenna as in Figure 4.14.

Inmarsat—*geosynchronous ("geostationary") orbits* (GEO)—Inmarsat satellites orbit the Earth directly over the equator at an altitude of about 22,236 miles. The period of their orbits exactly matches the rotational period of the Earth and so they appear to be stationary over one spot on the Earth at all times. GEO satellites always have a consistent view of the Earth below, and with control of onboard antennas they can increase capacity in some areas while reducing it in others. Most voyagers would never reach the polar limits of Inmarsat's coverage.

FIGURE 4.14 Iridium sat phone with external antenna.

Because GEO satellites are so far away, the antennas here on Earth need to be high-gain and highly directional. For this reason, dishes are the most common form of antenna and for use on boats these dishes need to be actively steered to always point at the selected satellite. As a result, the antennas tend to be larger, heavier, and more power hungry than most 40-foot boats care to carry.

Iridium—*low earth orbit* (LEO)—Iridium satellites orbit the Earth at an altitude of only 485 miles, and so lower power and omni-directional antennas can be used on board. Also, any given satellite will only be in view for a short period from a fixed point on the Earth's surface. Therefore, the Iridium space-borne fleet consists of 66 satellites in six different planes of orbit—in this way one or more satellites should be visible anywhere on Earth at any given time. (In the original concept, there would have been seven orbital planes with 11 satellites each. The name Iridium reflects the 77-component concept—the atomic number of iridium is 77. Later it was determined that six orbital planes were sufficient. The name was never changed to Dysprosium.) The Iridium network offers true global coverage because of its ability to relay data from satellite to satellite in real time. Since a satellite 485 miles above the Pacific Ocean is not going to be able to "see" a ground station at the same time, to establish its connection from boat to the Internet, it needs to relay data to another satellite nearby that can communicate with a ground station. Iridium phone calls and data connections can be reliably established from anywhere on Earth, but they tend drop out after a short period. Iridium modems operate at low speed of 10–20 kb/sec.

Globalstar—*LEO "bent pipe" satellites*—Globalstar satellites are also in low Earth orbit at an altitude of about 850 miles. Unlike Iridium satellites, the Globalstar birds cannot relay messages from satellite to satellite. Therefore, to establish a connection from user to network, the satellite must be simultaneously in view of the user and a Globalstar ground station. For this reason, Globalstar's coverage, while far-reaching, does not include the world's oceans, or indeed even southern Africa. Globalstar does offer competitive rates for North and South America, and in between. Like Iridium, its modems are smaller and the antennas are omni-directional.

4.9.3 Voice Communication on Ship

Presently voice communication on ship is provided with the help of an application called *Voice over IP* (VOIP) which uses a narrow bandwidth over the Internet or an IP network. Internet access on ship will not only facilitate business but also help crew members to stay in touch with their family members in a speedier and cost-effective way.

More affordable services are provided by clubbing land-based technology and satellite technology. The service utilizes *Internet protocol virtual private networks* (IP VPN), which prevents hackers and thus helps companies to protect their vital data. The system consists of a signal relaying device that resembles radar, a satellite earth station, and a satellite. This system allows unlimited and seamless broadband connectivity with speed up to 512K bps which enables voice calling, e-mail transferring, GPS system, and also web browsing [Figure 4.15].

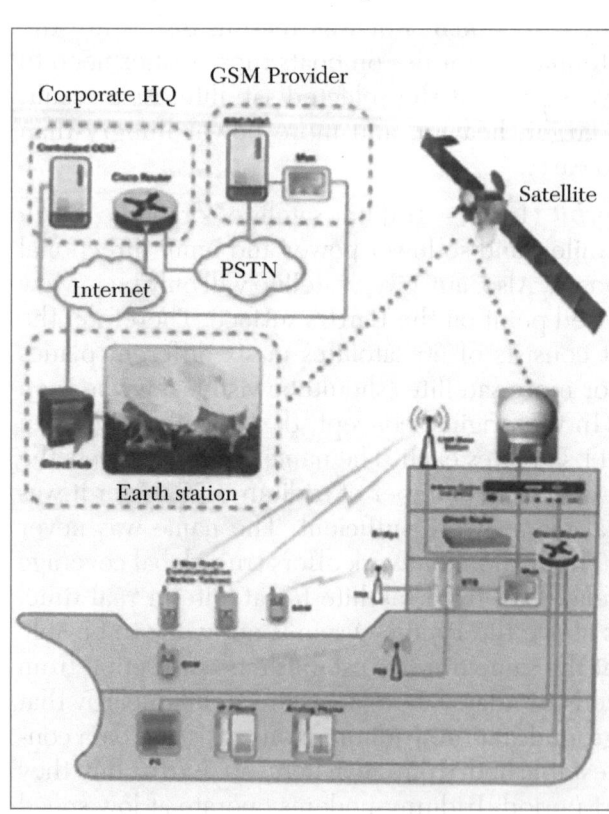

FIGURE 4.15 Internet on ship.

4.9.4. Submarine Communications Cable

Figure 4.16 shows a submarine communication cable cross section. The parts are 1) polyethylene, 2) Mylar tape, 3) stranded steel wires, 4) aluminum water barrier, 5) polycarbonate, 6) copper or aluminum tube, 7) petroleum jelly, 8) optical fibers.

A *submarine communications cable* is a cable laid on the sea bed between land-based stations to carry telecommu-

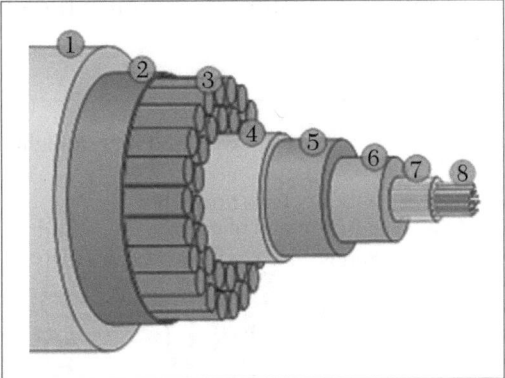

FIGURE 4.16 A cross section of a modern submarine communications cable.

nication signals across stretches of ocean. The first submarine communications cables carried telegraphy traffic. Subsequent generations of cables carried telephone traffic, then data communications traffic. Modern cables use optical fiber technology to carry digital data, which includes telephone, Internet, and private data traffic. Modern cables are typically 69 mm (2.7 in) in diameter and weigh around 10 kg/m (7 lb/ft), although thinner and lighter cables are used for deep-water sections. As of 2010, submarine cables link all the world's continents except Antarctica.

4.9.5 Optical Submarine Cable Repeaters

Optical fiber repeaters use a solid-state optical amplifier, usually an Erbium-doped fiber amplifier [Figure 4.17]. Each repeater contains separate equipment for each fiber. These comprise signal reforming, error measurement, and controls. A solid-state laser dispatches the signal into the next length of fiber. The solid-state laser excites a short length of doped fiber that itself acts as a laser amplifier as the light passes through it. This system also permits wavelength-division multiplexing, which dramatically increases the capacity of the fiber. Repeaters are powered by a constant direct current passed down the conductor near the center of the cable, so all repeaters in a cable are in series. Power feed equipment is installed at the terminal stations. Typically, both ends share the current generation, with one end providing a positive voltage and the other a negative voltage. A virtual earth point exists roughly halfway along the cable under normal operation. The amplifiers or repeaters derive their power from the potential difference across them. The optic fiber used in undersea cables is chosen for its exceptional clarity, permitting runs of more than 100 km between repeaters to minimize the number of amplifiers and the distortion they cause.

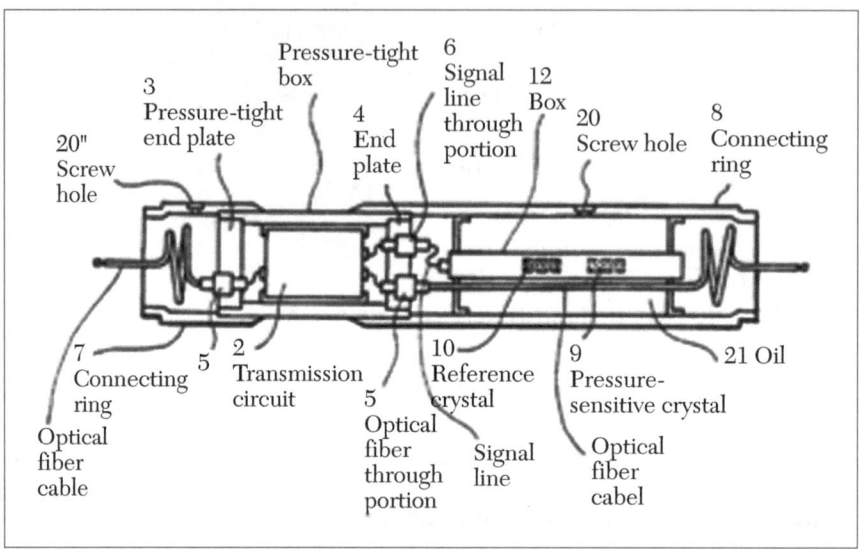

FIGURE 4.17 The optical submarine cable repeater.

4.9.6 Antarctica

Antarctica is the only continent yet to be reached by a submarine telecommunications cable. All phone, video, and e-mail traffic must be relayed to the rest of the world via satellite, which is still quite unreliable. Bases on the continent itself communicate with one another via radio, but this is only a local network. To be a viable alternative, a fiber-optic cable would have to be able to withstand temperatures of –80° C as well as massive strain from ice flowing up to 10 meters per year. Thus, plugging into the larger Internet backbone with the high bandwidth afforded by fiber-optic cable is still an infeasible economic and technical challenge in the Antarctic. Figure 4.18 shows the submarine cable map and Figure 4.19 shows the underwater cable.

4.9.7 Perfectly Secure Communication

Underwater vehicles are increasing in tactical importance. To achieve their full potential, underwater vehicles may be required to be virtually undetectable and be able to remain submerged for long periods of time.

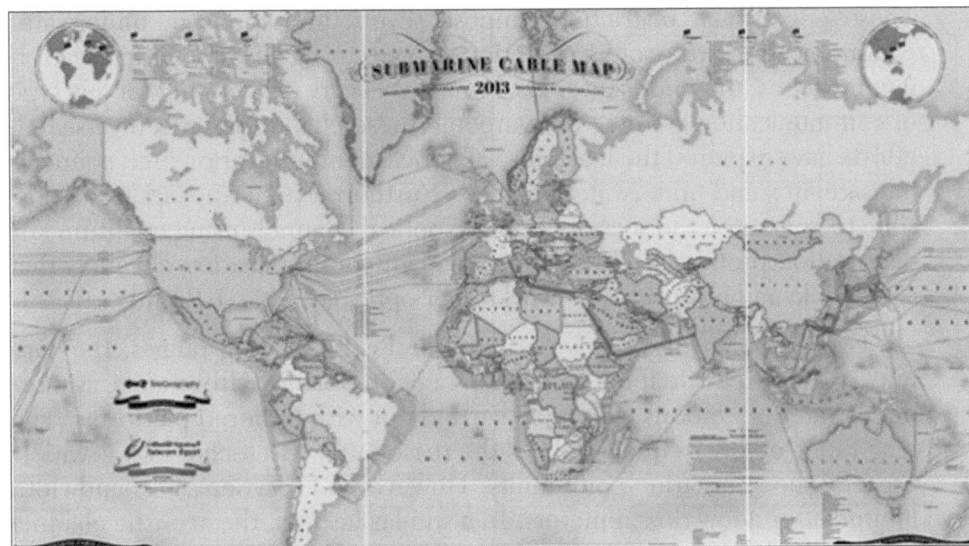

FIGURE 4.18 Submarine cable map.

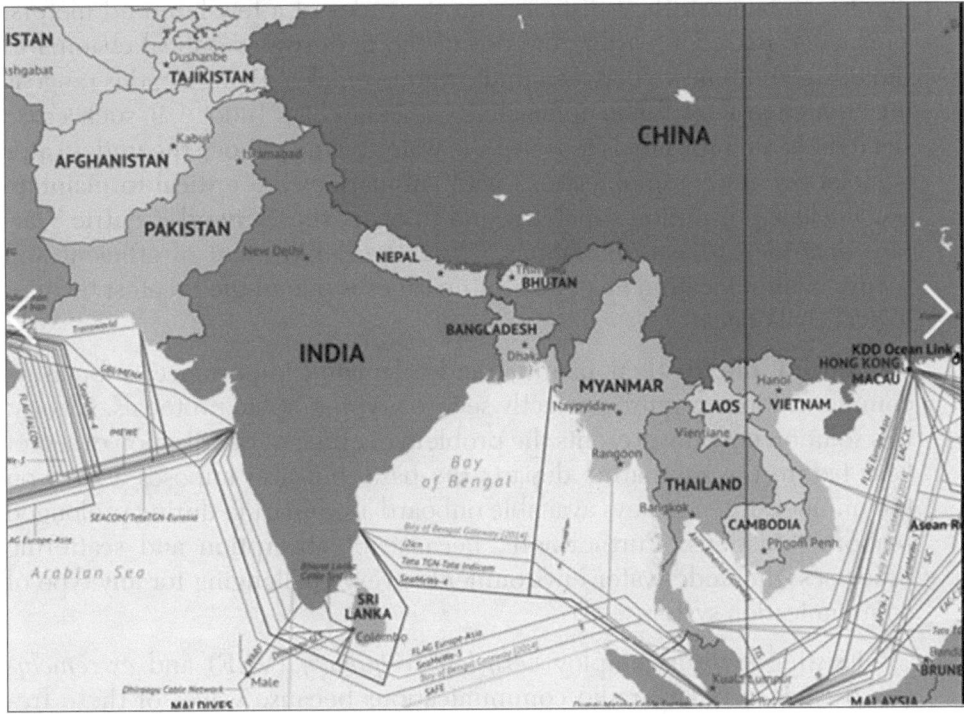

FIGURE 4.19 Undersea cables wiring the ends of the Earth.

These operational restrictions impose severe limitations to underwater communication protocols. In addition, due to strong absorption and scattering processes, the underwater environment is very challenging for any type of communication system. To overcome these challenges, recent research efforts have explored the feasibility of optical communication links connecting satellites and underwater vehicles. Furthermore, it appears to be possible to guarantee the security of these optical channels using quantum key distribution protocols. There are four main communication channels that can enable an underwater network: fiber optic, acoustic, RF, and optical.

Even though fibers provide high bandwidth at long ranges, they are infeasible for real life implementation in operational environments, as most underwater operations cannot be carried out with tethered platforms. The long range of low frequency RF is the reason why this technology is widely used by the submarine community. However, RF carriers are bandwidth limited after a few dozen meters. In a similar fashion, the acoustic channel has an even large range, but once again, bandwidth is severely limited after a few meters. On the other hand, the optical channel offers the possibility of high bandwidth at distances on the order of a few hundred meters. This alone makes infeasible the use of the underwater optical channel to directly communicate between underwater vehicles. However, this range is big enough to allow for an optical link with a satellite. Indeed, in such a case the light beam only needs to travel the water column above the underwater vehicle. Secure communications with submarines are critical to maintain our nuclear prevention capability and to enact the Network Centric Warfare doctrine of naval operations. Thus, the deployment of efficient and secure communication links with submarines is one of the greatest technological challenges.

Indeed, due to their planned and calculated importance, submarine communications require perfectly secure cryptographic protocols. Clearly, this solution not only presents the problem of efficient distribution of secret keys before the submarine departs the base, but also imposes a limit on the number of secret keys available onboard a submarine during prolonged seaborne missions. Furthermore, because of absorption and scattering processes, the underwater environment is very challenging for any type of communication systems.

Existing systems employ *very-low-frequency* (VLF) and *extremely-low-frequency* (ELF) radio communications because waves of these frequencies can partially penetrate a body of water. However, these systems impose severe operational limitations: these are extremely low bandwidth, one-way systems that require towed antennas or buoys, and submarines need to steer specific courses and reduce their speed. RF signals are attenuated greatly by water, so VLF is used for submarine communications but requires a huge antenna system. Therefore, Bluetooth and other RF-based systems won't be suitable. For a depth of 30 meters, acoustic modems are a practical solution. Optical communications would be another possibility. With optical it would depend a lot on the water quality and ambient light.

Submarine communication is restricted by the depth at which vessels can exchange information and the speed at which they can do so through the medium of water. Recently however, researchers have made impressive strides in solving this dilemma using a technique called *quantum key distribution* (QKD). QKD promises to guarantee secure communication through the principles of quantum mechanics, without sacrificing speed or forcing the submarine to rise nearer the surface. "Submarine communication is restricted by the depth at which vessels can exchange information." For a submarine to retain all its tactical advantage, it must remain submerged in the mixed layer, which is around 60–100 meters deep, below which surface sonars cannot detect them. Submarine communications are currently carried out while submerged using ELF or VLF radio waves because only very low or extremely low frequencies can penetrate the water at those depths.

Using ELF and VLF presents several disadvantages, however. The transmission sites must be very large, meaning the submarine must tow cumbersome antenna cables, plus it usually has to align on a specific orientation and reduce speed to obtain optimal reception. VLF and ELF frequencies only offer very low bandwidth: VLF supports a few hundred bits a second while ELF sustains just a few bits each minute. This prevents the transmission of complex data such as video. One potential solution is to carry out optical communications using a laser, a concept which has been around since the 1980s when experiments were carried out to demonstrate that it is possible to maintain an optical channel between a submarine and an airborne platform. The perfectly secure communication system requires two optical channels: a quantum channel to enable the generation of perfectly secure keys and a classical channel to transmit the encrypted information. Both channels are realized with lasers operating in the region of least attenuation (around 460–480 nm). In addition, both channels need to be operational at depths greater than the mixed layer (60–100m) so the underwater vehicle will not reveal its location to an active surface sonar. It is important to note that the proposed system does not require underwater optical communications over long distances.

Indeed, the communications link only needs to traverse the water column above an underwater vehicle traveling at a nominal depth of 100 meters. Of course, the optical beam also needs to travel across the atmospheric column to reach a low earth orbit satellite (between 160–2,000 km above Earth's surface). The proposed optical channel overcomes the biggest limitations of current VLF and ELF communication systems. Indeed, this is a two-way communication system which allows the underwater vehicle to receive and transmit messages; the transmitter does not have the vulnerabilities of the large ELF and VLF antenna sites; the underwater vehicle is not required to alter its course or reduce its speed; the optical carrier allows for higher data bandwidth; and finally, quantum encryption allows for perfectly secure communications without the need of using a trusted courier to distribute the cryptographic keys before the underwater vehicles leaves the base. In all fairness, the optical channel introduces its own share of potential disadvantages. The most important is the possibility that the optical beam could be detected by the opponent. If the adversary manages to locate two points in the laser beam, then the position of the underwater vehicle will be revealed.

There are some considerations and mitigation strategies that could overcome this problem. On one side, lasers are not highly susceptible to passive detection and interception because they are highly directional narrow beams. In principle, a satellite could detect any eavesdropper on the line of sight to the underwater vehicle and act accordingly (e.g., stop the transmission). On the other hand, it is known that if enough light is recovered, *scattered signal reconstruction* (SSR) techniques could reconstruct the original signal from the light that has been scattered out of the laser beam. This means that an optical channel, even if it is highly directional, requires some level of encryption. QKD is a protocol which uses quantum information to generate a pair of perfectly secure keys.

"Quantum information is different from classical information, because in classical information the unit is the bit and it can have the value of zero or one." "The unit of quantum information is the qubit, which is a quantum state of a photon. It can be on zero, one or any superposition of zero and one. It's more of a concept of information than the classical one." Quantum information has two important properties for securing communications. It cannot be copied which means it cannot be forged, and every time a quantum state is measured by an observer it gets collapsed, which means its properties are very difficult to detect. "QKD promises to guarantee secure communication through the principles of quantum mechanics, without sacrificing speed." Combined in QKD, these properties can be used to generate perfectly secure keys because the secrecy of the keys is guaranteed by the laws of physics.

4.9.8 Environmental Impact

The main point of interaction of cables with marine life is in the benthic zone of the oceans where most cable lies. Studies in 2003 and 2006 have indicated that cables pose minimal impact on life in these environments. In sampling sediment cores around cables and in areas removed from cables, there were few statistically significant differences in organism diversity or abundance. The main difference was that the cables provided an attachment point for anemones that typically could not grow in soft sediment areas.

FURTHER READINGS

1. OFDM for Underwater Acoustic Communications by Shengli Zhou and Zhaohui Wang

2. Advanced optical wireless communication systems by Shlomi Amon at al.

3. Wireless communication by S.K.Kataria.

PART A QUESTIONS: (ANSWER IN A WORD OR A SENTENCE)

1. What are the advantages of electromagnetic wave communication?

2. What are the limitations of electromagnetic wave communication?

3. List one advantage and one limitation of optical wave communication.

4. What is the main advantage of acoustic wave communication?

5. What is the speed of sound in water?

6. What are the range, data rate, and efficiency of acoustics?

7. What are the range, data rate, and efficiency of optical?

8. Compare optical and acoustic communication.

9. _____ lights are used in optical communication as sources.

10. What is the window effect of blue-green laser?

11. What are QAM, OFDM, and FSK anagrams of?

12. _____ are low earth orbit satellites.

13. Geosynchronous orbit satellites are _____.

14. Globalstar satellites are _____.

PART B QUESTIONS: (ANSWER IN A PAGE)

1. What are the properties of acoustic waves?

2. What is acoustic communication?

3. Write about the optical communication.

4. Compare acoustic, electromagnetic, and optical waves in a seawater environment.

5. What are the limitations of underwater acoustic communications?

6. Write about wireless underwater optical communication.

7. Write about underwater laser sensor architecture.

8. Compare underwater laser sensor networks with underwater acoustic sensor networks.

9. Explain the operation of wireless laser communication systems.

10. What is the role of MEMS in underwater communication?

11. What is frequency shift keying?

12. What is direct sequence spread spectrum?

13. What is multicarrier modulation?

14. Write a note on SeaNet.

15. Write a short note about GPS in ships.

16. What are the parts of the submarine communication cable?

PART C QUESTIONS: (ANSWER WITHIN 2 OR 3 PAGES)

1. What are the three different carriers used for underwater communication? Explain each.

2. Write about underwater laser communication in detail.

3. Explain underwater mobile communication.

4. Write about the different types of modulation for UAV.

5. How can you achieve Internet access on ship?

REFERENCES

http://www.brighthubengineering.com/

http://opticalengineering.org/

CHAPTER 5

OCEANOGRAPHIC WIRELESS SENSOR NETWORKS

This chapter discusses wireless sensor networks, oceanographic WSN, WSN architecture, WSN network topologies, WSN applications, and underwater network technology. It also discusses wireless underwater acoustic sensor networks, 2-dimensional underwater sensor network architecture, 3-dimensional underwater sensor network architecture, and sensor network architecture with AUV.

5.1 INTRODUCTION TO WIRELESS SENSOR NETWORKS

The ocean must be continuously observed to detect climate changes or pollution of the environment, which effects human and animal habitats. A *wireless sensor network* (WSN) consists of thousands of sensor nodes, each node containing a processing unit, a transceiver, memory, a battery, and sensors. Current terrestrial applications include the prevention of wood fires or the detection of leakages along a dike. The new approach is to verify the advantages of WSNs in oceanography to enhance hydrography, eutrophication, and detection of pollution.

A wireless sensor network (WSN) consists of dedicated sensor nodes with sensing and computing capabilities, which can sense and monitor the physical parameters and transmit the collected data to a central location using wireless communication technologies. A WSN has several inherent characteristics, including uncontrollable environments, topological constraints, and limited node resources for energy and computational power. Generally, a WSN deploys more sensors than the optimal placement to improve system reliability and fault tolerance. Marine environment systems are particularly vulnerable to the effects of human activities related to industry, tourism, and urban development. Traditionally, oceanographic research vessels or ships were used to monitor marine environments, which is a very expensive and time-consuming process that has a low resolution both in time and space. In a WSN-based marine environment monitoring system, various kinds of sensors are used to monitor and measure different physical and chemical parameters such as water temperature, pressure, wind direction, wind speed, salinity, turbidity, pH, oxygen density, and chlorophyll levels.

5.1.1 Terrestrial vs. Oceanographic WSNs

Extensive research has been conducted on *terrestrial wireless sensor networks* (t.WSNs). In t.WSNs the high miniaturization of each node is demanded, in *oceanographic WSNs* (o.WSNs) the size of a node is less important because high density is not required. Table 5.1 gives the difference between the two different networks.

	Terrestrial WSN	**Oceanographic WSN**
Network-Density	Fine grained (small area, many nodes)	Coarse grained (large area, few nodes)
Size of one Device	Required tiny	Small
Network-Mobility	Low	High (due to streams)
Energy Consumption	Low (radio frequency, short distances)	High (acoustic waves, high attenuation, long distances)
Prize per Node	Cheap	Expensive (modem, sensors)

TABLE 5.1 Basic Differences between Terrestrial and Oceanographic WSNs

Mobility in t.WSNs is not a problem in most scenarios. However, due to water streams, mobility in o.WSNs is very high and must be considered if nodes are not fixed. In both networks, minimized power consumption is imperative because, on the one hand, nodes in t.WSNs are very tiny with a small battery, while on the other hand, in o.WSNs underwater modems feature high power consumption (omni-directional transmission) and radio transceivers on the water surface have to transmit over large distances. Correspondingly, o.WSNs are very expensive, due to the need for acoustic modems, special sensors (e.g., chlorophyll), and water, press-proofed cases.

5.2 OCEANOGRAPHIC WSNS

At present, measurements of the ocean environment are mainly done by ships and with buoys fixed at the sea bottom. Particularly, research ships enable detailed measurements with a high variety of sensors (e.g., temperature, salinity, and oxygen). In contrast, area coverage is highly limited since the ships cruise only on limited routes. Additionally, *ship of opportunity sampling* is an inexpensive alternative, wherein regular ferries, freighter and other commercial ships take measurements on their routes. Moreover, satellites measure sea surface temperature and sea level (e.g., SeaSat, ERS-X, and IRS). The problems of classical methods are the high costs of navigation, the small period of data acquisition, the partly missing interaction between control systems ashore and in the water, and the rising financial costs. Moreover, the time expenditure is high due to possible device errors or failures that cannot be fixed immediately. Sometimes, more than months lie in between measurement and its examination. A new approach is to deploy large WSNs over and under the ocean surface. There are two different scenarios. First, short-term monitoring facilitates the acquisition of high resolution measurements in a small region (e.g., disaster prevention, pollution observation). Second, long-term monitoring of large areas is possible (e.g., improved weather forecasts, detection of climate change), which has been extremely difficult. Additionally, new observation techniques are imaginable, like the dynamic tracing of streams, where sensor nodes float with the stream (e.g., tracing the Gulf Stream). After spreading the sensor nodes over the ocean by plane, in a first phase the network is established in a self-organized manner and every node computes an initial position, as in Figure 5.1.

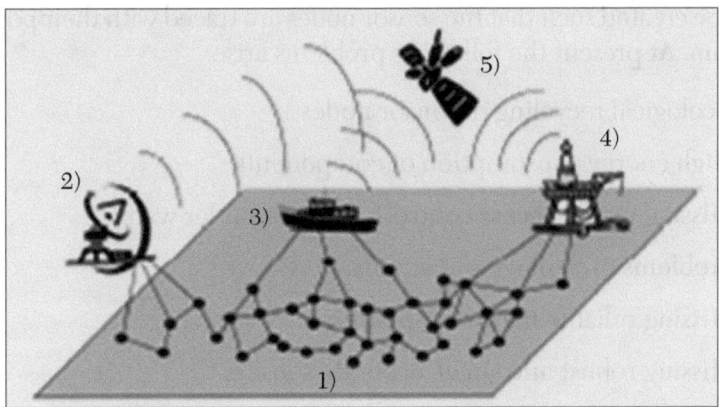

FIGURE 5.1.A Oceanographic wireless sensor network: 1) Sensor node with links; 2) on-shore base station; 3) ship; 4) fixed base station; 5) satellite

FIGURE 5.1.B Prototype of sensor node.

Then, the nodes start the observation phase by measuring the environment with specific sensors. At constant time intervals, nodes send sensor data and positions, hop-by-hop fashion, to one of the base stations in range. Either the base station is the endpoint (on-shore base station) and the information is analyzed instantly, or the data are forwarded to a satellite (from ship, fixed base station), where all endpoints are connected. Finally, over time a stream profile can be created such that the sensor nodes are traced with their positions in the stream. At present the following problems arise:

- Ecological recycling of sensor nodes
- High energy consumption of components
- Missing media access control-protocols under water
- Problems of deployment and fixation
- Missing reliable hardware-platforms
- Missing robust and small, accurate sensors
- High cost per sensor node

Encapsulated prototype modules (Figure 5.1.b) with integrated temperature sensors allow monitoring of the surface temperature. Wireless sensor networks enhance the quality of ocean monitoring. In addition to the higher resolution of measuring data, the higher coverage of surface, and long-term monitoring in real time, new applications are enabled, such as dynamic observation of streams.

5.2.1 Application Areas of Oceanographic WSN

WSN-based marine environment monitoring has broad coverage, including a number of application areas: water quality monitoring, ocean sensing and monitoring, coral reef monitoring, and marine fish farm monitoring. Different application areas require different WSN system architectures, communication technologies, and sensing technologies. A water quality monitoring system is usually developed to monitor water conditions and qualities including temperature, pH, turbidity, conductivity, and dissolved oxygen (DO) for ocean bays, lakes, rivers, and other water bodies. An ocean sensing and monitoring system is used to monitor ocean water conditions and other environmental parameters. A coral reef monitoring system is normally installed to monitor coral reef habitats using an autonomous, real-time, and in-situ wireless sensor network. A marine fish farm monitoring system is developed to monitor water conditions and qualities, including temperature and pH, and accurately quantify the amount of fecal waste and uneaten feed for a fish farm.

5.2.2 Common WSN Architecture

Figure 5.2 shows a common wireless sensor network architecture for monitoring marine environments, which consists of sensor nodes, sink nodes, a base station, a server, and user terminals. Sensor nodes can sense and monitor the in-situ environmental parameters such as water temperature, salinity, turbidity, pH, oxygen density, and chlorophyll levels and transmit the collected data to sink nodes via wireless communication using ZigBee or some other communication protocol. Communication between sensor nodes and a sink node is usually point-to-point. A sink node collects data from a group of sensor nodes, and transmits the collected data to the base station via the GPRS network. The server stores and processes the received data from the base station. The user terminals connect the server over the Internet.

The design and deployment of a lasting and scalable WSN for marine environment monitoring should carefully take into account the following factors: the hostile environment, the network topology, communication protocols, the number of nodes, buoys, mooring systems, oceanographic sensors, energy supply, and so on.

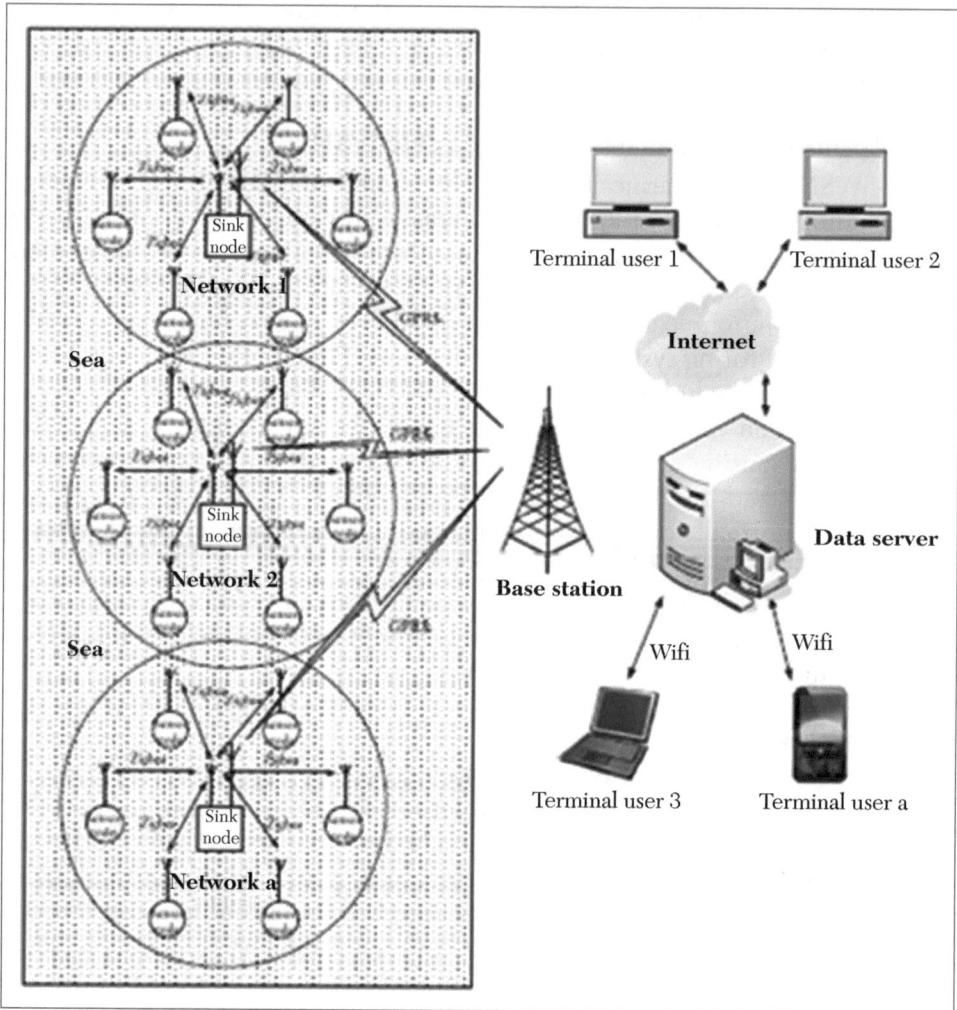

FIGURE 5.2 Common architecture of WSN-based marine monitoring systems.

5.2.3 General Sensor Node

Figure 5.3 shows the architecture of a general sensor node in a marine environment monitoring system. It usually includes a buoy device to protect nodes' electronic devices against water. A marine monitoring sensor node normally consists of the following four main modules:

1. A sensing module for data acquisition;

2. A central processing module for local data processing and storage;

3. A wireless transceiver module for wireless data communication;

4. A power supply module for energy supply.

A *sensing module* is usually composed of several probes and sensors (with associated amplifiers and A/D converters) to sense and monitor the physicochemical parameters of marine environment as mentioned above. A *central processing module* normally includes a CPU and memory to process and store the collected data. A *wireless transceiver module* mainly consists of a RF transceiver and an antenna to send the collected data and receive instructions from the sink node. A *power supply module* usually contains energy storage devices (rechargeable batteries), a power management system, and energy harvesting devices (solar panel, wind energy, tidal power, seawater generator, etc.). Finally, the buoy has an anchor device to prevent it from moving (due to waves, marine currents, wind, tide, etc.).

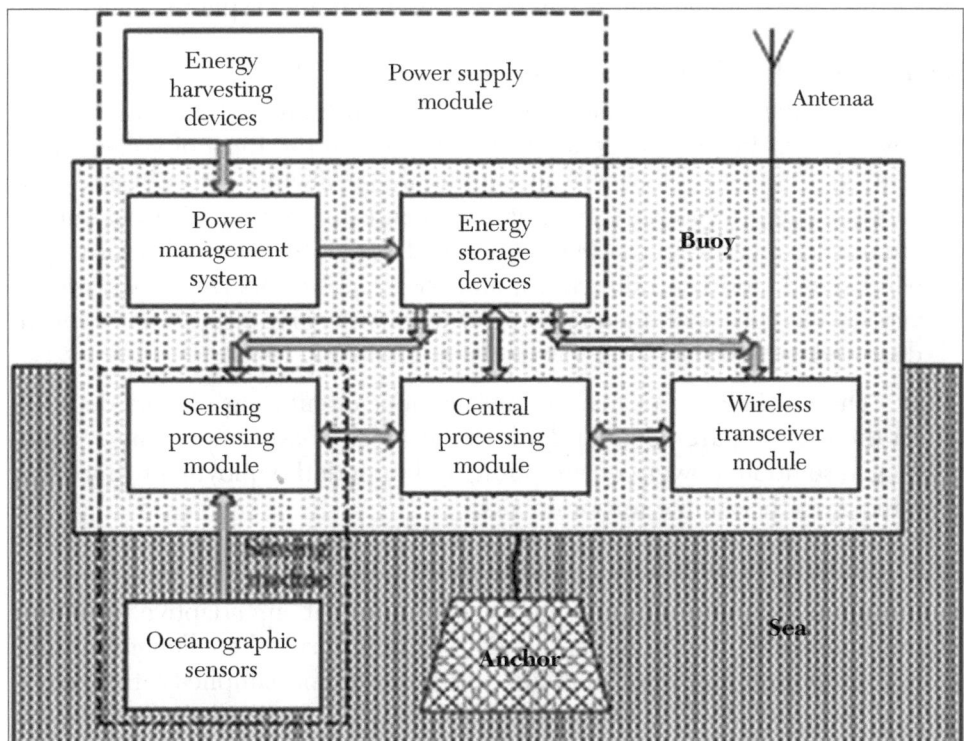

FIGURE 5.3 General architecture of an oceanographic sensor node.

The energy options for sensor nodes usually include batteries, capacitors, heat engines, fuel cells, and energy harvesting. Sensor nodes are battery powered in most application systems. However, the use of a battery in sensor nodes has disadvantages:

1. As sensor nodes increase in number and size, the replacement of depleted batteries is wasteful and time-consuming.

2. A battery has limited energy that cannot last through the long life of sensor nodes.

3. Batteries have environmental contamination and disposal issues since the chemical composition of a battery often involves toxic heavy metals.

It is therefore necessary to explore an alternative power supply for sensor nodes. Harvesting energy from the ambient environment is a promising power supply for sensor networks with lower cost and long life. Energy harvesting methods include photovoltaics, fluid flow, temperature gradients, pressure variations, and vibration harvesting. In terms of efficiencies and reliability, the most outstanding energy harvesting at the moment is photovoltaics.

5.2.4 Sensing Parameters and Sensors

The operating principle of sensors is to respond to changes in their environment by producing an electrical signal in the form of voltage, current, or frequency. Sensors can commonly be divided into physical sensors and chemical sensors. In a marine monitoring system, physical sensors are used to measure physical parameters (temperature, humidity, pressure, wind speed, and wind direction), and chemical sensors are used to sense various chemical parameters (salinity, turbidity, pH, nitrate, chlorophyll, dissolved oxygen [DO], etc.) as shown in Table 5.2.

The right choice of marine environment monitoring sensors depends on the user requirements of deployment area, measurement range, accuracy, resolution, power consumption, and intended deployment time. The table gives the sample sensor details.

5.2.5. Challenges in Oceanographic WSN

While the development and deployment of an adaptive, scalable, and self-healing WSN system needs to address critical challenges such as autonomy, scalability, adaptability, self-healing, and simplicity, the design and deployment of a lasting and scalable WSN for marine environment

Sensors	Monitoring parameters	Range	Accuracy	Power supply	Unit
SBE 16plusV2	Temperature	–5 to +35°C	+/–0.005°C	9–28V	°C
GT301	Pressure	0 to 60	<+/–0.5% of FRO	24V	Bar
SBE 16plusV2	Conductivity (salinity)	0–9	+/– 0.0005	9–28V	S/m
OBS-3+	Turbidity	Mud 5,000–10,000 mg/L Sand 50,000–100,000 mg/L	0.5NTU	15V	NTU
PS-2102	pH	0–14 pH	+/–0.1	N/A	pH
YSI 5025	Chlorophyll	0–400 μg/L	0.1 μg/L	6V	μg/L
ISUS V3	Nitrate	0.007–28 mg/L	+/–0.028 mg/L	0–18V	mg/L
SBE 63	Dissolved Oxygen(DO)	120% of surface saturation in all natural waters	0.1	0–24V; 35Ma	mg/L

TABLE 5.2 Common Marine Environment Monitoring Sensors

monitoring should take into account the following challenges, different from those on land:

Higher water resistance: Sensor nodes of a marine monitoring system require greater levels of water resistance.

Stronger robustness: A marine monitoring system needs stronger robustness, since the marine environment with waves, marine currents, tides, typhoons, vessels, etc., is aggressive and complex and causes movement of nodes.

Higher energy consumption: Energy consumption is higher due to long communication distances and an environment in constant motion.

More unstable line-of-sight: The oscillation of the radio antenna can cause a more unstable line-of-sight between transmitters and receivers.

Other problems: There are also some other problems, including the difficulty for deployment and maintenance of nodes, the need for buoy and mooring devices, sensor coverage problems, and possible acts of vandalism.

5.3 WIRELESS COMMUNICATION TECHNOLOGIES

WSN physical topology and density are entirely dependent on the applications, so the design and deployment of a WSN should consider its environment and application. Sensor nodes are densely deployed to improve data accuracy and achieve better system connectivity. However, a dense deployment of sensor nodes has some disadvantages: high energy consumption, data collisions, interferences, etc. WSN nodes normally have three typical kinds of network topologies: star topology, cluster/tree topology and mesh topology, as shown in Figure 5.4.

Star topology: A star topology is a point-to-point single-hop architecture in which each sensor node connects directly to a sink node. It potentially uses the least amount of power among the three topology architectures.

Mesh topology: A mesh topology is a one-to-many multi-hopping architecture in which each router node connects to multiple nodes. Its advantages over a star topology include a longer range of transmission, decreased loss of data, and higher self-healing communication ability. However, its disadvantages are at the cost of higher latency and higher power consumptions.

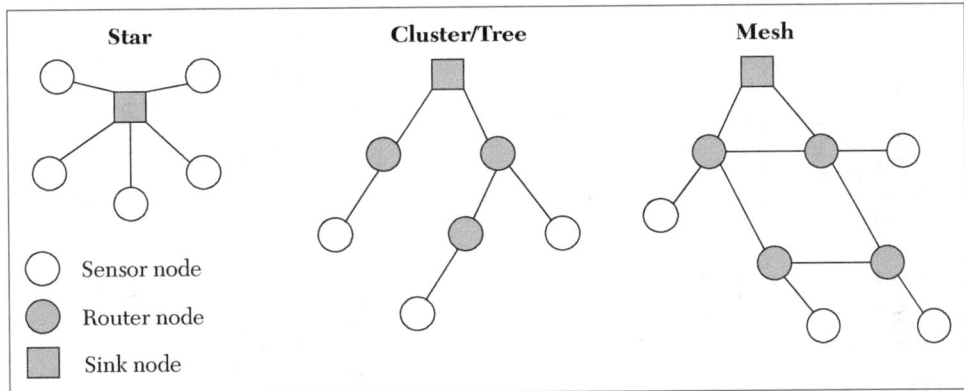

FIGURE 5.4 General WSN network topologies.

Cluster/tree topology: A cluster/tree topology is hybrid star-mesh architecture. It takes advantage of the low power consumption and simple architecture of a star topology, as well as the extended range and fault tolerance of a mesh one. However, there probably exists some latency.

The right and reasonable choice of network topology depends on the amount and frequency of data to be transmitted, transmission distance, battery life requirements, and the mobility of the sensor node. It should be noted that a WSN physical topology may change due to available energy, position variations of nodes, malfunction, reachability (due to noise, severe weathers, moving obstacles, etc.), and task details of sensor nodes. A sensor node normally incorporates a radio module for wireless communication. The transmitted distance of wireless communication can be anywhere between a few meters (Bluetooth, ZigBee, WiFi, etc.) and thousands of kilometers (GSM or GPRS radio communication). Wireless communication has various standards and technologies including Bluetooth, ZigBee, WiFi, GSM, GPRS and WiMAX. Table 5.3 provides a summary and brief comparison of these communication technologies. Usually, two or more wireless communication technologies are used in a real wireless sensor network. Underwater acoustic communication technologies can be a particularly good choice for data collection and exchange among underwater sensors.

Generally, the longer the range a radio module must transmit, the more energy consumption a radio module will have. The choice of wireless communication technology depends on the amount and frequency of the transmitted data, transmission distance, and amount of available energy.

Technology	Standard	Description	Throughput	Range	Frequency
WiFi	IEEE 802.11a; 802.11b/g/n	System of wireless data transmission over computational networks	11/54/300/ Mbps	<100m	5.8GHz 2.4GHz
Blue tooth	IEEE 802.15.1	Industrial specification for WPAN which enables voice and data transmission between different devices by means of a secure, globally free radio link	v.1.2: 1Mbps v.2.0: 3Mbps UWB: 53–480Mbps	Class 1: 100m Class 2: 15–20m Class 3: 1m	2.4GHz
Zig Bee	IEEE 802.15.4	Specification of a set of high level wireless communication protocols for use with low consumption digital radios, based on WPAN standard IEEE 802.13.4	250Kbps	<75m	2.4GHz
Wi Max	IEEE 802.16	Standard for data transmission using radio waves	<75Mbps	<10km	2–11GHz
GSM		Standard system for communication via mobile telephone incorporating digital technology	9.6Kbps	Dependent on service provider	850/900/1800/ 1900 MHz
GPRS		GSM extension for unswitched (or packaged) data transmission	56–144Kbps	Dependent on service provider	850/900/1800/ 1900 MHz

TABE 5.3 Wireless Communication Technologies

5.3.1 Oceanographic Sensors Protection

In marine environments, there are over 4,000 organisms related to fouling problems. Organisms can be classified by size into micro-organisms (or so-called biofilms, slimes, and micro-fouling) and macro-fouling. Biofouling development on a sensor surface is subject to several chemical, physical and biological factors such as pH, dissolved oxygen, temperature, light, location depth, conductivity, organic material, and hydrodynamic conditions. When oceanographic sensors are immersed in seawater, they are susceptible to biofouling problems, which often cause long-term accuracy issues in marine environmental sensor measurements. Since the marine environment is aggressive and seawater is corrosive, oceanographic sensors should take appropriate fouling protection measures.

Biofouling protection for oceanographic sensors may be divided into three techniques according to their different actions: wipers mechanisms, copper corrosion mechanisms, and chlorine evolution mechanisms.

Wiper mechanisms: A biofouling protection system based on wipers is a purely mechanical method. It is an effective biofouling protection technique as long as the sensor head has a suitable shape for wiper cleaning and the wipers are in good condition.

Copper corrosion mechanisms: A copper corrosion mechanism is an effective biofouling protection method to protect the sensitive sensor head, but the protection mechanism is not easy to apply to existing sensors and the cost is relatively high.

Chlorine evolution mechanisms: A biofouling protection system based on a chlorine evolution mechanism uses bleach or chlorine generation by seawater electrolysis. Moreover, this protection mechanism is easily adapted to existing sensors and the cost is relatively low.

Biofouling protection for oceanographic in-situ sensors is a very difficult problem. The ideal biofouling protection for oceanographic sensors should consider six aspects: low cost, low power consumption, easy installation on existing sensors, no or low impact on measurement precision and the environment, long lifetime, and robustness against aggressive conditions.

5.3.2 Advanced Buoy Design

Because the marine environment is aggressive and complex, it is crucial to design an advanced flotation device (buoy) for a marine environment monitoring system. A buoy normally consists of a wireless sensor network node (CPU, sensors, radio, and batteries), an energy harvesting module, underwater sensors, and a mooring system. The design and deployment of an advanced buoy for marine wireless sensor networks should address the following requirements: low cost, watertightness, strong stability, energy harvesting, and mooring system.

Low cost: A marine environment monitoring system using wireless sensor networks is usually composed of many sensor nodes. Therefore, each buoy device needs to be low cost.

Watertightness: To protect the stability of the marine environment monitoring system and prolong its lifetime, its electronic devices must be in a waterproof housing to avoid water damage.

Strong stability: As the marine environment is aggressive and complex, the monitoring system should have strong stability against adverse atmospheric conditions.

Energy harvesting: Since it is not convenient to replace the batteries deployed on the marine surface and the sensor nodes, which are far away from the land and are power-hungry, it is necessary to consider the use of energy harvesting to reduce system maintenance requirements.

Mooring system: Due to tides, waves, marine currents, wind, etc., an anchor is required on the seabed to avoid the movement of the buoy devices.

Besides these requirements, the buoy mechanic design should meet a number of requirements, including buoy visibility with bright yellow color and a warning light for maritime traffic, the use of environmentally friendly materials, the connection of several sensors, and a reasonable antenna height for better communication propagation.

5.3.3 Energy Harvesting System Design

The energy supply of a wireless sensor network is generally provided by batteries which have limited energy. In addition, in marine environment monitoring systems, wireless sensor nodes are often deployed in unapproachable sea surface areas, and they are mostly planned for long-time operation, therefore, it is not convenient to replace the sensor batteries. Moreover, marine sensor nodes (*sink nodes*) have high energy consumption due to the use of *long-range wireless communication protocol* (GPRS).

To reduce system maintenance requirements effectively, there is a clear need to design an energy harvesting system which uses renewable energy sources such as solar power, tidal power, or wind energy. To design an advanced energy harvesting system for marine environment monitoring, we should consider the following three aspects: energy harvesting devices, power management system, and energy storage devices.

Energy harvesting devices: An energy harvesting device is responsible for harvesting energy from the ambient environment. According to the characteristics of available ambient energies, we should choose appropriate energy harvesting devices and consider how to install the energy harvesting devices.

Power management system: A power management system can intelligently manage the batteries to be charged and discharged at separate intervals of time. An ideal power management system can prolong the lifetime of batteries and easily store more energy for the system.

Energy storage devices: Energy storage devices normally use rechargeable batteries. Usually, the energy capacity of rechargeable batteries is larger than the system's daily energy consumption and daily harvesting stores energy and permits the system to supply power even in bad weather.

Given the aggressive and hostile marine environment, to harvest and use more reliable renewable energies, we can envision a hybrid harvesting energy system for marine environment monitoring in the future, which can use several renewable power sources such as solar power, tidal power, seawater generation, and wind energy.

5.3.4 System Stability and Reliability

Considering the aggressive and complex environment, it is very important to analyze reliability in a marine environmental monitoring system using wireless sensor networks. Therefore, research on the reliability of a WSN-based marine environment monitoring system should consider the following aspects.

Battery life issues: As mentioned above, marine sensor nodes (sink nodes) consume more energy than other kinds of wireless sensor nodes. Therefore, battery life always affects system reliability.

Communication relay issues: Communication relay dramatically affects system reliability, when some nodes fail or simply disappear.

Severe environment conditions: The marine environment always has external interference from ships, fish, and birds and has severe weather conditions such as waves, marine currents, tides, and typhoons. Such severe environment conditions further influence system reliability.

5.4 WIRELESS UNDERWATER SENSOR NETWORK

Applications of underwater sensing range from the oil industry to aquaculture, and include instrument monitoring, pollution control, climate recording, prediction of natural disturbances, search and survey missions, and study of marine life. Underwater wireless sensing systems are envisioned for stand-alone applications and control of autonomous underwater vehicles (AUVs), and as an addition to cabled systems. For example, cabled ocean observatories are being built on submarine cables to deploy an extensive fiber optic network of sensors (cameras, wave sensors, and seismometers) covering miles of ocean floor. These cables can support communication access points, very much as cellular base stations are connected to the telephone network, allowing users to move and communicate from places cables cannot reach. Another example is cabled submersibles, also known as *remotely operated vehicles* (ROVs). These vehicles, which may weigh more than 10 metric tons, are connected to the mother ship by a cable that can extend over several kilometers and deliver high power to the remote end, along with high-speed communication signals. A popular example of an ROV/AUV tandem is the Alvin/Jason pair of vehicles deployed by the Woods Hole Oceanographic Institution (WHOI) in 1985 to discover the *Titanic*. Such vehicles were also instrumental in the discovery of hydro-thermal vents, sources of extremely hot water on the bottom of deep ocean, which revealed forms of life different from any others previously known. The first vents were found in the late 1970s, and new ones are still being discovered. The importance of such discoveries is comparable only to space missions, and so is the technology that supports them.

Today, both vehicle technology and sensor technology are mature enough to motivate the idea of underwater sensor networks. To turn this idea into reality, however, one must face the problem of communications. Underwater communication systems today mostly use acoustic technology. Complementary communication techniques, such as optical and radio frequency or even electrostatic communication, have been proposed for short-range links (typically 1–10 m), where their very high bandwidth (MHz or more) can be exploited.

These signals attenuate very rapidly, within a few meters (radio) or tens of meters (optical), requiring either high-power or large antennas. Acoustic communications offer longer ranges, but are constrained by three factors: limited and distance-dependent bandwidth, time-varying multi-path propagation, and low speed of sound. Together, these constraints result in a communication channel of poor quality and high latency, thus combining the worst aspects of terrestrial mobile and satellite radio channels into a communication medium of extreme difficulty.

Among the first underwater acoustic systems was the submarine communication system during the end of World War II, which used analogue modulation in the 8–11 kHz band (single-side band amplitude modulation). Research has since advanced, pushing digital modulation-detection techniques into the forefront of modern acoustic communications. At present, several types of acoustic modems are available commercially, typically offering up to a few kilobits per second (kbps) over distances up to a few kilometers.

The major challenges were identified over the past decade, pointing once again to the fundamental differences between acoustic and radio propagation. For example, acoustic signals propagate at 1,500 m/s, causing propagation delays as long as a few seconds over a few kilometers. With bit rates of the order of 1,000 bps, propagation delays are not negligible with respect to typical packet durations—a situation very different from that found in radio based networks. Moreover, acoustic modems are typically limited to half duplex operation. These constraints imply that acoustic-conscious protocol design can provide better efficiencies than direct application of protocols developed for terrestrial networks (e.g., 802.11 or *transmission control protocol* [TCP]). In addition, for anchored sensor networks, energy efficiency will be as important as in terrestrial networks, since battery recharging hundreds of meters below the sea surface is difficult and expensive. Finally, underwater instruments (sensors, robots, modems, and batteries) are neither cheap nor disposable.

5.4.1 Underwater Sensing Applications

The need to sense the underwater world drives the development of underwater sensor networks. Applications can have very different requirements: fixed or mobile, short or long lived, best effort or life or death; these requirements can result in different designs. The different kinds of deployments, classes of applications and several specific examples, both current and speculative are explained below.

Deployments: Mobility and density are two parameters that vary over different types of deployments of underwater sensor networks. Here, the focus is on wireless underwater networks, although there is significant work in cabled underwater observatories, from the sound surveillance system military networks in the 1950s to the recent Ocean Observatories Initiative. Figure 5.5 illustrates several ways to deploy an underwater sensor network. Underwater networks are often static: individual nodes attached to docks, anchored buoys, or the seafloor (as in the cabled or wireless seafloor sensors in Figure 5.5). Alternatively, semi-mobile underwater networks can be suspended from buoys that are deployed by a ship and used temporarily, but then left in place for hours or days. (The moored sensors in Figure 5.5 may be short-term deployments.) The topologies of these networks are static for long durations, allowing engineering of the network topology to promote connectivity. However, network connectivity still may change owing to small-scale movement or to water dynamics (as currents, surface waves, or other effects change). When battery powered, static deployments may be energy constrained.

Underwater networks may also be mobile, with sensors attached to AUVs, low power gliders, or unpowered drifters. Mobility is useful to maximize sensor coverage with limited hardware, but it raises challenges for localization and maintaining a connected network. Energy for communications is plentiful in AUVs, but it is a concern for gliders or drifters.

As with surface sensor networks, network density, coverage and number of nodes are interrelated parameters that characterize a deployment. Underwater deployments to date are generally less dense, have longer range and employ significantly fewer nodes than terrestrial sensor networks. For example, the Seaweb deployment in 2000 involved 17 nodes spread over a 16km^2 area, with a median of five neighbours per node. Finally, as with remote terrestrial networks, connectivity to the Internet is important and can be difficult. Figure 5.5 shows several options, including underwater cables, point-to-point wireless and satellite.

Application domains of underwater networks: Applications of underwater networks fall into similar categories as for terrestrial sensor networks. Scientific applications observe the environment: from geological processes on the ocean floor, to water characteristics (temperature, salinity, oxygen levels, bacterial and other pollutant content, dissolved matter, etc.) to counting or imaging animal life (micro-organisms, fish, or mammals). Industrial applications monitor and control commercial activities, such as underwater

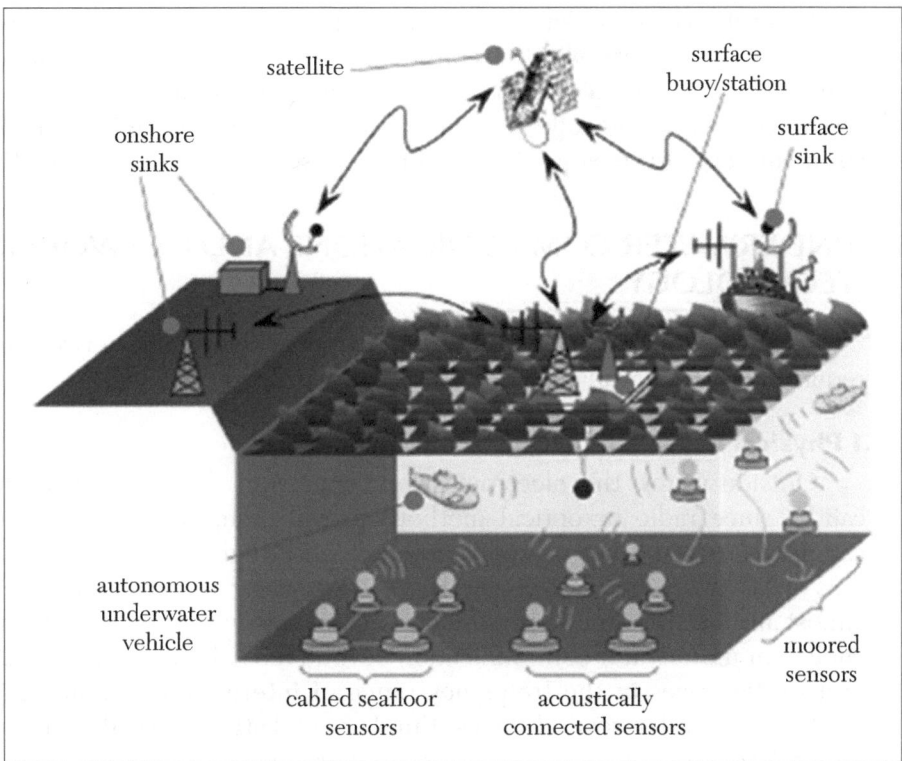

FIGURE 5.5 Deployments can be cabled, fixed, and moored wireless, mobile (on AUVs), and can have different links to shore.

equipment related to oil or mineral extraction, underwater pipelines, or commercial fisheries. Industrial applications often involve control and actuation components as well. Military and homeland security applications involve securing or monitoring port facilities or ships in foreign harbors, de-mining, and communication with submarines and divers.

While the classes of applications are similar, underwater activities have traditionally been much more resource-intensive than terrestrial sensing. Finally, underwater sensing deployments occur over shorter periods (several hours), rather than the days to months or years common in terrestrial sensing. Primary reasons are deployment cost, coupled with a large area of interest, and battery limitations. Underwater deployments can be harsher than surface sensing, with biofouling requiring periodic maintenance. Powered or glider-based AUVs may be coupled with buoys or anchored deployments.

Motivations for underwater sensor networks are similar to those for terrestrial sensor nets: wireless communications reduce deployment costs; interactive data indicate whether sensing is operational or prompts corrective actions during collection; and data analysis during collection allows attendant scientists to adjust sensing in response to interesting observations.

5.5 UNDERWATER COMMUNICATIONS AND NETWORKING TECHNOLOGY

In this section, physical layer, medium access layer, and network layer requirements and available techniques are discussed.

5.5.1 Physical Layer

Outside water, the electromagnetic spectrum dominates communication, since radio or optical methods provide long-distance communication (meters to hundreds of kilometers) with high bandwidths (kHz to tens of MHz), even at low power. In contrast, water absorbs and disperses almost all electromagnetic frequencies, making acoustic waves a preferred choice for underwater communication beyond tens of meters. Propagation of acoustic waves in the frequency range of interest for communication can be described in several stages. Fundamental attenuation describes the power loss that a tone at frequency f experiences as it travels from one location to another. The first (basic) stage takes into account this fundamental loss that occurs over a transmission distance d. The second stage takes into account the site-specific loss due to surface bottom reflections and refraction that occurs as sound speed changes with depth and provides a more detailed prediction of the acoustic field around a given transmitter. The third stage addresses the apparently random changes in the large-scale received power (averaged over some local interval of time) that are caused by slow variations in the propagation medium (e.g., tides). These phenomena are relevant for determining the transmission power needed to close a given link. A separate stage of modeling is required to address the small-scale, fast variations of the instantaneous signal power.

Multi-path propagation creates signal echoes that arrive with varying delays. Delay spreading depends on the system location and can range from a few milliseconds to several hundreds of milliseconds. In a wideband system, this leads to a frequency selective channel transfer function as different frequency components may exhibit substantially different attenuation. The channel response and the instantaneous power often exhibit small-scale, fast variations, typically caused by scattering and the rapid motion of the sea surface (waves) or of the system itself. While large-scale variations influence power control at the transmitter, small-scale variations influence the design of adaptive signal processing algorithms at the receiver. Directional motion causes additional time variation in the form of Doppler effect. A typical AUV velocity is on the order of a few meters per second, while freely suspended platforms can drift with currents at similar speeds. Because sound propagates slowly, the ratio of the relative transmitter/receiver velocity to the speed of sound can be as high as 0.1%—an extreme value that implies the need for dedicated synchronization. This situation is in stark contrast with radio systems, where corresponding values are orders of magnitude smaller, and typically only the center frequency shifting needs to be taken into account.

To avoid the long delay spread and time-varying phase distortion, early systems focused on frequency modulation (frequency shift keying) and noncoherent (energy) detection. Although these methods do not make efficient use of the bandwidth, they are favored for robust communication at low bit rates (typically of the order of 100 bps over a few kilometers), and are used in both commercial modems. The development of bandwidth-efficient communication methods that use amplitude or phase modulation (quadrature amplitude modulation, phase-shift keying) gained momentum in the 1990s, after coherent detection was shown to be feasible on acoustic channels. The research focused on adaptive equalization and synchronization for single carrier wideband systems, leading to real time implementations that today provide high speed communications at several kbps over varying link configurations (horizontal, vertical), as well as with AUVs.

Research on the physical layer is extremely active. Single carrier modulation/detection is being improved using powerful coding and turbo equalization, while multi-carrier modulation/detection is considered as an alternative. Both types of systems are being extended to multi-input multi-output configurations that provide spatial multiplexing (the ability to send parallel data streams from multiple transmitters), and bit rates of several tens of kbps have been demonstrated experimentally. Respecting the physical aspects of acoustic propagation is crucial for successful signal processing; understanding its implications is essential for proper network design. A greater bandwidth yields a greater bit rate and shorter packets—as measured in seconds for a fixed number of bits per packet. While shorter bits imply less energy per bit, shorter packets imply fewer chances of collision on links with different, non-negligible delays. Both facts have beneficial implications on the network performance (and lifetime), if the interference can be managed.

These characteristics of the physical layer influence medium access and higher layer protocol design. For example, the same network protocol may perform differently under a different frequency allocation—moving to a higher frequency region will cause more attenuation to the desired signal, but the interference will attenuate more as well, possibly boosting the overall performance. Also, propagation delay and packet duration matter, since a channel that is sensed to be free may nonetheless contain interfering packets; their length will affect the probability of collisions and the efficiency of retransmission (*throughput*). Finally, power control, coupled with intelligent routing, can greatly help us to limit interference.

5.5.2 Medium Access Control and Resource Sharing

Multi-user systems need an effective means to share the communication resources among the participating nodes. In wireless networks, the frequency spectrum is inherently shared and interference needs to be properly managed. Several techniques have been developed to provide rules to allow different stations to effectively share the resource and separate the signals that coexist in a common medium. In designing resource sharing schemes for underwater networks, one needs to keep in mind the peculiar characteristics of the acoustic channel. Most relevant are long delays, frequency dependent attenuation and the relatively long reach of acoustic signals. In addition, the bandwidth constraints of acoustic hardware (and the transducer in particular) must also be considered. Signals can be deterministically separated in time (*time division multiple access*; TDMA) or frequency (FDMA). In the first case, users take turns accessing the medium, so that signals do not overlap in time and therefore interference is avoided. In FDMA, instead, signal separation is achieved in the frequency domain; although they may overlap in time, signals occupy disjoint parts of the spectrum. These techniques are extensively used in most communications systems, and have been considered for underwater networks as well. For example, owing to acoustic modem limitations, in FDMA use of guard bands for channel separation leads to some inefficiency and this type of frequency channel allocation has very little flexibility (e.g., to accommodate varying transmission rates). TDMA can be more flexible, but requires synchronization among all users to make sure they access disjoint time slots. Many schemes and protocols are based on such an underlying time division structure, which however needs some coordination and some guard times to compensate for inconsistencies in dealing with propagation delays.

Another quasi-deterministic technique for signal separation is *code division multiple access* (CDMA), in which signals that coexist in both time and frequency can be separated using specifically designed codes in combination with signal processing techniques. The price to pay in this case is a bandwidth expansion, especially acute with the narrow bandwidth of the acoustic channel (20 kHz or less for typical hardware). CDMA-based medium access protocols with power control have been proposed for underwater networks and have the advantages of not requiring slot synchronization and being robust to multi-path fading. While these deterministic techniques can be used directly in multi-user systems, data communication nodes typically use contention-based protocols that prescribe the rules by which nodes decide when to transmit on a shared channel. In the simplest protocol, ALOHA, nodes just transmit whenever they need to (random access), and end-terminals recover from errors owing to overlapping signals (called collisions) with retransmission. More advanced schemes implement *carrier-sense multiple access* (CSMA), a listen-before-transmit approach, with or without *collision avoidance* (CA) mechanisms, with the goal of avoiding transmission on an already occupied channel. While CSMA/CA has been very successful in radio networks, the latencies encountered underwater (up to several seconds) make it very inefficient underwater (even worse than ALOHA). In fact, while ALOHA is rarely considered in radio systems owing to its poor throughput, it is a potential candidate for underwater networks when combined with simple CSMA features. Two examples of protocols specifically designed for underwater networks following the CSMA/CA approach are *distance aware collision avoidance protocol* (DACAP) and *tone-Lohi* (T-Lohi). DACAP is based on an initial signaling exchange to reserve the channel, thereby decreasing the probability of collision. T-Lohi exploits CA tones, whereby nodes that want to transmit signal their intention by sending narrowband signals and proceed with data transmission if they do not hear tones sent by other nodes, providing lightweight signaling at the cost of greater sensitivity to the hidden-terminal problem. T-Lohi also exploits high acoustic latency to count contenders in ways impossible with radios, allowing very rapid convergence.

While unsynchronized protocols are simpler, explicit coordination can improve the performance at the price of acquiring and maintaining a time reference. Although long propagation still causes inefficiency, synchronization allows protocols to exploit the space time volume, intentionally overlapping packets in time while they remain distinct in space.

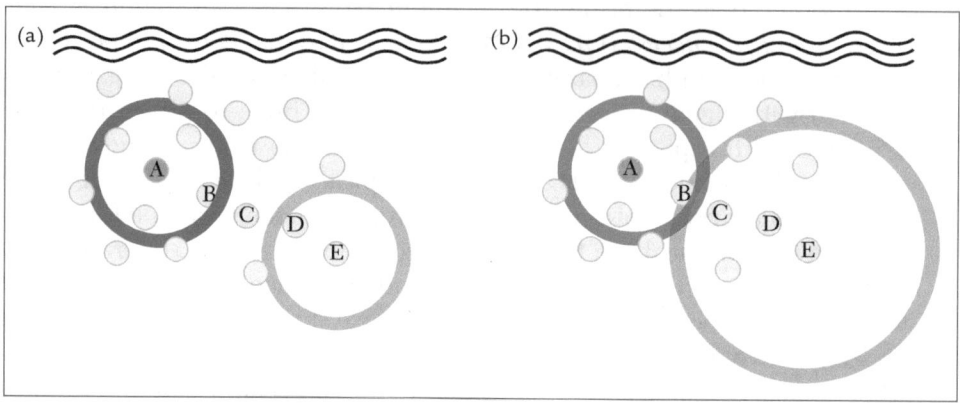

FIGURE 5.6 Illustration of space-time volume: long acoustic latencies mean that packets from A and E are successfully received at B and D in part (*a*), even though they are sent concurrently, while in part (*b*), packets collide at B even though they are sent at different times. (*a*) Same transmission time, no collision; (*b*) different transmission time but collision at B.

Figure 5.6 gives an example of this principle, where unlike in near instant radio communications, long acoustic latencies mean concurrent packets can be received successfully (Figure 5.6a) and packets sent at different times may collide (Figure 5.6b). Even though, in most cases, it is very difficult to operate such protocols in large networks, local synchronization can be achieved and used to improve efficiency. Several protocols have been proposed that assume a common slotted structure accessed by the various nodes in the system. Early work exploited this effect, using centralized scheduling instead of random access to completely avoid collisions, although for static topologies and with additional signaling. Slotted *floor acquisition multiple access* (FAMA) is a decentralized, CSMA-based protocol that uses synchronization to reduce the probability of collision, but is also subject to longer delays due to guard times. The underwater wireless acoustic networks media access protocol is another such protocol that is designed to minimize energy consumption through sleep modes and local synchronization. Several hybrid schemes have also been developed in which two or more of the earlier mentioned techniques are combined.

5.5.3. Network Layer, Routing, and Transport

In large networks, it is unlikely that any pair of nodes can communicate directly, and multi-hop operation, by which intermediate nodes are used to forward messages towards the final destination, is typically used. In addition, multi-hop operation is beneficial in view of distance bandwidth dependence. In this case, routing protocols are used to determine a variable route that a packet should follow through a topology. The design of transport protocols in underwater acoustic networks is another critical issue. Protocols such as TCP (*transmission control protocol*) are designed for low to moderate latencies, not the large fractions of a second commonly encountered in underwater networks, and limited bandwidth and high loss suggest that end-to-end retransmission will perform poorly. Work on higher layer data dissemination protocols underwater has been sparse, with each deployment typically using a custom solution. Finally, an important issue is that of topology control, where nodes sleep to reduce energy while maintaining network connectivity. With this feature, it is possible to wake up nodes on demand and to obtain a virtually perfect topology control mechanism. The *sensor networks for undersea seismic experimentation* (SNUSE) modem implements such as a low-power wake-up circuit, which has been integrated into the *media access protocol* (MAC) layer.

5.5.4 Network Services

Of the many network services possible, localization and time synchronization have seen significant research because of their applicability to many scenarios. Localization and time synchronization are, in a sense, duals of each other: localization often estimates communication time of flight, assuming accurate clocks, and time synchronization estimates clock skew, modeling slowly varying communication delays. Underwater, both pose the challenge of coping with long communications latency and noisy, time-varying channels. Time synchronization in wired networks dates to the network time protocol in the 1990s; wireless sensor networks prompted a resurgence of research a decade later, with an emphasis on message and energy conservation through one-to-many or many-to-many synchronization and integration with hardware to reduce jitter. Underwater time synchronization has built upon these ideas, revised to address challenges in slow acoustic propagation. Time synchronization for high latency networks showed that clock drift during message propagation dominates the error for acoustic channels longer than 500 meters. More recently, D-Sync incorporated Doppler shift estimation to account for error due to node mobility or water currents. Localization, too, has a history in wired and radio-based wireless networks, where node-to-node ranging (based on communications time of flight) and beacon proximity (reachability due to attenuation) are the two fundamental methods used to locate devices. As with time synchronization, localization protocols are often pairwise, or a beacon may broadcast to many potential receivers. Slow acoustic propagation improves localization, since each microsecond error in timing only corresponds to a 15 mm error in location. However, bandwidth limitations make reducing message counts even more important than for radio networks.

Two underwater-specific localization systems with experimental validation are *sufficient distance map estimation* (SDME). SDME exploits post facto localization (analogous to post facto time synchronization of reference broadcast synchronization) to reduce message counts using an otherwise standard scheme based on all pairs, broadcast-based, inter-station ranging. They observe localization accuracy of about 1 meter at ranges of 139 meters. Their localization scheme is based on acoustic ranging between vehicles with synchronized, high precision clocks, combined with AUV location estimates from inertial navigation, combined post facto with an extended Kalman filter. In sea trials tracking an AUV at 4,000 meter depths, their scheme estimates position with a standard deviation of about 10–14 meters.

5.5.5 Sensing and Application Techniques

Some types of underwater sensors are easy and inexpensive, but many rapidly become difficult and expensive from a few hundreds to thousands of dollars or more. Inexpensive sensors include pressure sensing, which can give approximate depth, and photo diodes and thermistors that measure ambient light and temperature. More specialized sensors include fluorimeters that estimate concentrations of chlorophyll, devices to measure water CO_2 concentrations or turbidity, and sonar to detect objects underwater. Such specialized sensors can be much more expensive than more basic sensors. Traditional biology and oceanography rely on samples taken in the environment and returned to the laboratory for analysis. As traditional underwater research has assumed personnel on site, the cost of sample return is relatively small compared with the cost of getting the scientist to the site.

Algorithms for managing underwater sensing, sensor fusion, and coordinated and adaptive sensing are just beginning to develop. Sonar has been used for over more than 60 years for processing single sensors and sensor array data, and today, offline pre-mission planning of AUVs has become routine. As the field matures, in future the work involves online, adaptive sampling using communicating AUVs.

5.5.6 Hardware Platforms

Many hardware platforms for acoustic communication have been developed over the years, with commercial, military, and research success. These platforms are essential to support testing and field use. Teledyne/Benthos modems are widely used commercial devices. They have been extensively used, but their firmware is not accessible to general users, limiting their use for new physical layer and MAC research. Evologics S2C modems may provide some additional flexibility in that they support the transmission of short packets, which are completely customizable by users and can be transmitted instantly without any medium access protocol rule. By using such packets, there is some room for implementing and testing protocols, even though the level of reprogrammability of commercial devices remains rather limited. The data rates supported by these modems range from a few hundred bps to a few kbps in various bands of the tens of kHz frequency range, over distances up to a few tens of kilometers and with power consumptions of tens of watts. While there is no universal development environment or operating system for underwater research, platforms are generally large enough that traditional embedded systems operating environments are feasible. Many groups use embedded variants of Linux, for example.

5.5.7 Test Beds, Simulators, and Models

The breadth of interest in underwater networks has resulted in a great deal of work in the laboratory and simulation, but field experiments remain difficult, and the cost and time of boat rental and offshore deployment are high. Unlike in radio frequency wireless sensor networks, where experimentation is comparatively accessible and affordable, underwater hardware is expensive and costly to deploy, so alternatives are important. Also important is the need for rapid and controlled, reproducible testing over a wide range of conditions. Simulation and modeling is ideal to address both problems. Unfortunately, in many instances, the accuracy of networking simulators in modeling the physical layer and the propagation effects is poor, limiting the predictive value of such tools. Many researchers develop custom simulators to address their specific question, and others develop personal extensions to existing tools such as the network simulator (ns-2, ns-3). However, distribution and generality of these tools is often minimal. Several recent efforts have approached the goal of building underwater simulation tools for the general research community, particularly striving to capture, in sufficient detail, the key properties of acoustic propagation.

A complementary approach also under consideration is to connect a simulator directly to acoustic modems (instead of simulating propagation and physical layers), combining simulation and hardware to emulate a complete system. Several sophisticated modeling tools (including both analytical and computational approaches, e.g., ray tracing) have been developed to study acoustic propagation. However, in most cases, the complexity of such models makes them unsuitable for use in the analysis of communication systems and networks, where the time scales involved require lightweight channel/error models and where many lower level details may have a lesser effect on the overall performance. For this reason, there is currently a strong interest in the development of alternative models, designed to be used in analytical or simulation systems studies.

5.6 WIRELESS UNDERWATER ACOUSTIC SENSOR NETWORK

Underwater acoustic sensor networks (UW-ASN) consist of a variable number of sensors and vehicles that are deployed to perform collaborative monitoring tasks over a given area. To achieve this objective, sensors and vehicles self-organize in an autonomous network that can adapt to the characteristics of the ocean environment. Underwater networking is a rather unexplored area, although underwater communications have been experimented since World War II when, in 1945, an underwater telephone was developed in the United States to communicate with submarines. Acoustic communications are the typical physical layer technology in underwater networks. In fact, radio waves propagate at long distances through conductive sea water only at extra low frequencies (30–300 Hz), which require large antennae and high transmission power. Optical waves do not suffer from such high attenuation but are affected by scattering. Moreover, transmission of optical signals requires high precision in pointing the narrow laser beams. Thus, links in underwater networks are based on acoustic wireless communications. The traditional approach for ocean-bottom or ocean column monitoring is to deploy underwater sensors that record data during the monitoring mission, and then recover the instruments. This approach has the following disadvantages:

1. Real-time monitoring is not possible. This is critical especially in surveillance or in environmental monitoring applications such as seismic monitoring. The recorded data cannot be accessed until the instruments are recovered, which may happen several months after the beginning of the monitoring mission.

2. No interaction is possible between onshore control systems and the monitoring instruments. This impedes any adaptive tuning of the instruments, nor is it possible to reconfigure the system after particular events occur.

3. If failures or misconfigurations occur, it may not be possible to detect them before the instruments are recovered. This can easily lead to the complete failure of a monitoring mission.

4. The amount of data that can be recorded during the monitoring mission by every sensor is limited by the capacity of the onboard storage devices (memories, hard disks, etc.).

Therefore, there is a need to deploy underwater networks that will enable real-time monitoring of selected ocean areas, remote configuration, and interaction with onshore human operators. This can be obtained by connecting underwater instruments by means of wireless links based on acoustic communication. Major challenges in the design of underwater acoustic networks are:

1. Battery power is limited and usually batteries cannot be recharged, also because solar energy cannot be exploited

2. Available bandwidth is severely limited

3. Channel characteristics, including long and variable propagation delays, multi-path and fading problems

4. High bit error rates and temporary losses of connectivity (shadow zones)

5. Underwater sensors are prone to failures because of fouling, corrosion, etc.

6. Propagation delay is five orders of magnitude higher than in Radio Frequency (RF) terrestrial channels

5.6.1 Underwater Acoustic Sensor Network vs. Terrestrial Network

1. Communication method: Terrestrial sensor networks employ electromagnetic waves but in underwater network is relied on physical means like acoustic sounds to transmit the signal.

2. Protocols: Due to distinct network dynamics, existing communication protocols for terrestrial networks are not suitable for underwater environment.

3. Cost: Terrestrial networks are becoming inexpensive but underwater sensors are still expensive devices.

4. Deployment: While terrestrial sensor networks are densely deployed, in underwater, the deployment is generally sparser.

5. Power: The power needed for acoustic underwater communications is higher than in terrestrial radio communications.

6. Memory: Underwater sensors need to have large memory compared to terrestrial sensors.

7. Node mobility: In terrestrial networks, nodes mobility can be predicted whereas in underwater networks, prediction of mobility of the node is difficult.

8. Spatial correlation: Readings taken from terrestrial networks with sensors are often correlated but this is not the case in underwater networks.

5.6.2 Unique Characteristics of Underwater Acoustic Sensor Network

1. Communication media: Acoustic communication is the most versatile and widely used technique in underwater due to low attenuation in water.

2. Transmission loss: Attenuation is mainly provoked by absorption due to conversion of acoustic energy into heat. The geometric spreading refers to the spreading of sound energy due to the expansion of wave fronts.

3. Noise: Man-made noise is mainly caused by machinery noise and shipping activity, while the ambient noise is related to hydrodynamics and to seismic and biological phenomena

4. Multipath propagation: Doppler spreading generates two effects: a simple frequency translation and a continuous spreading of frequencies.

5. Doppler spread: Multi-path propagation may be responsible for severe degradation of the acoustic communication signal, since it generates ISI.

6. High delay: The propagation speed in the UW-A channel is five orders of magnitude lower than in the radio channel. Large propagation delay of 0.67 s/km.

5.6.3 Underwater Acoustic Sensor Network Architecture

There are several different architectures for Underwater Acoustic Sensor Networks, depending on the application:

1. Two-dimensional UW-ASNs for ocean bottom monitoring: These are constituted by sensor nodes that are anchored to the bottom of the ocean. Typical applications may be environmental monitoring, or monitoring of underwater plates in tectonics.

2. Three-dimensional UW-ASNs for ocean column monitoring: These include networks of sensors whose depth can be controlled and may be used for surveillance applications or monitoring of ocean phenomena (ocean bio-geo-chemical processes, water streams, pollution, etc.).

3. Three-dimensional networks of Autonomous Underwater Vehicles (AUVs): These networks include fixed portions composed of anchored sensors and mobile portions constituted by autonomous vehicles.

5.6.4 Two-Dimensional Underwater Sensor Network Architecture

The reference architecture for 2-dimensional underwater networks is shown in the Figure 5.7. A group of sensor nodes are anchored to the bottom of the ocean with deep ocean anchors. By means of wireless acoustic links, underwater sensor nodes are interconnected to one or more *underwater sinks* (uw-sinks), which are network devices in charge of relaying data from the ocean bottom network to a surface station. To achieve this objective, uw-sinks are equipped with two acoustic transceivers, namely a vertical and a horizontal transceiver. The horizontal transceiver is used by the uw-sink to communicate with the sensor nodes to: 1) send commands and configuration data to the sensors (uw-sink to sensors); 2) collect monitored data (sensors to uw-sink). The vertical link is used by the uw-sinks to relay data to a surface station. Vertical transceivers must be long-range transceivers for deep water applications, as the ocean can be as deep as 10 km. The surface station is equipped with an acoustic transceiver that can handle multiple parallel communications with the deployed uw-sinks.

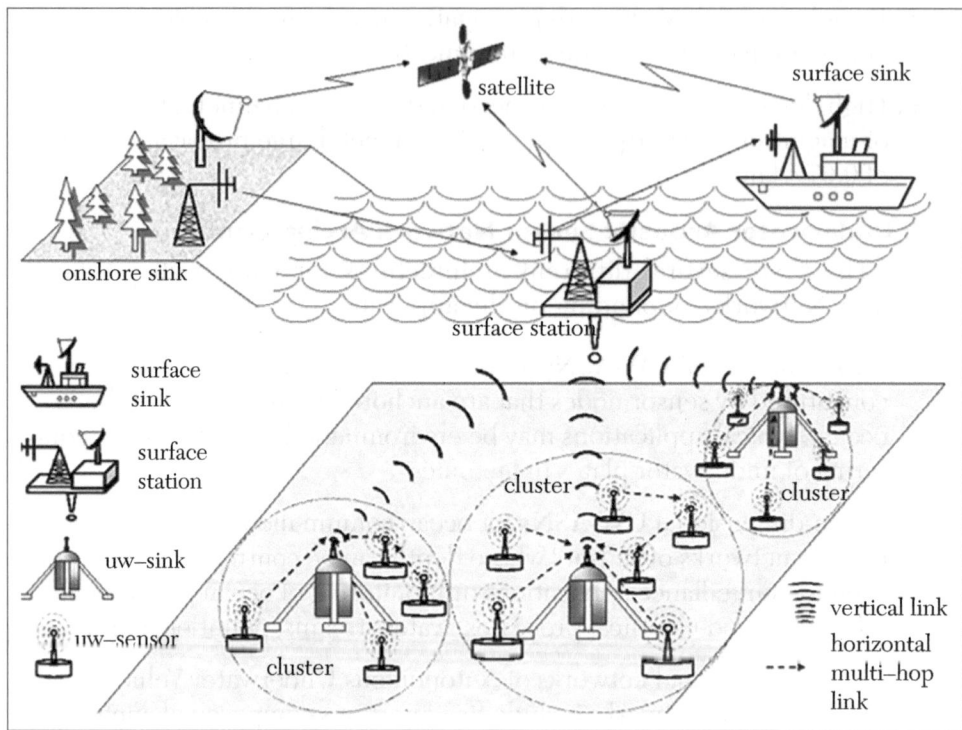

FIGURE 5.7 Two-dimensional underwater sensor networks.

It is also endowed with a long- range RF and/or satellite transmitter to communicate with the *onshore sink* (os-sink) and/or to a *surface sink* (s-sink).

Sensors can be connected to uw-sinks via direct links or through multi-hop paths. In the former case, each sensor directly sends the gathered data to the selected uw-sink. This is the simplest way to network sensors, but it may not be the most energy efficient, since the sink may be far from the node and the power necessary to transmit may decay with powers greater than two of the distance. Furthermore, direct links are very likely to reduce the network throughput because of increased acoustic interference due to high transmission power. In case of multi-hop paths, as in terrestrial sensor networks, the data produced by a source sensor is relayed by intermediate sensors until it reaches the uw-sink. This results in energy savings and increased network capacity, but increases the complexity of the routing functionality as well. In fact, every network device usually takes part in a collaborative process whose objective is to diffuse topology information such that efficient and loop free routing decisions can be made at each intermediate node. This process involves signaling and computation. Since energy and capacity are precious resources in underwater environments, in UW-ASNs the objective is to deliver event features by exploiting multi-hop paths and minimizing the signaling overhead necessary to construct underwater paths at the same time.

5.6.5 Three-Dimensional Underwater Sensor Network Architecture

Three-dimensional underwater networks are used to detect and observe phenomena that cannot be adequately observed by means of ocean bottom sensor nodes, i.e., to perform cooperative sampling of the 3D ocean environment. In 3-dimensional underwater networks, sensor nodes float at different depths in order to observe a given phenomenon. One possible solution would be to attach each uw-sensor node to a surface buoy, by means of wires whose length can be regulated to adjust the depth of each sensor node. However, although this solution allows easy and quick deployment of the sensor network, multiple floating buoys may obstruct ships navigating on the surface, or they can be easily detected and deactivated by enemies in military settings.

For these reasons, a different approach can be to anchor sensor devices to the bottom of the ocean. In this architecture, depicted in the figure 5.8 above, each sensor is anchored to the ocean bottom and equipped with a floating buoy that can be inflated by a pump. The buoy pushes the sensor towards the ocean surface. The depth of the sensor can then be regulated by adjusting the length of the wire that connects the sensor to the anchor, by means of an electronically controlled engine that resides on the sensor.

Many challenges arise with such an architecture, that needs to be solved to enable 3D monitoring, including:

1. Sensing coverage: Sensors should collaboratively regulate their depth to achieve full column coverage, according to their sensing ranges. Hence, it must be possible to obtain sampling of the desired phenomenon at all depths.

2. Communication coverage: Since in 3D underwater networks there is no notion of uw-sink, sensors should be able to relay information to the surface station via multi-hop paths. Thus, network devices should coordinate their depths such a way that the network topology is always connected, i.e., at least one path from every sensor to the surface station always exists.

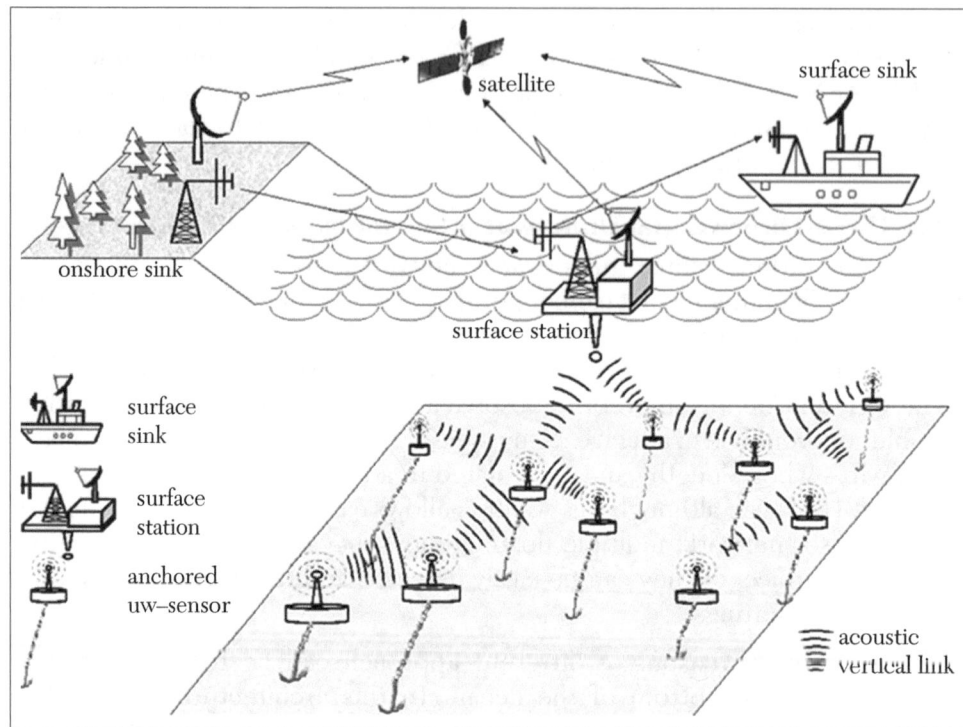

FIGURE 5.8 Three-dimensional underwater sensor networks.

5.6.6 Sensor Networks Architecture with AUVs

AUVs can function without tethers, cables, or remote control, and thus have a multitude of applications in oceanography, environmental monitoring, and underwater resource study. Inexpensive AUV submarines equipped with multiple underwater sensors can reach any depth in the ocean. Hence, they can be used to enhance the capabilities of underwater sensor networks in many ways. The integration and enhancement of fixed sensor networks with AUVs is an almost unexplored research area which requires new network coordination algorithms, such as:

Adaptive sampling: This includes control strategies to command the mobile vehicles to places where their data will be most useful. This approach is also known as adaptive sampling and has been proposed in pioneering monitoring missions. For example, the density of sensor nodes can be adaptively increased in an area when a higher sampling rate is needed for a given monitored phenomenon.

Self-configuration: This includes control procedures to automatically detect connectivity holes due to node failures and request the intervention of an AUV. AUVs can either be used to deploy new sensors or as relay nodes to restore connectivity.

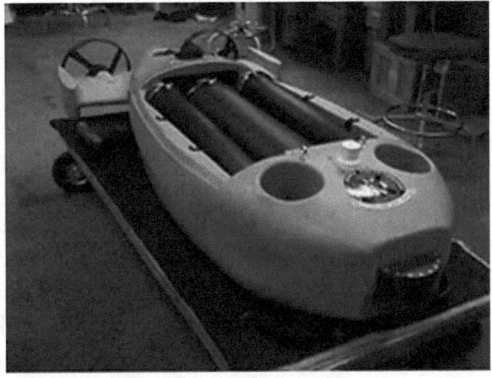

FIGURE 5.9 Different sensors are integrated in AUV.

Figure 5.9 shows an Automatic Underwater Vehicle with sensors of Doppler velocity logs, Acoustic doppler, current profilers, conductivity and temperature, fluorimeter, Li-Cor PAR sensor, inertial navigation systems, attitude heading reference, marine global positioning systems, and depth gauges.

5.7. NETWORKING CHALLENGES FOR UNDERWATER ACOUSTIC SENSOR NETWORKS

Due to the unique characteristics of underwater acoustic channels (long latency and low bandwidth) and the harsh underwater environments (resulting in high channel dynamics), technology used in terrestrial radio networks could not be applied to underwater acoustic networks.

Medium access control: Due to the dense deployment of sensors in UWSNs, it is necessary to design an efficient *medium access control* (MAC) protocol to coordinate communication among sensors. This is a largely unexplored challenge in the communication/networking community. On the one hand, there is no need for MAC protocols in existing small-scale acoustic networks, since in such networks, sensors are sparsely separated from each other and point-to-point communication is sufficient. On the other hand, most existing MAC protocols in radio-based networks assume that the signal propagation delay between neighbor nodes is negligible. In UWSNs, the propagation delay of sound in water is five magnitudes higher than that of radio in air. Moreover, the bandwidth capacities of acoustic channels are very low compared with those of RF channels. While ALOHA-type random access protocols used in satellite networks address the long delay issue to some extent, medium access control handling both long propagation delay and low bandwidth is relatively uninvestigated.

Furthermore, the energy efficiency of MAC protocols in satellite networks is usually not a major concern. In short, a viable MAC solution for UWSNs should take long propagation delay, low available bandwidth, energy efficiency (for long-term applications), and node mobility (for mobile UWSNs) into account. So far, various approaches have been explored. Among the scheduling based protocols—including *time division multiple access* (TDMA), *frequency division multiple access* (FDMA), and *code division multiple access* (CDMA)—CDMA is considered a promising technique for underwater sensor networks. For contention-based protocols (where nodes compete for a shared channel, resulting in probabilistic coordination), the applicability of random access methods and RTS/CTS-based approaches in underwater sensor networks has been used and are suitable for dense underwater sensor networks with high traffic rate.

Multi-hop routing: Forwarding data from source nodes to command/control stations efficiently is very challenging in UWSNs, especially in mobile UWSNs for long-term applications. In such networks, saving energy is a major concern. At the same time, routing should be able to handle node mobility. This requirement makes most existing energy efficient routing protocols unsuitable for UWSNs. In mobile UWSNs, however, most sensor nodes are mobile, and the "network topology" changes very rapidly. The frequent maintenance and recovery of forwarding paths is very expensive in highly dynamic networks, and even more expensive in dense 3-dimensional UWSNs. Geographic routing is considered promising for mobile UWSNs. Another critical issue challenge for routing in UWSNs is link outrage due to water turbulence, currents, obstacles (e.g., ships), etc., as these may cause intermittent network partitioning (that is, some nodes are disconnected from the other nodes). There may be situations where no connected path exists at any given time between the source and the destination.

Reliable data transfer: Reliable data transfer is important in UWSNs, especially for those aquatic exploration applications requiring reliable information. There are typically two approaches to reliable data transfer: end-to-end and hop-by-hop. The most common end-to-end solution is TCP (*transmission control protocol*). In UWSNs, due to the high and dynamic channel error rates and the long propagation delay, TCP's performance will be problematic. There are several techniques that can be used to render TCP's performance more efficient. Another type of approach for reliable data transfer is hop-by-hop. The hop-to-hop approach is favored in wireless and error-prone networks and is believed to be more suitable for sensor networks. The protocol that mainly takes the hop-by-hop approach is ARQ. Due to the long propagation delay of acoustic signals, conventional ARQ will cause very low channel utilization in underwater environments. Thus, new approaches are desired for efficient reliable data transfer in UWSNs. One possible direction to solve the reliable data transfer problem in UWSNs is to investigate coding schemes, including erasure coding and network coding, which, though introducing additional computational and packet overhead, can avoid retransmission delay and significantly enhance the network robustness.

Localization: Localization of mobile sensor nodes is indispensable for UWSNs. Some applications, such as aquatic monitoring, demand high precision localization, while other applications, such as surveillance networks, require a localization solution that can scale to a large number of nodes. However, underwater acoustic propagation characteristics and sensor mobility pose great challenges on high-precision and scalable localization solutions in that:

1. Underwater acoustic channels are highly dispersive, and time delay of arrival (TDOA) estimation is hampered by dense multipath.

2. Acoustic signal does not travel on a straight path due to the stratification effect.

3. Underwater acoustic channels have extremely low bandwidth that renders any approach based on frequent message exchange not appealing.

4. Large-scale sensor deployment prevents centralized solutions.

5. Sensor mobility entails dynamic network topology change. To effectively handle the channel effects, high-precision localization usually involves advanced signal processing algorithms.

5.7.1 Research Challenges and Opportunities

From the above discussions (though the problem list is far from complete), we conclude that, although acoustic waves are practical for underwater acoustic sensor networks from the physics and communication point of view, a tremendous amount of work is demanded from the networking perspective. So far, wireless sensor networks have been widely applied to terrestrial areas, and some of these deployments have achieved satisfactory performance. However, the application of WSNs in marine environment monitoring is still in its infancy, and most WSN-based systems are purely experimental. Few challenges of wireless sensor networks for marine environment monitoring including oceanographic sensors protection, advanced buoy design, energy harvesting system design, network protocol, MAC protocol, simulator tools, hardware platforms, and system stability and reliability.

EXERCISES: PART A (ANSWER IN A WORD OR A SENTENCE)

1. What is meant by "wireless sensor network"?
2. List the challenges in oceanographic WSN.
3. List some application area in WSN marine.
4. What is meant by "star topology"?
5. List the disadvantages of mesh topology.
6. What are the advantages of tree topology?
7. What are the components of a buoy?
8. What are CDMA, TDMA, and FDMA acronyms for?
9. What is TDMA?
10. Define FDMA.
11. What is the meaning of "CDMA"?
12. What is UW-ASN?

EXERCISES: PART B (ANSWER IN A PAGE)

1. What are the differences between land and water WSNs?
2. Write briefly about sensing parameters and sensors.
3. Explain WSN network topologies.
4. What are the different wireless communication technologies available for underwater sensors?
5. Write a note on oceanographic sensor protection.
6. What are the design requirements of buoy design?

7. Compare underwater acoustic sensor networks and terrestrial networks.

8. What are the characteristics of underwater acoustic sensor networks?

EXERCISES: PART C (ANSWER WITHIN 2 OR 3 PAGES)

1. Explain WSN architecture for marine monitoring.

2. Write in detail about underwater communication and network technologies.

3. What are the three different architectures of underwater acoustic sensor network? Explain any one of the architectures in detail.

4. Explain in detail about the 2-dimensional underwater sensor networks.

5. With the help of diagram, write in detail about 3- dimensional underwater sensor network architecture.

6. Write about your own network challenges in WSN.

WEB LINKS

http://rsta.royalsocietypublishing.org/

http://www.ece.gatech.edu/research/labs/bwn/UWASN/

CHAPTER 6

IMAGE PROCESSING FOR THE OCEAN

This chapter deals with analog and digital images, EM spectrum, multi-layer images, spectral response patterns, multi-spectral images, multi-spectral remote sensing, superspectral images, hyperspectral images, hyperspectral remote sensing, sensor/platform systems, special resolution, pixel size, radiometric resolution, data volume, infrared remote sensing, black body radiation, microwave remote sensing, digital image processing, and software for ocean color and algorithms.

6.1 ANALOG AND DIGITAL IMAGES

An image is a 2-dimensional representation of objects in a real scene. Remote sensing images are representations of parts of the Earth's surface as seen from space. The images may be *analog* or *digital*. Aerial photographs are examples of analog images while satellite images acquired using electronic sensors are examples of digital images. A digital image is a 2-dimensional array of pixels. Each pixel has an intensity value (represented by a digital number) and a location address (referenced by its row and column numbers). A digital image comprises a 2-dimensional array of individual picture elements called *pixels* arranged in columns and rows, as in Figure 6.1. Each pixel represents

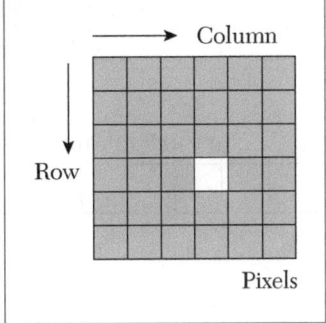

FIGURE 6.1 The representation of a digital image.

an area on the Earth's surface. A pixel has an intensity value and a location address in the 2-dimensional image.

The intensity value represents a measured physical quantity such as the solar radiance in a wavelength band reflected from the ground, emitted infrared radiation, or backscattered radar intensity. This value is normally the average value for the whole ground area covered by the pixel. The intensity of a pixel is digitized and recorded as a digital number. Due to finite storage capacity, a digital number is stored with a finite number of *bits* (binary digits). The number of bits determines the radiometric resolution of the image. For example, an 8-bit digital number ranges from 0 to 255 (i.e., 2^8–1), while a 11-bit digital number ranges from 0 to 2047. The detected intensity value needs to be scaled and quantized to fit within this range of value. In a *radiometrically calibrated image*, the actual intensity value can be derived from the pixel digital number. The address of a pixel is denoted by its row and column coordinates in the 2-dimensional image. There is a one-to-one correspondence between the column-row address of a pixel and the geographical coordinates (e.g., longitude, latitude) of the imaged location. To be useful, the exact geographical location of each pixel on the ground must be derivable from its row and column indices, given the imaging geometry and the satellite orbit parameters.

6.2 THE ELECTROMAGNETIC SPECTRUM

Most remote sensing devices make use of electromagnetic energy. However, the *electromagnetic spectrum* is very broad and not all wavelengths are equally effective for remote sensing purposes. Figure 6.2 illustrates the

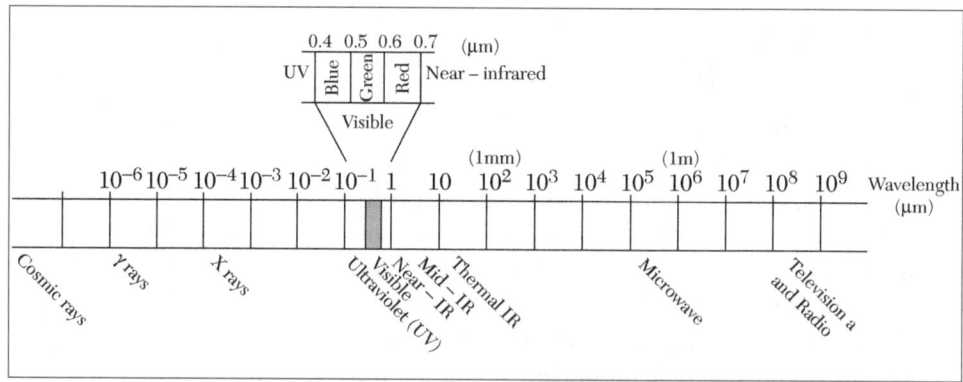

FIGURE 6.2 The electromagnetic spectrum.

electromagnetic spectrum. The atmosphere itself causes significant absorption and/or scattering of the very shortest wavelengths. In addition, the glass lenses of many sensors also cause significant absorption of shorter wavelengths such as the *ultraviolet* (UV). Thus the first significant window (i.e., a region in which energy can significantly pass through the atmosphere) opens up in the visible wavelengths. Even here, the blue wavelengths undergo substantial attenuation by atmospheric scattering, and are thus often left out in remotely sensed images. However, the green, red, and near-*infrared* (IR) wavelengths all provide good opportunities for gauging Earth surface interactions without significant interference by the atmosphere. In addition, these regions provide important clues to the nature of many Earth surface materials. Chlorophyll, for example, is a very strong absorber of red visible wavelengths, while the near-infrared wavelengths provide important clues to the structures of plant leaves. Thus the bulk of remotely sensed images used in GIS-related applications are taken in these regions. Extending into the middle and thermal infrared regions, a variety of good windows can be found. The longer of the middle infrared wavelengths have proven to be useful in many geological applications. The thermal regions have proven to be very useful for monitoring not only the obvious cases of the spatial distribution of heat from industrial activity, but a broad set of applications ranging from fire monitoring to animal distribution studies to soil moisture conditions.

After the thermal IR, the next area of major significance in environmental remote sensing is in the microwave region. Many important windows exist in this region and are of particular importance for the use of active radar imaging. The texture of the Earth's surface materials causes significant interactions with several microwave wavelength regions. This can thus be used as a supplement to information gained in other wavelengths, and offers the significant advantage of being usable at night (because, as an active system, it is independent of solar radiation) and in regions of persistent cloud cover (since radar wavelengths are not significantly affected by clouds).

6.2.1 A Push Broom Scanner

This type of imaging system is commonly used in optical remote sensing satellites such as *SPOT*. The imaging system has a linear detector array (usually of the CCD type) consisting of many detector elements (6,000 elements in SPOT HRV). Each detector element projects an *instantaneous field of view* (IFOV) on the ground. The signal recorded by a detector

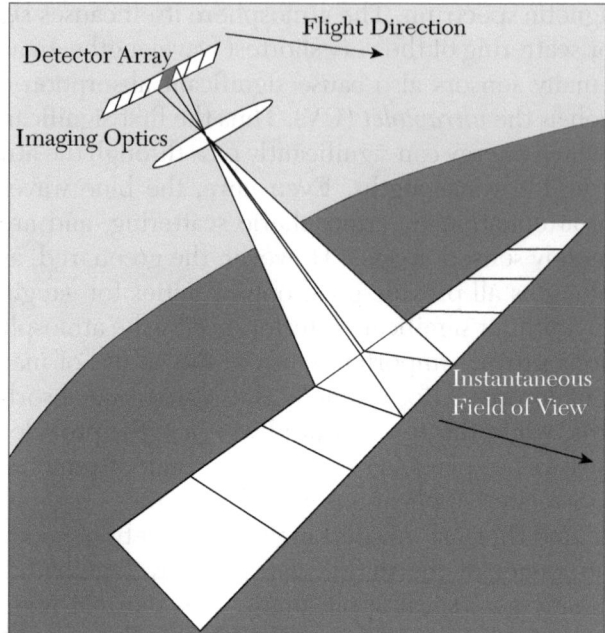

FIGURE 6.3 Push-broom scanner.

element is proportional to the total radiation collected within its IFOV [Figure 6.3]. At any instant, a row of pixels is formed. As the detector array flies along its track, the row of pixels sweeps along to generate a 2-dimensional image.

6.3 MULTI-LAYER IMAGE

Several types of measurement may be made from the ground area covered by a single pixel. Each type of measurement forms images which carry some specific information about the area. By "stacking" these images from the same area together, a multi-layer image is formed. Each component image is a layer in the multi-layer image. Multi-layer images can also be formed by combining images obtained from different sensors, and other subsidiary data. For example, a multi-layer image may consist of three layers from a SPOT multi-spectral image, a layer of ERS synthetic aperture radar image, and perhaps a layer consisting of the digital elevation map of the area being studied. The multi-layer image consisting of five component layers is as shown in Figure 6.4.

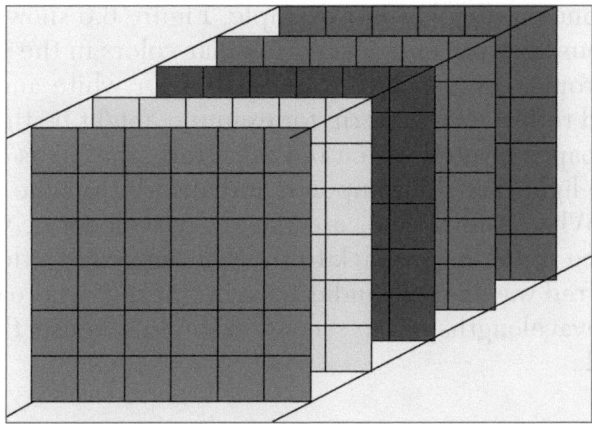

FIGURE 6.4 Multi-layer image.

6.3.1 Interaction Mechanisms

When electromagnetic energy strikes a material, three types of interaction can follow: reflection, absorption and/or transmission (as in Figure 6.5). The main concern is with the reflected portion, since it is usually this which is returned to the sensor system. Exactly how much is reflected will vary and will depend upon the nature of the material and where in the electromagnetic spectrum the measurement is being taken. The nature of this reflected component over a range of wavelengths characterizes the result as a spectral response pattern.

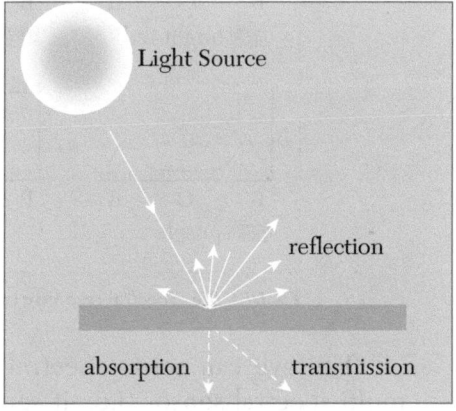

FIGURE 6.5 Interaction process.

6.4 SPECTRAL RESPONSE PATTERNS

A spectral response pattern is sometimes called a *signature*. It is a description (often in the form of a graph) of the degree to which energy is reflected in different regions of the spectrum. Most humans are very familiar with spectral response patterns since they are equivalent to the human concept of color. For example, Figure 6.6 shows idealized spectral response patterns for several familiar colors in the visible portion of the electromagnetic spectrum, as well as for white and dark grey. The bright red reflectance pattern, for example, might be that produced by a piece of paper printed with a red ink. Here, the ink is designed to alter the white light that shines upon it and absorb the blue and green wavelengths. What is left, then, are the red wavelengths, which reflect off the surface of the paper back to the sensing system (the eye). The high return of red wavelengths indicates a bright red, whereas the low return of green wavelengths in the second example suggests that it will appear quite dark.

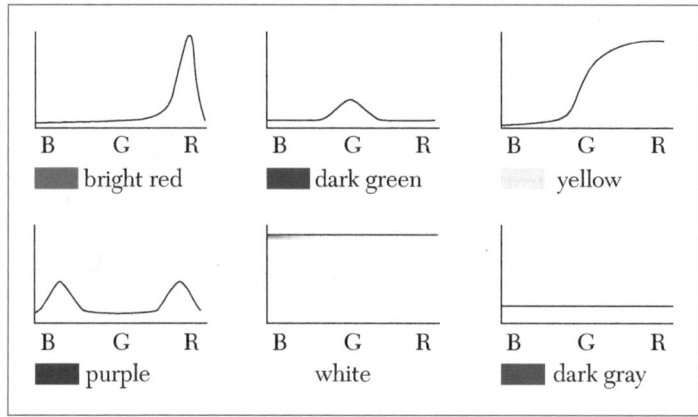

FIGURE 6.6 Spectral response patterns.

The eye can sense spectral response patterns because it is truly a multi-spectral sensor (i.e., it senses in more than one place in the spectrum). Although the actual functioning of the eye is quite complex, it does in fact have three separate types of detectors that can usefully be thought of as responding to the red, green, and blue wavelength regions. These are the additive primary colors, and the eye responds to mixtures of these three to yield a sensation of other hues. For example, the color perceived by the third spectral response pattern in Figure 6.6 would be a yellow—the result of mixing a red and a green. However, it is important to recognize that this is simply our phenomenological perception of a spectral response pattern. Consider, for example, the fourth curve. Here we have reflectance in both the blue and red regions of the visible spectrum. This is a bimodal distribution, and thus technically not a specific hue in the spectrum. However, we would perceive this to be a purple! Purple (a color between violet and red) does not exist in nature (i.e., as a hue—a distinctive dominant wavelength). It is very real in our perception, however. Purple is simply our perception of a bimodal pattern involving a nonadjacent pair of primary hues.

Figure 6.7 shows an idealized spectral response pattern for vegetation along with those of water and dry bare soil. The strong absorption by leaf pigments (particularly chlorophyll for purposes of photosynthesis) in the blue and red regions of the visible portion of the spectrum leads to the characteristic green appearance of healthy vegetation. However, while this signature is distinctively different from most nonvegetated surfaces, it is not very capable of distinguishing between species of vegetation—most will have a similar color of green at full maturation. In the near infrared, however, we find a much higher return from vegetated surfaces because of scattering within the fleshy mesophyllic layer of the leaves. Plant pigments do not absorb energy in this region, and thus the scattering, combined with the multiplying effect of a full canopy of leaves, leads to high reflectance in this region of the spectrum. However, the extent of this reflectance will depend highly on the internal structure of leaves (e.g., broadleaf versus needle). Significant differences between species can thereby often be detected in this region.

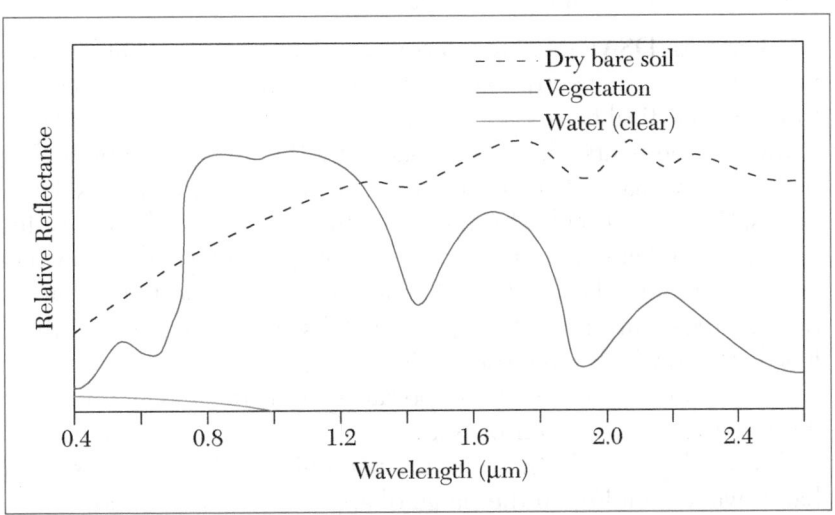

FIGURE 6.7 Spectral response pattern for different items.

6.5 MULTI-SPECTRAL IMAGE

A multi-spectral image consists of a few image layers; each layer represents an image acquired at a specific wavelength band. For example, the SPOT HRV (high resolution visible) sensor operating in the multi-spectral mode detects radiations in three wavelength bands: the green (500–590 nm), red (610–680 nm) and near-infrared (790–890 nm) bands. A single SPOT multi-spectral scene consists of three intensity images in the three wavelength bands. In this case, each pixel of the scene has three intensity values corresponding to the three bands. A multi-spectral IKONOS image consists of four bands: blue, green, red, and near infrared, while a LANDSAT TM (*thematic mapper*) multi-spectral image consists of seven bands: blue, green, red, and near-IR bands, two SWIR bands, and a thermal IR band.

6.5.1 Multi-Spectral Remote Sensing

In the visual interpretation of remotely sensed images, a variety of image characteristics are brought into consideration: color (or tone in the case of panchromatic images), texture, size, shape, pattern, context, and the like. However, with computer-assisted interpretation, it is most often simply color (i.e., the spectral response pattern) that is used. It is for this reason that a strong emphasis is placed on the use of multi-spectral sensors (sensors that, like the eye, look at more than one place in the spectrum and thus can gauge spectral response patterns), and the number and specific placement of these spectral bands.

The LANDSAT satellite is a commercial system providing multi-spectral imagery in seven spectral bands at a 30-meter resolution. It can be shown through analytical techniques such as *principal components analysis*, that in many environments, the bands that carry the greatest amount of information about the natural environment are the near-infrared and red wavelength bands. Water is strongly absorbed by infrared wavelengths and is thus highly distinctive in that region. In addition, plant species typically show their greatest differentiation here. The red area is also very important because it is the primary region in which chlorophyll absorbs energy for photosynthesis. Thus, it is this band which can most readily distinguish between vegetated and non-vegetated surfaces. Given this importance of the red and near-infrared bands, it is not surprising that sensor systems designed for earth resource monitoring will invariably include these in any particular multi-spectral system. Other bands will depend upon the range of applications envisioned. Many include the green visible band since it can be used, along with the other two, to produce a traditional false color composite—a full-color image derived from the green, red, and infrared bands (as opposed to the blue, green, and red bands of natural color images). This format became common with the advent of color infrared photography and is familiar to many specialists in the remote sensing field. In addition, the combination of these three bands works well in the interpretation of the cultural landscape as well as natural and vegetated surfaces. However, it is increasingly common to include other bands that are more specifically targeted to the differentiation of surface materials. For example, LANDSAT TM Band 5 is placed between two water absorption bands and has thus proven very useful in determining soil and leaf moisture differences. Similarly, LANDSAT TM Band 7 targets the detection of hydro-thermal alteration zones in bare rock surfaces. By contrast, the AVHRR system on the NOAA series satellites includes several thermal channels for the sensing of cloud temperature characteristics.

6.5.2 Superspectral Image

The more recent satellite sensors can acquire images at many more wavelength bands. For example, the MODIS sensor onboard NASA's TERRA satellite consists of 36 spectral bands, covering the wavelength regions ranging from the visible, near infrared, and short-wave infrared to the thermal infrared. The bands have narrower bandwidths, enabling the finer spectral characteristics of the targets to be captured by the sensor. The term *super-spectral* has been coined to describe such sensors.

6.6 HYPERSPECTRAL IMAGE

A *hyperspectral* image consists of about a hundred or more contiguous spectral bands. The characteristic spectrum of the target pixel is acquired in a hyperspectral image. The precise spectral information contained in a hyper-spectral image enables better characterization and identification of targets. Hyperspectral images have potential applications in such fields as precision agriculture (e.g., monitoring the types, health, moisture status, and maturity of crops), coastal management (e.g., monitoring of phytoplanktons, pollution, and bathymetry changes). Currently, hyperspectral imagery is not commercially available from satellites. There are experimental satellite-sensors that acquire hyperspectral imagery for scientific investigation (e.g., NASA's Hyperion sensor onboard the EO1 satellite, the CHRIS sensor onboard ESA's PRABO satellite). An illustration of a hyperspectral image cube is shown in Figure 6.8. The hyperspectral image data usually consists of over a hundred contiguous spectral bands, forming a three-dimensional (two spatial dimensions and one spectral dimension) image cube. Each pixel is associated with a complete spectrum of the imaged area. The high spectral resolution of hyperspectral images enables better identification of land cover.

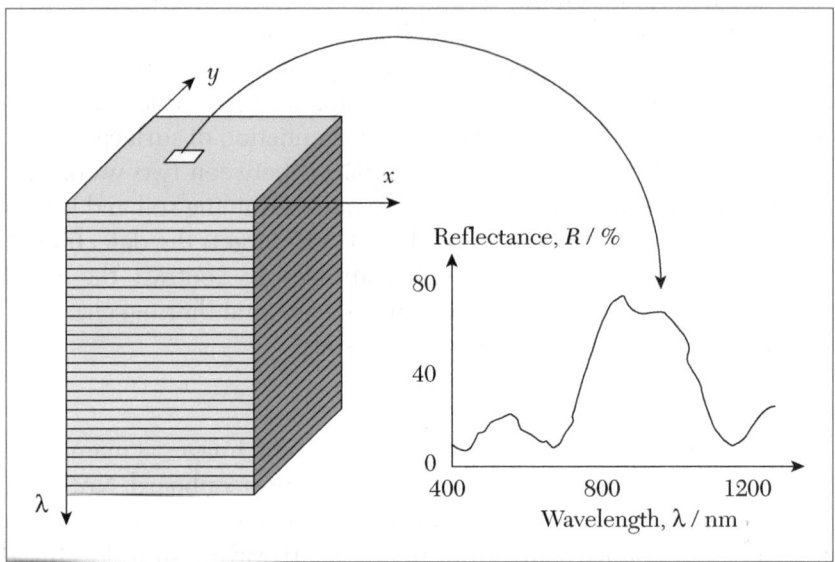

FIGURE 6.8 Hyperspectral image.

6.6.1 Hyperspectral Remote Sensing

In addition to traditional multi-spectral imagery, some new and experimental systems such as AVIRIS and MODIS can capture hyperspectral data. These systems cover a similar wavelength range to multi-spectral systems, but in much narrower bands. This dramatically increases the number of bands (and thus precision) available for image classification (typically tens and even hundreds of very narrow bands). Moreover, hyperspectral signature libraries have been created in lab conditions and contain hundreds of signatures for different types of land covers, including many minerals and other earth materials. Thus, it should be possible to match signatures to surface materials with great precision.

6.7 SENSOR/PLATFORM SYSTEMS

Given recent developments in sensors, a variety of platforms are now available for the capture of remotely sensed data. Here we review some of the major sensor/platform combinations that are typically available to the GIS user community.

6.7.1 Aerial Photography

Aerial photography is the method in which the cameras mounted in light aircraft flying between 200 and 15,000 meters capture a large quantity of detailed information. Aerial photos provide an instant visual inventory of a portion of the Earth's surface and can be used to create detailed maps. Aerial photographs commonly are taken by commercial aerial photography firms which own and operate specially modified aircraft equipped with large format (23 cm × 23 cm) mapping-quality cameras. Aerial photos can also be taken using small format cameras (35 mm and 70 mm), handheld or mounted in unmodified light aircraft. Camera and platform configurations can be grouped in terms of oblique and vertical. Oblique aerial photography is taken at an angle to the ground. The resulting images give a view as if the observer is looking out an airplane window. These images are easier to interpret than vertical photographs, but it is difficult to locate and measure features on them for mapping purposes. Vertical aerial photography is taken with the camera pointed straight down. The resulting images depict ground features in plan form and are easily compared with maps. Vertical aerial photos are always highly desirable, but are particularly useful for resource surveys in areas where no maps are available. Aerial photos depict features such as field patterns and vegetation which are often omitted on maps. Comparison of old and new aerial photos can also capture changes within an area over time. Vertical aerial photos contain subtle displacements due to relief, tip and tilt of the aircraft, and lens distortion. Vertical images may be taken with overlap, typically about 60% along the flight line and at least 20% between lines. Overlapping images can be viewed with a stereoscope.

6.7.2 Aerial Videography

Light, portable, inexpensive video cameras and recorders can be carried in chartered aircraft. In addition, many smaller aerial mapping companies offer videography as an output option. By using several cameras simultaneously, each with a filter designed to isolate a specific wavelength range; it is possible to isolate multi-spectral image bands that can be used individually or in combination in the form of a color composite. For use in digital analysis, special graphics hardware boards known as frame grabbers can be used to freeze any frame within a continuous video sequence and convert it to digital format, usually in one of the more popular exchange formats such as TIF or TARGA.

6.7.3 Satellite-Based Scanning Systems

Photography has proven to be an important input to visual interpretation and the production of analog maps. However, the development of satellite platforms, the associated need to telemeter imagery in digital form, and the desire for highly consistent digital imagery have given rise to the development of solid-state scanners as a major format for the capture of remotely sensed data. The specific features of systems vary (including, in some cases, the removal of a true scanning mechanism). However, in the discussion which follows, an idealized scanning system is presented that is highly representative of current systems in use. The basic logic of a scanning sensor is the use of a mechanism to sweep a small field of view (known as an instantaneous field of view [IFOV] as in Figure 6.3) in a west-to-east direction at the same time the satellite is moving in a north-to-south direction. Together, this movement provides the means of composing a complete raster image of the environment. A simple scanning technique is to use a rotating mirror that can sweep the field of view in a consistent west-to-east fashion. The field of view is then intercepted with a prism that can spread the energy contained within the IFOV into its spectral components. Photoelectric detectors (of the same nature as those found in the exposure meters of commonly available photographic cameras) are then arranged in the path of this spectrum to provide electrical measurements of the amount of energy detected in various parts of the electromagnetic spectrum. As the scan moves from west to east, these detectors are polled to get a set of readings along the east-west scan. These form the columns along one row of a set of raster images, one for each detector. Movement of the satellite from north to south then positions the system to detect the next row, ultimately leading to the production of a set of raster images as a record of reflectance over a range of spectral bands.

There are several satellite systems in operation today that collect imagery that is subsequently distributed to users. Each type of satellite data offers specific characteristics that make it appropriate for a particular application. In general, there are two characteristics that may help guide the choice of satellite data: spatial resolution and spectral resolution. The spatial resolution refers to the size of the ground area that is summarized by one data value in the imagery. This is the instantaneous field of view (IFOV) described earlier. Spectral resolution refers to the number and width of the spectral bands that the satellite sensor detects. In addition, issues of cost and imagery availability must also be considered.

6.8 SPATIAL RESOLUTION

Spatial resolution refers to the size of the smallest object that can be resolved on the ground. In a digital image, the resolution is limited by the pixel size, i.e., the smallest resolvable object cannot be smaller than the pixel size. The intrinsic resolution of an imaging system is determined primarily by the instantaneous field of view (IFOV) of the sensor, which is a measure of the ground area viewed by a single detector element in a given instant in time. However, this intrinsic resolution can often be degraded by other factors which introduce blurring of the image, such as improper focusing, atmospheric scattering, and target motion. The pixel size is determined by the sampling distance. A *high-resolution* image refers to one with a small resolution size. Fine details can be seen in a high-resolution image. On the other hand, a *low-resolution* image is one with a large resolution size, i.e., only coarse features can be observed in the image. A low-resolution MODIS scene with a wide coverage is shown in Figure 6.9. The intrinsic resolution of the image was approximately 1 km, but the image shown here has been resampled to a resolution of about 4 km. The coverage is more than 1,000 km from east to west. A large part of Indochina, Peninsular Malaysia, Singapore, and Sumatra can be seen in the image.

A browse image of a high-resolution SPOT scene is shown in Figure 6.10. The multi-spectral SPOT scene has a resolution of 20 meters and covers an area of 60 km by 60 km. The browse image has been resampled to 120 meter pixel size, and hence the resolution has been reduced. This scene shows Singapore and part of the Johor State of Malaysia.

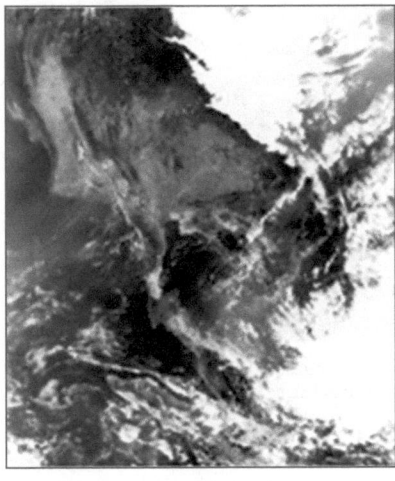

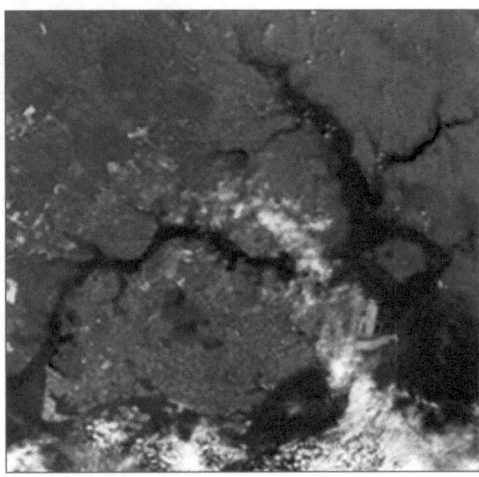

FIGURE 6.9 Low-resolution image. *FIGURE 6.10* High-resolution image.

6.8.1 Spatial Resolution and Pixel Size

Image resolution and pixel size are often used interchangeably, but in reality, they are not equivalent. An image sampled at a small pixel size does not necessarily have a high resolution. The three images in Figure 6.11 illustrate this point. The first image is a SPOT image of 10-meter pixel size. It was derived by merging a SPOT panchromatic image of 10-meter resolution with a SPOT multi-spectral image of 20 meters resolution. The merging procedure "colors" the panchromatic image using the colors derived from the multi-spectral image. The effective resolution is thus determined by the resolution of the panchromatic image, which is 10 meters. This image is further processed to degrade the resolution while

FIGURE 6.11 Different resolution, same pixel size images: L–R: 10 meter resolution, 10 meter pixel size; 30 meter resolution, 10 meter pixel size; 80 meter resolution, 10 meter pixel size.

maintaining the same pixel size. The next two images are the blurred versions of the image with larger resolution size, but still digitized at the same pixel size of 10 meters. Even though they have the same pixel size as the first image, they do not have the same resolution.

The images in Figure 6.12 illustrate the effect of pixel size on the visual appearance of an area. The first image is a SPOT image of 10 meter pixel size derived by merging a SPOT panchromatic image with a SPOT multispectral image. The subsequent images show the effects of digitizing the same area with larger pixel sizes.

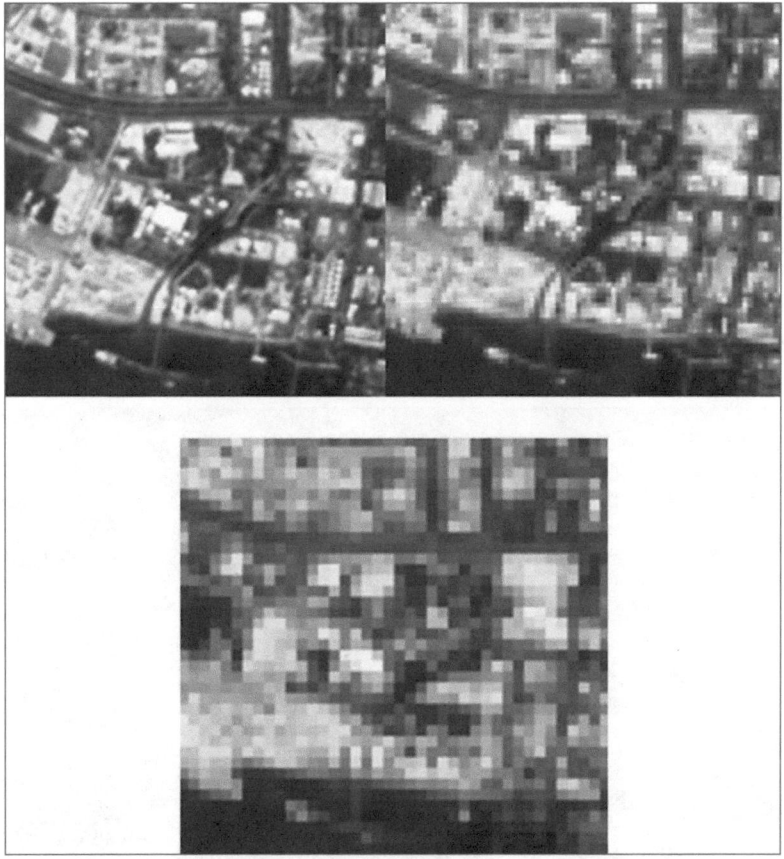

FIGURE 6.12 Effect of pixel size in the images; L–R: pixel size = 10 meters, image width, height = 160 pixels; pixel size = 20 meters, image width, height = 80 pixels; pixel size = 80 meters, image width, height = 20 pixels.

6.9 RADIOMETRIC RESOLUTION

Radiometric resolution refers to the smallest change in intensity level that can be detected by the sensing system. The intrinsic radiometric resolution of a sensing system depends on the signal-to-noise ratio of the detector. In a digital image, the radiometric resolution is limited by the number of discrete quantization levels used to digitize the continuous intensity value. The images in Figure 6.13 illustrate the effects of the number of quantization levels on the digital image. The first image is a SPOT panchromatic image quantized at 8 bits (i.e., 256 levels) per pixel. The subsequent images show the effects of degrading the radiometric resolution by using fewer quantization levels.

Digitization using a small number of quantization levels does not affect very much the visual quality of the image. Even 4-bit quantization (16 levels) seems acceptable in the examples shown. However, if the image is to be subjected to numerical analysis, the accuracy of analysis will be compromised if few quantization levels are used.

6.9.1 Data Volume

The volume of the digital data can potentially be large for multi-spectral data, as a given area is covered in many different wavelength bands.

8-bit quantization (256 levels).   6-bit quantization (64 levels).

FIGURE 6.13 Continued on next page

4-bit quantization (16 levels).   3-bit quantization (8 levels).

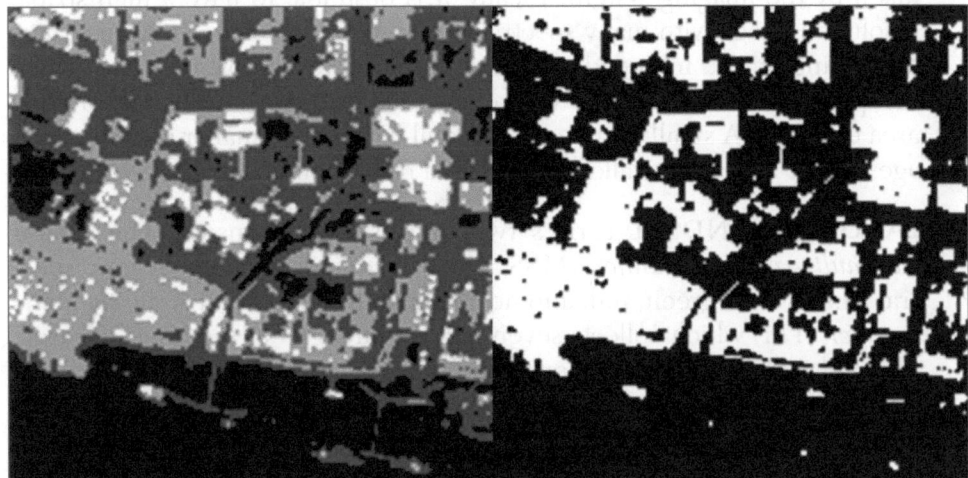

2-bit quantization (4 levels).   1-bit quantization (2 levels).

FIGURE 6.13 Different radiometric resolution images.

For example, a 3-band multi-spectral SPOT image covers an area of about 60×60 km^2 on the ground with a pixel separation of 20 meters, so there are about $3,000 \times 3,000$ pixels per image. Each pixel intensity in each band is coded using an 8 bit (i.e., 1 byte) digital number, giving a total of about 27 million bytes per image. In comparison, panchromatic data has only one band. Thus, panchromatic systems are normally designed to give a higher spatial resolution than the multi-spectral system. For example, a SPOT panchromatic scene has the same coverage of about 60 × 60 km² but the pixel size is 10 meters, giving about 6,000 × 6,000 pixels and a total of about 36 million bytes per image. If a multi-spectral SPOT scene is digitized also at a 10-meter pixel size, the data volume will be 108 million bytes. For very high spatial resolution imagery, such as the one acquired by the IKONOS satellite, the data volume is even more significant. For example, an IKONOS 4-band multi-spectral image at 4-meter pixel size covering an area of 10 km by 10 km, digitized at 11 bits (stored at 16 bits), has a data volume of 4 × 2,500 × 2,500 × 2 bytes, or 50 million bytes per image. A 1-meter resolution panchromatic image covering the same area would have a data volume of 200 million bytes per image. The images taken by a remote sensing satellite is transmitted to Earth through telecommunication. The bandwidth of the telecommunication channel sets a limit to the data volume for a scene taken by the imaging system. Ideally, it is desirable to have a high spatial resolution image with many spectral bands covering a wide area. In reality, depending on the intended application, spatial resolution may have to be compromised to accommodate a larger number of spectral bands, or wide area coverage. A small number of spectral bands or a smaller area of coverage may be accepted to allow high spatial resolution imaging.

LANDSAT: LANDSAT carries two multi-spectral sensors. The first is the *multi-spectral scanner* (MSS) which acquires imagery in four spectral bands: blue, green, red, and near-infrared. The second is the *thematic mapper* (TM) which collects seven bands: blue, green, red, near-infrared, two mid-infrared, and one thermal infrared. The MSS has a spatial resolution of 80 meters, while that of the TM is 30 meters. Both sensors image a 185-km wide swath, passing over each day at 09:45 local time, and returning every 16 days. With LANDSAT 7, support for TM imagery is to be continued with the addition of a coregistered 15-meter panchromatic band.

SPOT: SPOT satellites carry two *high resolution visible* (HRV) pushbroom sensors which operate in multi-spectral or panchromatic mode. The multi-spectral images have 20-meter spatial resolution while the panchromatic images have 10-meter resolution. All SPOT images cover a swath 60 kilometers wide. The SPOT sensor may be pointed to image along adjacent paths. This allows the instrument to acquire repeat imagery of any area 12 times during its 26-day orbital period. The pointing capability makes SPOT the only satellite system which can acquire useful stereo satellite imagery.

IRS: The Indian Space Research Organization currently has five satellites in the IRS system. IRS-1D satellites that together provide continuing global coverage with the following sensors:

- IRS-Pan: 5.8 meter panchromatic

- IRS-LISS38: 23.5 meter multi-spectral in the following bands: green (0.52–0.59), red (0.62–0.68), near-infrared (0.77–0.86), shortwave infrared (1.55–1.7)

- IRS-WiFS9: 180 meter multi-spectral in the following bands: red (0.62–0.68), near-infrared (0.77–0.86)

6.10 INFRARED REMOTE SENSING

Infrared remote sensing makes use of infrared sensors to detect infrared radiation emitted from the Earth's surface. The *middle-wave infrared* (MWIR) and *long-wave infrared* (LWIR) are within the thermal infrared region. These radiations are emitted from warm objects such as the Earth's surface. They are used in satellite remote sensing for measurements of the Earth's land and sea surface temperature. Thermal infrared remote sensing is also often used for detection of forest fires. This is shown in Figure 6.14.

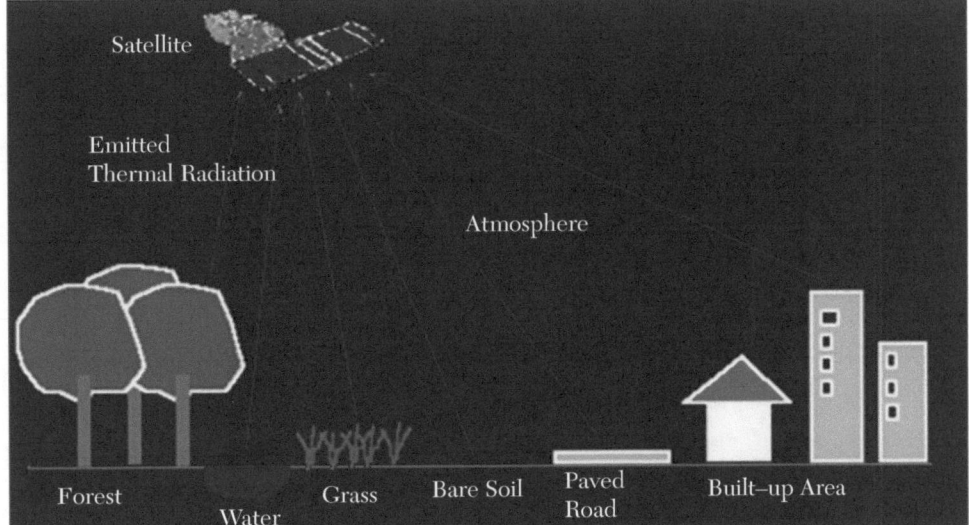

FIGURE 6.14 Infrared remote sensing.

6.10.1 Black Body Radiation

The amount of thermal radiation emitted at a wavelength from a warm object depends on its temperature. If the Earth's surface is regarded as a blackbody emitter, its apparent temperature (known as the *brightness temperature*) and spectral radiance are related by Planck's blackbody equation, plotted in Figure 6.15 for several temperatures. For a surface at a brightness temperature around 300 K, spectral radiance peaks at a wavelength around 10 µm. The peak wavelength decreases as the brightness temperature increases. For this reason, most satellite sensors for measurement of Earth's surface temperature have a band detecting infrared radiation around 10 µm. Besides the measurement of regular surface temperature, infrared sensors can be used for detection of forest fires or other warm/hot objects. For typical fire temperatures from about 500 K (smoldering fire) to over 1,000 K (flaming fire), the radiance versus wavelength curves peak at around 3.8 µm. Sensors such as the NOAA-AVHRR, ERS-ATSR and TERRA-MODIS are equipped with this band that can be used for detection of fire hot spots.

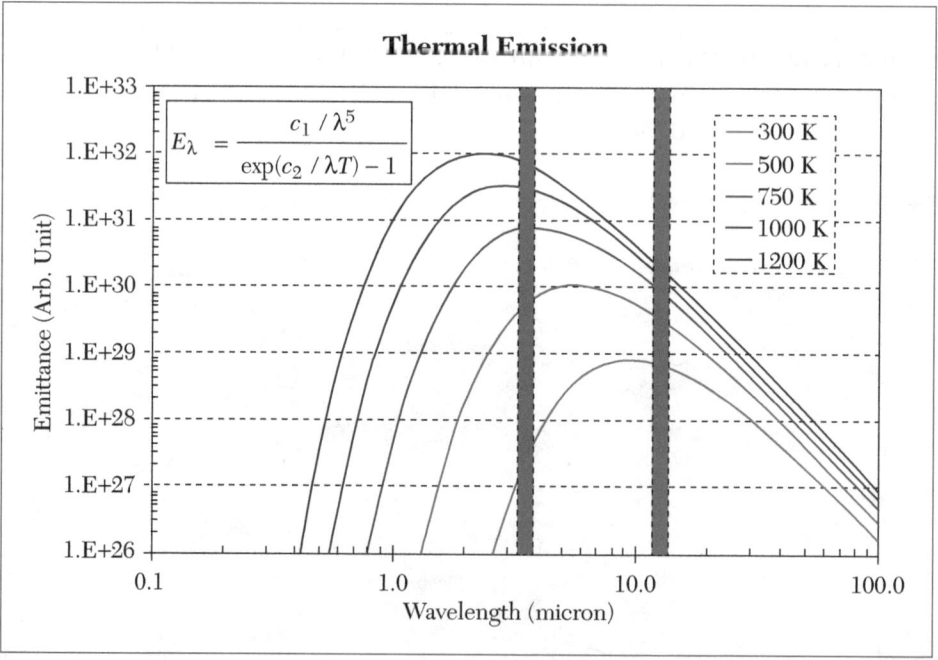

FIGURE 6.15 Thermal emission from a surface at various temperatures, modeled by Planck's equation for an ideal blackbody.

The two bands around 3.8 μm (e.g., AVHRR band 3) and 10 μm (e.g. AVHRR band 4) commonly available in infrared remote sensing satellite sensors are marked in Figure 6.15. A 50-km resolution global *sea surface temperature* (SST) field for the period 11 to 24 September 2014 derived from NOAA-AVHRR thermal infrared data is shown in Figure 6.16. Occurrence of abnormal climatic conditions such as the El Nino can be predicted by observations of the SST anomaly, i.e., the deviation of the daily SST from the mean SST.

FIGURE 6.16 Sea surface temperature data.

6.11 MICROWAVE REMOTE SENSING

Electromagnetic radiation in the microwave wavelength region is used in remote sensing to provide useful information about the Earth's atmosphere, land, and ocean. A microwave radiometer is a passive device which records the natural microwave emission from the Earth. It can be used to measure the total water content of the atmosphere within its field of view. A radar altimeter sends out pulses of microwave signals and record the signal scattered back from the Earth's surface. The height of the surface can be measured from the time delay of the return signals. A wind scatterometer can be used to measure wind speed and direction over the ocean's surface. It sends out pulses of microwaves along several directions and records the magnitude of the signals backscattered from the ocean surface. The magnitude of the backscattered signal is related to the ocean's surface roughness, which in turns is dependent on the sea surface wind condition, and hence wind speed and direction can be derived.

Synthetic aperture radar (SAR): In *synthetic aperture radar* (SAR) imaging, microwave pulses are transmitted by an antenna towards the earth surface. The microwave energy scattered back to the spacecraft is measured. The SAR makes use of the radar principle to form an image by utilizing the time delay of the backscattered signals.

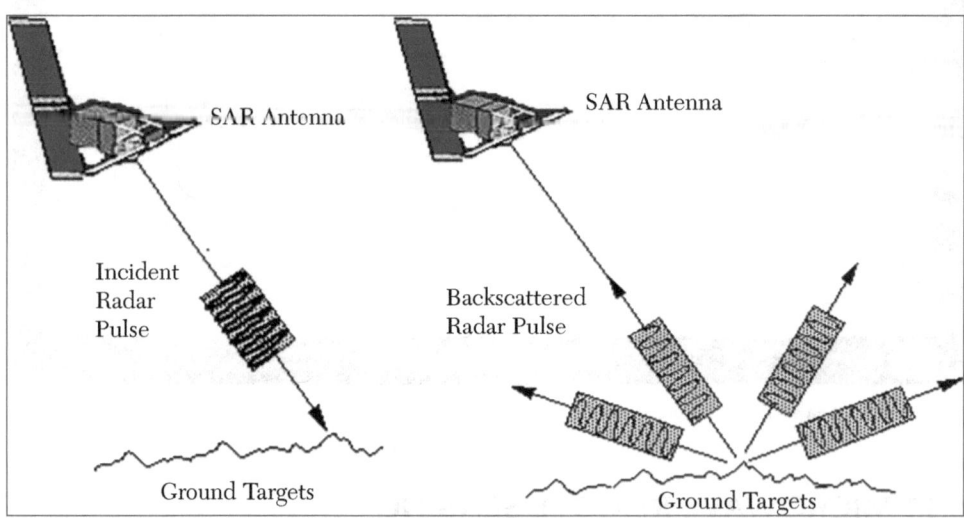

FIGURE 6.17 SAR using microwave remote sensing.

A radar pulse is transmitted from the antenna to the ground. The radar pulse is scattered by the ground targets back to the antenna as in Figure 6.17. In real aperture radar imaging, the ground resolution is limited by the size of the microwave beam sent out from the antenna. Finer details on the ground can be resolved by using a narrower beam. The beam width is inversely proportional to the size of the antenna, i.e., the longer the

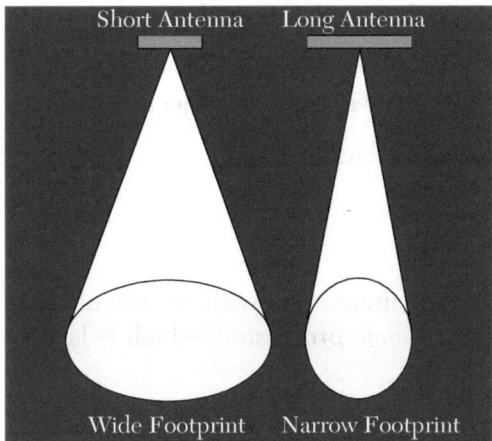

FIGURE 6.18 The relationship between antenna size and footprint.

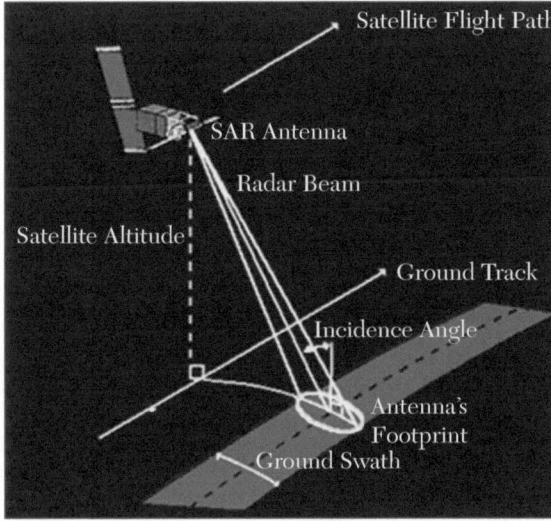

FIGURE 6.19 SAR imaging system.

antenna, the narrower the beam, as in Figure 6.18.

The microwave beam sent out by the antenna illuminates an area on the ground (known as the antenna's *footprint*). In radar imaging, the recorded signal strength depends on the microwave energy backscattered from the ground targets inside this footprint. Increasing the length of the antenna will decrease the width of the footprint. It is not feasible for a spacecraft to carry the very long antenna that is required for high-resolution imaging of the Earth's surface. To overcome this limitation, SAR capitalizes on the motion of the spacecraft to emulate a large antenna (about 4 km for the ERS SAR) from the small antenna (10 meters on the ERS satellite) it actually carries on board. Imaging geometry for a typical strip-mapping synthetic aperture radar imaging system is shown in Figure 6.19. The antenna's footprint sweeps out a strip parallel to the direction of the satellite's ground track.

6.11.1 Interaction between Microwaves and Earth's Surface

When microwaves strike a surface, the proportion of energy scattered back to the sensor depends on many factors:

- Physical factors such as the dielectric constant of the surface materials which also depends strongly on the moisture content

- Geometric factors such as surface roughness, slopes, and orientation of the objects relative to the radar beam direction

- The types of land cover (soil, vegetation, or man-made objects)

- Microwave frequency, polarization, and incident angle

6.12 DIGITAL IMAGE PROCESSING

Digital data analysis techniques to the image data with the use of computer is the field of study called *digital image processing*, which is largely concerned with four basic operations:

1. Image restoration or preprocessing

2. Image enhancement

3. Image classification

4. Image transformation

Image restoration is concerned with the correction and calibration of images to achieve as faithful a representation of the Earth's surface as possible—a fundamental consideration for all applications. *Image enhancement* is predominantly concerned with the modification of images to optimize their appearance to the visual system. Visual analysis is a key element, even in digital image processing, and the effects of these techniques can be dramatic. *Image classification* refers to the computer assisted interpretation of images, an operation that is vital to GIS. Finally, *image transformation* refers to the derivation of new imagery through mathematical treatment of the raw image bands. To undertake the operations listed, it is necessary to have access to image processing software. IDRISI is one such system. While it is known primarily as a GIS software system, it also offers a full suite of image processing capabilities.

6.12.1 Image Preprocessing

Prior to data analysis, initial processing on the raw data is usually carried out to correct for any distortion due to the characteristics of the imaging system and imaging conditions. Depending on the user's requirement, some standard correction procedures may be carried out by the ground station operators before the data is delivered to the end user. These procedures include radiometric correction to correct for uneven sensor response over the whole image and geometric correction to correct for geometric distortion due to Earth's rotation and other imaging conditions (such as oblique viewing). The image may also be transformed to conform to a specific map projection system. Furthermore, if accurate geographical location of an area on the image needs to be known, *ground control points* (GCPs) are used to register the image to a precise map (*georeferencing*).

Remotely sensed images of the environment are typically taken at a great distance from the Earth's surface. As a result, there is a substantial atmospheric path that electromagnetic energy must pass through before it reaches the sensor. Depending upon the wavelengths involved and atmospheric conditions (such as particulate matter, moisture content, and turbulence), the incoming energy may be substantially modified. The sensor itself may then modify the character of that data since it may combine a variety of mechanical, optical, and electrical components that serve to modify or mask the measured radiant energy. In addition, during the time the image is being scanned, the satellite is following a path that is subject to minor variations as the earth is moving underneath. The geometry of the image is thus in constant flux. Finally, the signal needs to be telemetered back to Earth, and processed to yield the final data we receive. Consequently, a variety of systematic and apparently random disturbances can combine to degrade the final quality of the image. Image restoration seeks to remove these degradation effects. Broadly, image restoration can be broken down into the subareas of radiometric restoration and geometric restoration.

Radiometric restoration: Radiometric restoration refers to the removal or diminishment of distortions in the degree of electromagnetic energy registered by each detector. A variety of agents can cause distortion in the values recorded for image cells. Some of the most common distortions for which correction procedures exist include: uniformly elevated values, due to atmospheric haze, which preferentially scatters short wavelength bands (particularly the blue wavelengths); striping, due to detectors going out of calibration; random noise, due to unpredictable and unsystematic performance of the sensor or transmission of the data; and scan line drop out, due to signal loss from specific detectors. It is also appropriate to include here procedures that are used to convert the raw, unitless relative reflectance values (known as *digital numbers,* or DN) of the original bands into true measures of reflective power (radiance).

Geometric restoration: For mapping purposes, it is essential that any form of remotely sensed imagery be accurately registered to the proposed map base. With satellite imagery, the very high altitude of the sensing platform results in minimal image displacements due to relief and registration can usually be achieved through the use of a systematic rubber sheet transformation process that gently warps an image (through the use of polynomial equations) based on the known positions of a set of widely dispersed control points. This capability is provided in IDRISI through the module RESAMPLE. With aerial photographs, however, the process is more complex. In these instances, it is necessary to use photogrammetric rectification to remove these distortions and provide accurate map measurements. Failing this, the central portions of high altitude photographs can be resampled with some success. RESAMPLE is a module of major importance, and it is essential that one learn to use it effectively. Doing so also requires a thorough understanding of reference systems and their associated parameters such as datums and projections.

6.12.2 Image Enhancement

Image enhancement is concerned with the modification of images to make them more suited to the capabilities of human vision. Regardless of the extent of digital intervention, visual analysis invariably plays a very strong role in all aspects of remote sensing. While the range of image enhancement techniques is broad, the following fundamental issues form the backbone of this area:

Contrast stretch: Digital sensors have a wide range of output values to accommodate the strongly varying reflectance values that can be found in different environments. However, in any single environment, it is often the case that only a narrow range of values will occur over most areas. Grey level distributions thus tend to be much skewed. Contrast manipulation procedures are thus essential to most visual analyses. Figure 6.20 shows TM band 3 (visible red) and its histogram. Note that the values of the image are quite skewed. The right image of the figure shows the same image band after a linear stretch between values 12 and 60 has been applied. In IDRISI, this type of contrast enhancement may be performed interactively through *composer's layer properties* while the image is displayed. This is normally used for visual analysis only—original data values are used in numeric analyses. New images with stretched values are produced with the module STRETCH.

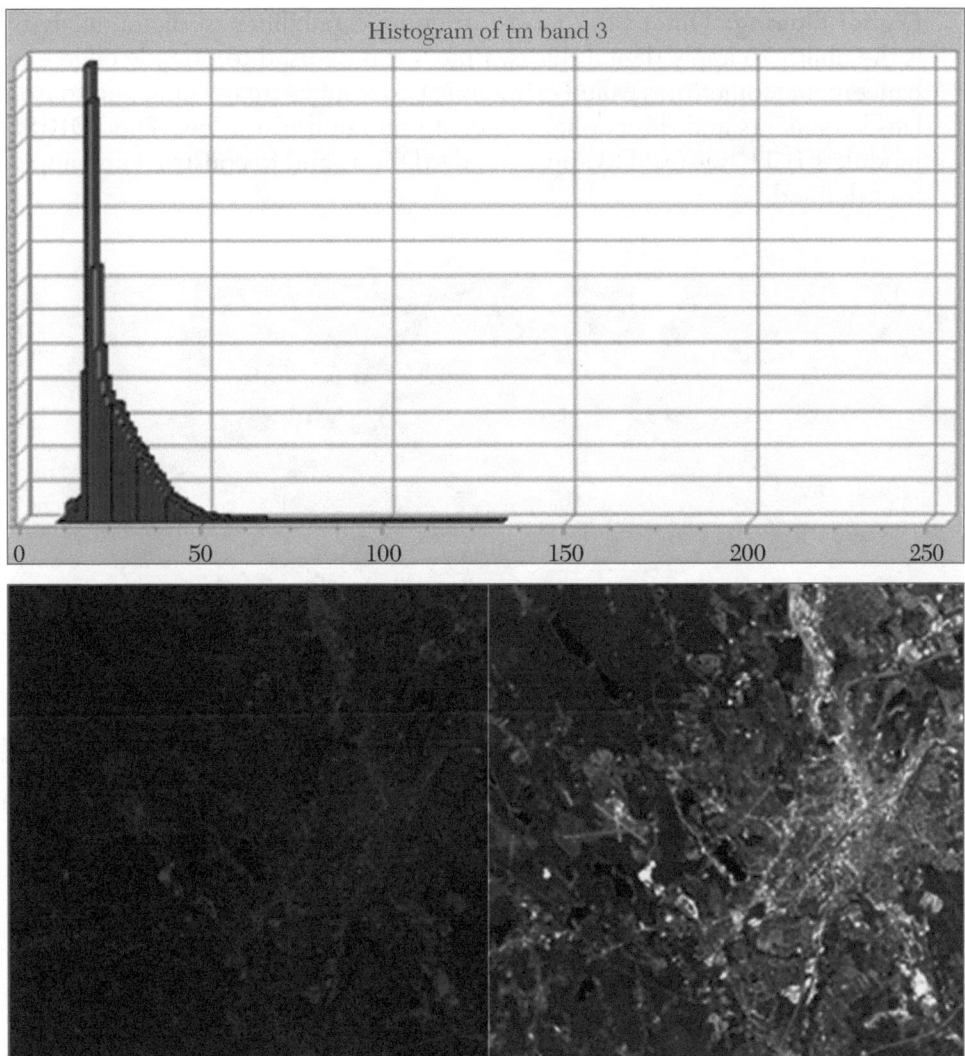

FIGURE 6.20 Linear stretch.

Composite generation: For visual analysis, color composites make fullest use of the capabilities of the human eye. Depending upon the graphics system in use, composite generation ranges from simply selecting the bands to use, to more involved procedures of band combination and associated contrast stretch. Figure 6.21 shows several composites made with different band combinations from the same set of TM images. (See Figure 6.20 for TM band definitions.) The IDRISI module COMPOSITE is used to construct three-band composite images.

Digital filtering: One of the most intriguing capabilities of digital analysis is the ability to apply digital filters. Filters can be used to provide edge enhancement (sometimes called *crispening*), to remove image blur, and to isolate lineaments and directional trends, to mention just a few. The IDRISI module FILTER is used to apply standard filters and to construct and apply user defined.

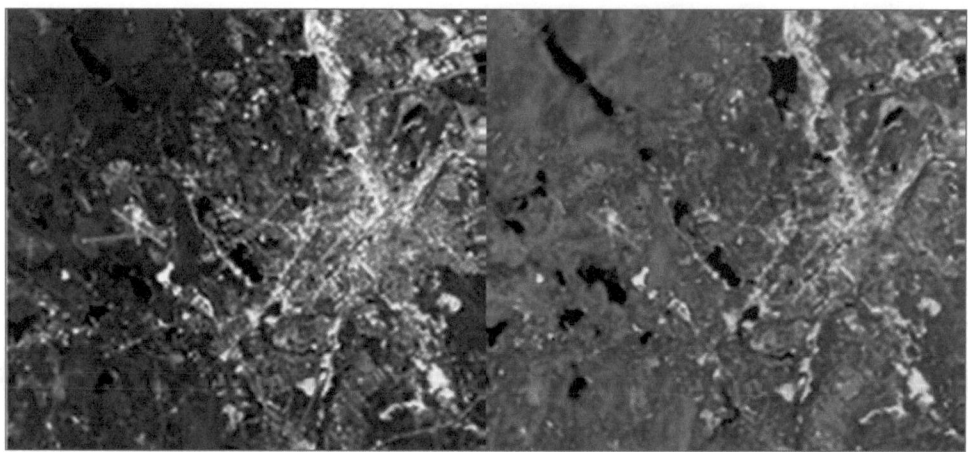

RGB = bands 3, 2, 1.   RGB = bands 4, 3, 2.

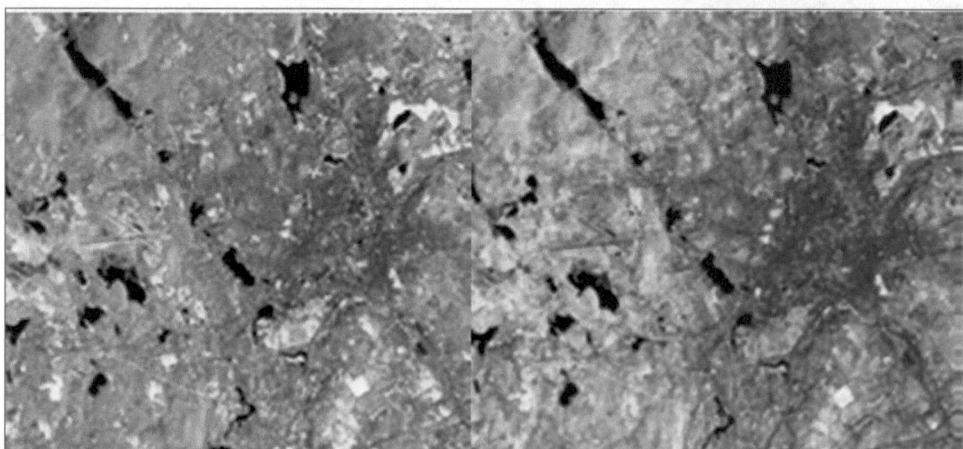

RGB = bands 4, 5, 3.   RGB = bands 7, 4, 2.

FIGURE 6.21 Several composites made with different band combinations from the same (Figure 6.20) set of TM images.

To aid visual interpretation, the visual appearance of the objects in the image can be improved by image enhancement techniques such as grey level stretching to improve the contrast and spatial filtering for enhancing the edges. An example of an enhancement procedure is shown here.

FIGURE 6.22 Unenhanced images.

A multi-spectral SPOT image is shown in Figure 6.22. Radiometric and geometric corrections have been done. The image has also been transformed to conform to a certain map projection (UTM projection). This image is displayed without any further enhancement. In the above unenhanced image, a bluish tint can be seen all over the image, producing a hazy appearance. This hazy appearance is due to scattering of sunlight by atmosphere into the field of view of the sensor. This effect also degrades the contrast between different land covers. It is useful to examine the image histograms before performing any image enhancement. The X-axis of the histogram is the range of the available digital numbers, i.e., 0–255. The Y-axis is the number of pixels in the image having a given digital number. The histograms of the three bands of this image are shown in Figure 6.23: XS3 (near-infrared) band (displayed in red), XS2 (red) band (displayed in green), and XS1 (green) band (displayed in blue).

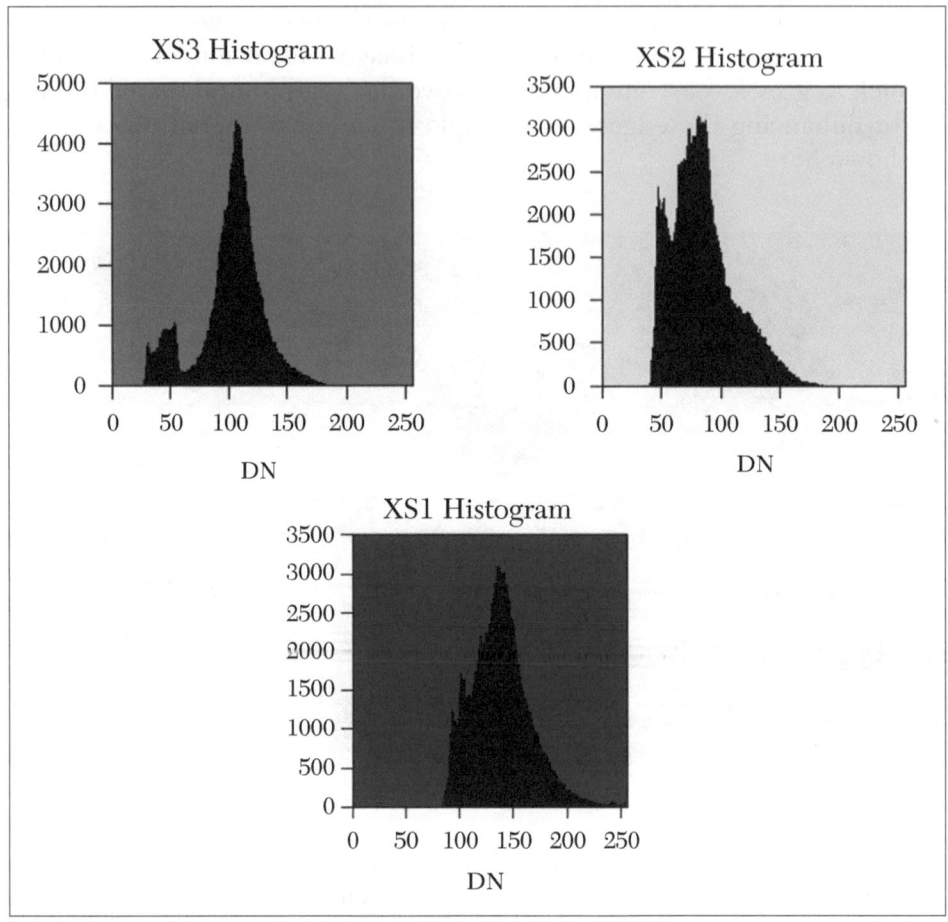

FIGURE 6.23 Histograms of the three bands' images.

Note that the minimum digital number for each band is not zero. Each histogram is shifted to the right by a certain amount. This shift is due to the atmospheric scattering component adding to the actual radiation reflected from the ground. The shift is particularly large for the XS1 band compared to the other two bands due to the higher contribution from Rayleigh scattering for the shorter wavelength. The maximum digital number of each band is also not 255. The sensor's gain factor has been adjusted to anticipate any possibility of encountering a very bright object. Hence, most of the pixels in the image have digital numbers well below the maximum value of 255. The image can be enhanced by a simple linear grey-level

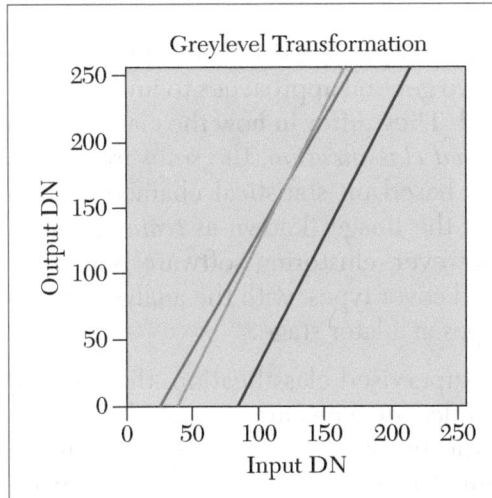

FIGURE 6.24 Grey-level transformation table for performing linear grey level stretching of the three bands of the image. Red line: XS3 band; Green line: XS2 band; Blue line: XS1 band.

stretching. In this method, a level threshold value is chosen so that all pixel values below this threshold are mapped to zero. An upper threshold value is also chosen so that all pixel values above this threshold are mapped to 255. All other pixel values are linearly interpolated to lie between 0 and 255. The lower and upper thresholds are usually chosen to be values close to the minimum and maximum pixel values of the image. The grey-level transformation table is shown in Figure 6.24.

The result of applying the linear stretch is shown in Figure 6.25. Note that the hazy appearance has generally been removed, except for some parts near the top of the image. The contrast between different features has been improved.

FIGURE 6.25 Multi-spectral SPOT image after enhancement by a simple linear grey-level stretching.

6.12.3 Image Classification

Image classification refers to the computer-assisted interpretation of remotely sensed images. There are two general approaches to image classification: supervised and unsupervised. They differ in how the classification is performed. In the case of *supervised classification*, the software system delineates specific land cover types based on statistical characterization data drawn from known examples in the image (known as *training sites*). With *unsupervised classification*, however, clustering software is used to uncover the commonly occurring land cover types, with the analyst providing interpretations of those cover types at a later stage.

Supervised classification: In supervised classification, the spectral features of some areas of known landcover types are extracted from the image. These areas are known as the training areas. Every pixel in the whole image is classified as belonging to one of the classes depending on how close its spectral features are to the spectral features of the training areas. The software system is then used to develop a statistical characterization of the reflectance for each information class. This stage is often called *signature analysis* and may involve developing a characterization as simple as the mean or the range of reflectance on each band, or as complex as detailed analyses of the mean, variances and covariances over all bands. Once a statistical characterization has been achieved for each information class, the image is then classified by examining the reflectance for each pixel and deciding which of the signatures it resembles most. There are several techniques for making these decisions, called *classifiers*. Most image processing software will offer several, based on varying decision rules. IDRISI offers a wide range of options falling into three groups depending upon the nature of the output desired and the nature of the input bands.

Hard classifiers: The distinguishing characteristic of hard classifiers is that they all make a definitive decision about the land cover class to which any pixel belongs. IDRISI offers three supervised classifiers in this group: *parallelepiped* (PIPED), *minimum distance to means* (MINDIST), and *maximum likelihood* (MAXLIKE). They differ only in how they develop and use a statistical characterization of the training site data. Of the three, the maximum likelihood procedure is the most sophisticated and is unquestionably the most widely used classifier in the classification of remotely sensed imagery.

Soft classifiers: In contrast to hard classifiers, soft classifiers do not make a definitive decision about the land cover class to which each pixel belongs. Rather, they develop statements of the degree to which each pixel belongs to each of the land cover classes being considered. Thus, for example, a soft classifier might indicate that a pixel has a 0.72 probability of being forest, a 0.24 probability of being pasture, and a 0.04 probability of being bare ground. A hard classifier would resolve this uncertainty by concluding that the pixel was forest. However, a soft classifier makes this uncertainty explicitly available, for any of a variety of reasons. For example, the analyst might conclude that the uncertainty arises because the pixel contains more than one cover type and could use the probabilities as indications of the relative proportion of each. This is known as subpixel classification. Alternatively, the analyst may conclude that the uncertainty arises because of unrepresentative training site data and therefore may wish to combine these probabilities with other evidence before hardening the decision to a final conclusion. IDRISI offers three soft classifiers (BAYCLASS, BELCLASS, and FUZCLASS) and three corresponding hardeners (MAXBAY, MAXBEL, and MAXFUZ). The difference between them relates to the logic by which uncertainty is specified: Bayesian, Dempster-Shafer, and Fuzzy Sets, respectively. In addition, the system supplies a variety of additional tools specifically designed for the analysis of subpixel mixtures (e.g., UNMIX, FUZSIG, MIXCALC, and MAXSET).

Hyperspectral classifiers: All the classifiers mentioned above operate on multi-spectral imagery—images where several spectral bands have been captured simultaneously as independently accessible image components. Extending this logic to many bands produces what has come to be known as hyperspectral imagery. Although there is essentially no difference between hyperspectral and multi-spectral imagery (i.e., they differ only in degree), the volume of data and high spectral resolution of hyperspectral images does lead to differences in the way that they are handled. IDRISI provides special facilities for creating hyperspectral signatures either from training sites or from libraries of spectral response patterns developed under lab conditions (HYPERSIG) and an automated hyperspectral signature extraction routine (HYPERAUTOSIG). These signatures can then be applied to any of several hyperspectral classifiers: spectral angle mapper (HYPERSAM), minimum distance to means (HYPERMIN), linear spectral unmixing (HYPERUNMIX), orthogonal subspace projection (HYPEROSP), and absorption area analysis (HYPERABSORB). An unsupervised classifier for hyperspectral imagery (HYPERUSP) is also available.

Unsupervised classification: In unsupervised classification, the computer program automatically groups the pixels in the image into separate clusters, depending on their spectral features. Each cluster will then be assigned a landcover type by the analyst. Each class of landcover is referred to as a *theme* and the product of classification is known as a *thematic map*. In contrast to supervised classification, where the system about the character (i.e., signature) of the information classes are looking for, unsupervised classification requires no advance information about the classes of interest. Rather, it examines the data and breaks it into the most prevalent natural spectral groupings, or clusters, present in the data. The analyst then identifies these clusters as land cover classes through a combination of familiarity with the region and ground truth visits.

The logic by which unsupervised classification works is known as *cluster analysis*, and is provided in IDRISI primarily by the CLUSTER module. CLUSTER performs classification of composite images (created with COMPOSITE) that combine the most useful information bands. It is important to recognize, however, that the clusters unsupervised classification produces are not information classes, but spectral classes (i.e., they group together features [pixels] with similar reflectance patterns). It is thus usually the case that the analyst needs to reclassify spectral classes into information classes. For example, the system might identify classes for asphalt and cement which the analyst might later group together, creating an information class called pavement. While attractive conceptually, unsupervised classification has traditionally been hampered by very slow algorithms. However, the clustering procedure provided in IDRISI is extraordinarily fast (unquestionably the fastest on the market) and can thus be used iteratively in conjunction with ground truth data to arrive at a very strong classification. With suitable ground truth and accuracy assessment procedures, this tool can provide a remarkably rapid means of producing quality land cover data on a continuing basis.

In addition to these techniques, two modules bridge both supervised and unsupervised classifications. ISOCLUST uses a procedure known as *self-organizing cluster analysis* to classify up to 7 raw bands with the user specifying the number of clusters to process. The procedure uses the CLUSTER module to initiate a set of clusters that seed an iterative application of the MAXLIKE procedure, each stage using the results of the previous stage as the training sites for this supervised procedure. The result is an unsupervised classification that converges on a final set of stable members using a supervised approach (hence the notion of "self-organizing"). MAXSET is also, at its core, a supervised procedure. However, while the procedure starts with training sites that characterize individual classes, it results in a classification that includes not only these specific classes, but also significant (but unknown) mixtures that might exist. Thus, the end result has much the character of that of an unsupervised approach.

Accuracy assessment: A vital step in the classification process, whether supervised or unsupervised, is the assessment of the accuracy of the final images produced. This involves identifying a set of sample locations (such as with the SAMPLE module) that are visited in the field. The land cover found in the field is then compared to that which was mapped in the image for the same location. Statistical assessments of accuracy may then be derived for the entire study area, as well as for individual classes (using ERRMAT). In an iterative approach, the error matrix produced (sometimes referred to as a *confusion matrix*), may be used to identify cover types for which errors are in excess of that desired. The information in the matrix about which covers are being mistakenly included in a class (errors of commission) and those that are being mistakenly excluded (errors of omission) from that class can be used to refine the classification approach.

Other transformations: As mentioned earlier, IDRISI offers a variety of other transformations. These include *color space transformations* (COLSPACE), *texture calculations* (TEXTURE), *blackbody thermal transformations* (THERMAL), and a wide variety of ad hoc transformations (such as *image ratioing*) that can be most effectively accomplished with the image calculator utility. The availability of this data, coupled with the computer software necessary to analyze it, provides opportunities for environmental scholars and planners, particularly in the areas of land use mapping and change detection that would have been unheard of only a few decades ago. The inherent raster structure of remotely sensed data makes it readily compatible with raster GIS. Thus, while IDRISI provides a wide suite of image processing tools, they are completely integrated with the broader set of raster GIS tools the system provides. A plausible assignment of landcover types to the thematic classes is shown in Table 6.1. The accuracy of the thematic map derived from remote sensing images should be verified by field observation.

Class No. (Color in Map)	Land Cover Type
1 (black)	Clear water
2 (green)	Dense forest with closed canopy
3 (yellow)	Shrubs, less-dense forest
4 (orange)	Grass
5 (cyan)	Bare soil, built-up areas
6 (blue)	Turbid water, bare soil, built-up areas
7 (red)	bare soil, built-up areas
8 (white)	bare soil, built-up areas

TABLE 6.1 Land Cover Types to the Thematic Class

The spectral features of these land cover classes can be exhibited in two graphs shown in Figure 6.26. The first graph is a plot of the mean pixel values of the XS3 (near-infrared) band versus the XS2 (red) band for each class. The second graph is a plot of the mean pixel values of the XS2 (red) versus XS1 bands. The standard deviations of the pixel values for each class are also shown.

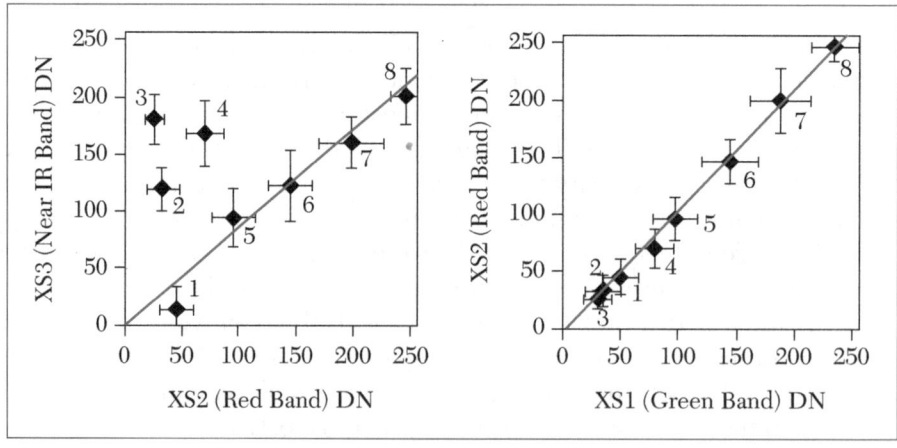

FIGURE 6.26 Scatterplot of the mean pixel values for each land cover class.

In the scatterplot of the class means in the XS3 and XS2 bands, the data points for the nonvegetated landcover classes generally lie on a straight line passing through the origin. This line is called the *soil line*. The vegetated landcover classes lie above the soil line due to the higher reflectance in the near-infrared region (XS3 band) relative to the visible region. In the XS2 (visible red) versus XS1 (visible green) scatterplot, all the data points generally lie on a straight line. This plot shows that the two visible bands are very highly correlated. The vegetated areas and clear water are generally dark while the other nonvegetated landcover classes have varying brightness in the visible bands.

6.12.4 Image Transformation

Digital image processing offers a limitless range of possible transformations on remotely sensed data. Two are mentioned here specifically, because of their special significance in environmental monitoring applications.

Vegetation indices: There are a variety of vegetation indices that have been developed to help in the monitoring of vegetation. Most are based on the very different interactions between vegetation and electromagnetic energy in the red and near-infrared wavelengths. Reflectance in the red region (about 0.6–0.7µ) is low because of absorption by leaf pigments (principally chlorophyll). The infrared region (about 0.8–0.9µ), however, characteristically shows high reflectance because of scattering by the cell structure of the leaves. A very simple vegetation index can thus be achieved by comparing the measure of infrared reflectance to that of the red reflectance. Although several variants of this basic logic have been developed, the one which has received the most attention is the normalized difference vegetation index (NDVI). It is calculated in the following manner:

$$NDVI = (NIR - R) / (NIR + R)$$

Where,
NIR = Near Infrared and R = Red

Figure 6.27 shows NDVI calculated with TM bands 3 and 4 for the same area shown in Figures 6.20 and 6.21.

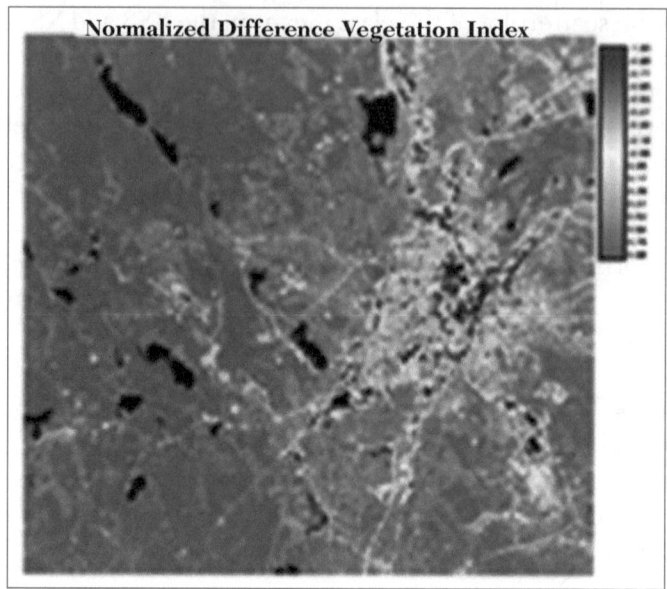

FIGURE 6.27 NDVI corrected image.

Principal components analysis: *Principal components analysis* (PCA) is a linear transformation technique related to factor analysis. Given a set of image bands, PCA produces a new set of images, known as components, that are uncorrelated with one another and are ordered in terms of the amount of variance they explain from the original band set. PCA has traditionally been used in remote sensing as a means of data compaction. For a typical multi-spectral image band set, it is common to find that the first two or three components can explain virtually all of the original variability in reflectance values. Later components thus tend to be dominated by noise effects. By rejecting these later components, the volume of data is reduced with no appreciable loss of information. Given that the later components are dominated by noise, it is also possible to use PCA as a noise removal technique. The output from the PCA module in IDRISI includes the coefficients of both the forward and backward transformations. By zeroing out the coefficients of the noise components in the reverse transformation, a new version of the original bands can be produced with these noise elements removed.

Recently, PCA has also been shown to have special application in environmental monitoring. In cases where multi-spectral images are available for two dates, the bands from both images are submitted to a PCA as if they all came from the same image. In these cases, changes between the two dates tend to emerge in the later components. More dramatically, if a time series of NDVI images (or a similar single-band index) is submitted to the analysis, a very detailed analysis of environmental changes and trends can be achieved. In this case, the first component will show the typical NDVI over the entire series while each successive component illustrates change events in an ordered sequence of importance. By examining these images, along with graphs of their correlation with the individual bands in the original series, important insights can be gained into the nature of changes and trends over the time series. The TSA (time series analysis) module in IDRISI is a specially tailored version of PCA to facilitate this process.

Measurement of biogeophysical parameters: Specific instruments carried onboard the satellites can be used to make measurements of the biogeo-physical parameters of the Earth. Some of the examples are: atmospheric water vapor content, stratospheric ozone, land and sea surface temperature, sea water chlorophyll concentration, forest biomass, sea surface wind field, tropospheric aerosol, etc. Specific satellite missions have been launched to continuously monitor the global variations of these environmental param-eters that may show the causes or the effects of global climate change and the impacts of human activities on the environment.

Geographical information system (GIS): Different forms of imagery such as optical and radar images provide complementary information about the landcover. More detailed information can be derived by combining several different types of images. For example, radar images can form one of the layers in combination with the visible and near-infrared layers when per-forming classification. The thematic information derived from the remote sensing images is often combined with other auxiliary data to form the basis for a geographic information system (GIS). A GIS is a database of different layers, where each layer contains information about a specific aspect of the same area which is used for analysis by the resource scientists.

6.13 SOFTWARE FOR OCEAN COLOR DATA

- ODESA provides users with a complete Level 2 processing environ-ment for the MERIS instrument: It includes MEGS 8.1 and associated auxiliary data files (ADF) used for the MERIS 3rd reprocessing in line with the current version of the IPF (6.0). Within this environment, users have several possibilities:

  - Access to MERIS data from remote processing facility
  - Download MERIS level 2 processor and its operational environment

- Download the ODESA analysis tools
- Access to the MERMAID algorithm validation facilities
- Access to the ODESA and MERIS forum

- BEAM Software: The Basic ERS and Envisat (A) ATSR and MERIS Toolbox (BEAM) is a collection of executable tools and an application programming interface (API) which has been developed to facilitate the utilization, viewing, and processing of ESA MERIS, (A)ATSR, and ASAR data. It can be downloaded free of charge.

- SeaDAS (SeaWiFS Data Analysis System) is a comprehensive image analysis package developed by NASA's Ocean Biology Processing Group (OBPG) for the processing, display, analysis, and quality control of all SeaWiFS data products. It is freely available for download.

- WASI (Water Color Simulator) is a tool for the simulation of optical properties and light field parameters of deep and shallow waters and for data analysis of instruments disposed above the water surface and submerged in the water. Examples of supported measurements are downwelling irradiance, upwelling radiance, irradiance reflectance, remote sensing reflectance, attenuation, and absorption. Data analysis is done by inverse modeling. The provided database, which covers the spectral range from 350 to 1000 nm in 1 nm intervals, can be exchanged easily to represent the studied area. The module WASI-2D extends the functionality towards image processing of atmospherically corrected data from airborne sensors and satellite instruments. WASI is free of charge.

- SeaBatch: If you work with ocean color data and utilize SeaDAS (see above), you likely need a way to batch process multiple files. SeaBatch can help. SeaBatch is a group of UNIX shell scripts that batch process ocean color data derived from NASA's MODIS (Aqua and Terra) and SeaWiFS sensors. SeaBatch is a powerful tool that will greatly assist you with your research. It is free, as are SeaDAS and Unix. If using runtime SeaDAS, an IDL license is not required. With SeaBatch you can:

  - Process MODIS Level-0 files (utilize high-resolution bands)
  - Process Level-1 files to Level-2
  - Spatially bin Level-2 files (.5, 1, 2, 4, 9, and 36 km)
  - Temporally bin Level-2 files (day, 7 day, 8 day, and month)
  - Output Level-3 files as ascii, flat, hdf, png, etc.

- Borstad Associates Satellite Image Toolbox. Free online tools and reassembled datasets to help generate and test ideas and to facilitate use of remotely sensed imagery in support of oceanographic and limnological research. Most of the datasets focus on the Canadian West Coast, but a very useful "temporal profiler" is available for the entire North West Hemisphere.

- ArcGIS and satellite data. Importing satellite data into ArcGIS just got easier. There is now an ArcGIS extension that allows users to browse THREDDS catalogs and connect directly to OPeNDAP servers to access large amounts of scientific data and ingest the data into ArcGIS desktop 9.3. This extension, called the environmental data connector (EDC), uses a Java-based browser and leverages existing components from Unidata and NOAA/PMEL libraries so that users can filter large amounts of data in space and time. The user has a choice of importing the data into ArcGIS in either raster or feature format. The time stamped data can then be animated using a TimeSlider extension which is built into the EDC. A stand-alone version is also available, which provides a GUI to browse THREDDS catalogs or OPeNDAP directories, to subset the selected data in space and time, and to download the data as a netcdf file.

- Software for Graphics and Data Analysis

- Software for the calculation of surface solar irradiance and PAR, using SeaWiFS data, written by Robert Frouin and John McPherson.

- WIM (Windows Image Manager) is a general-purpose image display and analysis program for various satellite images, including those from ocean color sensors (http://wimsoft.com). This is commercial software, but it is available for free evaluation. A major addition to the tools is the WIM Automation Module (WAM), which allows automating repetitive tasks by writing simple programs using WIM functions e.g., calculating primary production according to the Behrenfeld Falkowski model.

- ACRI-ST of France has developed several tools for the commissioning phase of the ENVISAT MERIS and GOMOS instruments. Details are available at *http://www.acri-st.fr/tools/*

- Additional useful links for analyzing satellite data can be found on the data processing and analysis webpage of the Marine Environmental Protection of the Northwest Pacific Region website.

Software for Ocean-color Algorithms

- Inversion of IOP based on Rrs and Remotely Retrieved Kd

- Over Constrained Linear Matrix Inversion

- Quasi-Analytical Algorithm

- Garver, Siegel, Maritorena Model (GSM-01) The updated IDL code files for this model can be downloaded from *http://www.icess.ucsb.edu/OCisD/*

- PML algorithm is an IOP algorithm

The PML IOP model is an analytical approach for determining the spectral inherent optical properties of the ocean which uses spectral slopes, derived from field measurements, at the central wavelengths of 490 and 510 nm (or 531 for MODIS). Once the absorption and backscatter are known at these wavelengths, based on the assertion of Morel (1980), then the absorption and backscatter across the spectrum can be determined if you assume a spectral shape for backscatter. Once the primary inherent optical properties of total absorption and backscatter have been determined the biogeochemical parameters can be determined using standard relationships and slopes for CDOM and phytoplankton.

Remote Sensing Software

Among the leading remote sensing software programs, the following names can be given:

ENVI *http://www.envi-sw.com/*

PCI *http://www.pcigeomatics.com/*

ERDAS *http://www.erdas.com/*

There are also some low-cost remote sensing-GIS raster softwares:

IDRISI *http://www.clarklabs.org/*

ILWIS *http://www.itc.nl/ilwis/*

And, there are some remote sensing-GIS raster freewares (downloadable):

GRASS *http://www.baylor.edu/~grass/*

SPRING *http://sputnik.dpi.inpe.br/spring/english/home.html*

PIT *http://priede.bf.lu.lv/GIS/Descriptions/Remote_Sensing/An_Online_Handbook/Appendix/nicktutor_A.html*

Other remote sensing links:

SPOT IMAGE *http://www.spotimage.fr/*

Airborne Laser Mapping *http://www.airbornelasermapping.com/ALMNews.html*

Canada Center for Remote Sensing *http://www.ccrs.nrcan.gc.ca*

Landsat 7 *http://landsat.gsfc.nasa.gov/*

Center for the Study of Earth from Space *http://cires.colorado.edu/cses/*

Table of Fundamental Physical Constants *http://www.fpl.uni-stuttgart.de/fpl/physchem/const.html*

European Space Agency *http://earth.esa.int/*

TOPEX/Poseidon *http://topex-www.jpl.nasa.gov/*

IKONOS *http://www.spaceimaging.com/*

RADARSAT *http://www.rsi.ca/*

EXERCISES: PART A (ANSWER IN A WORD OR A SENTENCE)

1. What is an image?
2. Give an example of an analog image.
3. Define "digital image."
4. What is a pixel?
5. What do you mean by "intensity value"?
6. What is the address of a pixel?
7. _____ is the strong absorber of red visible wavelength.
8. The _____ IR region is used for monitoring animal distribution studies and soil moisture conditions.

9. How is the formation of a multi-layer image achieved?

10. A spectral response pattern is sometimes also called _____.

11. _____, _____ and _____ are the primary colors.

12. _____ color is the result of mixing a red and a green.

13. Define "multi-spectral image."

14. What is a "superspectral image"?

15. _____ refers to the size of the smallest object that can be resolved on the ground.

16. _____ is the measure of the ground area viewed by a single detector element at a given instant in time.

17. What is the difference between a high resolution and low resolution image?

18. Define "radiometric resolution."

19. What is a microwave radiometer?

20. What is a radar altimeter?

21. _____ is used to measure wind speed and direction over the ocean surface.

22. Define "digital image processing."

23. Define "image restoration."

24. What is image enhancement?

25. What is image classification?

26. What is image transformation?

27. _____ correction corrects for uneven sensor response over the whole image.

28. _____ correction corrects for geometric distortion due to Earth's rotation.

29. Image restoration can be broken down into the two subareas of _____ and _____.

EXERCISES: PART B (ANSWER IN A PAGE)

1. In the context of the ocean, explain the EM spectrum.

2. Write about push-broom scanners.

3. Write a short note on spectral response patterns.

4. What is multi-spectral remote sensing?

5. Explain hyperspectral remote sensing.

6. Explain spatial resolution and pixel size.

7. Write about spatial resolution and radiometric resolution.

8. Write about data volume in the context of ocean image processing.

9. Write about infrared remote sensing.

10. Write about microwave remote sensing.

EXERCISES: PART C (ANSWER WITHIN 2 OR 3 PAGES)

1. Give an explanation of the variety of platforms available for the capture of remotely sensed data.

2. Write in detail about the four basic operations in digital image processing.

3. Write about image preprocessing in detail.

4. Write about image enhancement.

5. In detail, explain about image classification.

6. Explain in detail about image transformation.

WEB LINKS

http://www.crisp.nus.edu.sg/~research/tutorial/image.htm

http://www.ioccg.org/data/software.html

CHAPTER 7

OCEAN ENERGY

This chapter discusses ocean energy as renewable source of energy, wave energy, wave energy technologies, tidal power, ocean thermal energy conversion, ocean current energy, offshore wind energy, offshore wind energy technology, offshore solar energy, offshore solar energy technology, and concentrating solar power technology.

7.1 OCEAN ENERGY AS A RENEWABLE SOURCE OF ENERGY

To replace fossil fuels and nuclear energy, a renewable energy source must meet four basic criteria.

1. Near limitless supply of energy: Energy that will not run out in the next 1,000 years and comes from a source that has existed for the last 1,000 years.

2. A constant even supply of energy: No periods of nonproduction, no spikes in production, no movement in energy location.

3. High energy density: Energy can be collected at one location with realistic commitment of technology and resources.

4. Survivability: Energy supply and collection is not disrupted by storms, natural disaster, war, etc.

Ocean current energy is the only form of renewable nonpolluting energy that meets all these requirements and it is the only energy source with the potential to replace fossil fuel and nuclear energy. Ocean energy is a term used to describe all forms of renewable energy derived from the sea. The ocean can produce two types of energy: thermal energy from the sun's heat, and mechanical energy from tides and waves. Oceans cover more than 70% of Earth's surface, making them the world's largest solar collectors. The sun's heat warms the surface water much more than the deep ocean water, and this temperature difference creates thermal energy. Just a small portion of the heat trapped in the ocean could power the world. Ocean thermal energy is used for many applications, including electricity generation. There are three types of electricity conversion systems: *closed cycle*, *open cycle*, and *hybrid*.

Closed cycle systems: These use the ocean's warm surface water to vaporize a working fluid with a low boiling point, such as ammonia. The vapor expands and turns a turbine. The turbine then activates a generator to produce electricity, as shown in Figure 7.1.

- Pump warm surface sea water through a heat exchanger to vaporize a low boiling point fluid (such as ammonia) that then turns a generator

- Cold deep sea water pumped through a second exchanger condenses the vapor back to liquid

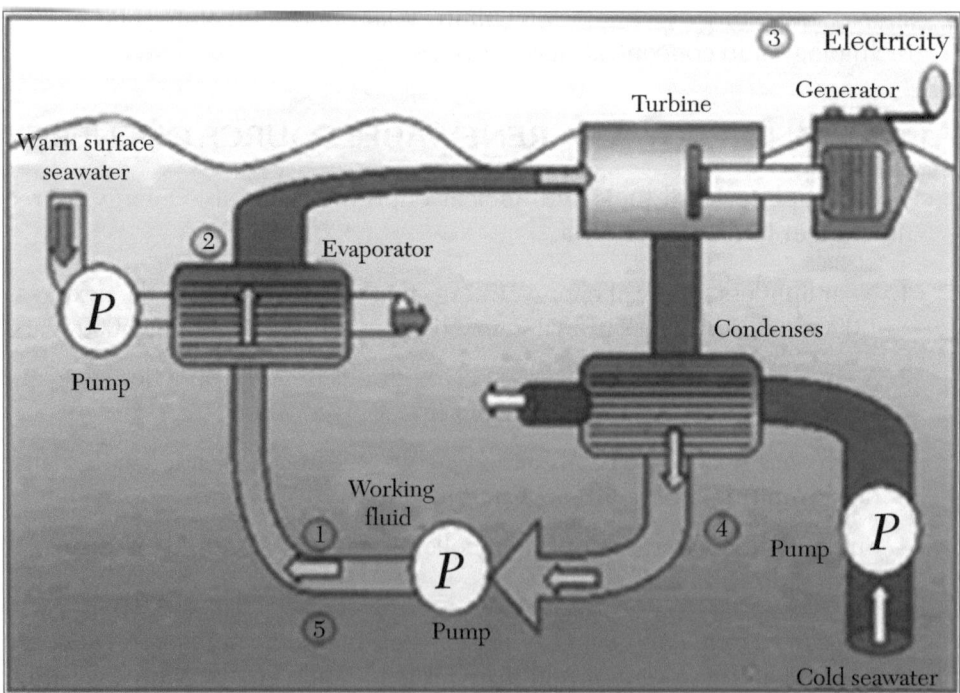

FIGURE 7.1 Closed cycle OTEC.

Open cycle systems: Open cycle systems boil the sea water by operating at low pressures. This produces steam that passes through a turbine/generator.

- Boil warm surface sea water to create steam that turns a low-pressure turbine

- Steam is turned back to liquid by exposure to cold temperatures from deep-ocean water

Hybrid systems: Hybrid systems combine both closed cycle and open cycle systems.

- Combine elements of open and closed cycle OTEC systems

- Warm sea water is flash-evaporated into steam (open cycle)

- Steam vaporizes a low-boiling-point fluid (closed cycle) that turns a turbine

A byproduct of open or hybrid cycle OTEC plants is the production of fresh water from sea water, known as *desalinization*.

Ocean mechanical energy is quite different from ocean thermal energy. Even though the sun affects all ocean activity, tides are driven primarily by the gravitational pull of the Moon, and waves are driven primarily by the winds. Thus, tides and waves are intermittent sources of energy, while ocean thermal energy is fairly constant. Also, unlike thermal energy, the electricity conversion of both tidal and wave energy usually involves mechanical devices. A barrage (dam) is typically used to convert tidal energy into electricity by forcing the water through turbines, activating a generator. For wave energy conversion, there are three basic systems: channel systems that funnel the waves into reservoirs; float systems that drive hydraulic pumps; and oscillating water column systems that use the waves to compress air within a container. The mechanical power created from these systems either directly activates a generator or is transferred to a working fluid, water, or air, which then drives a turbine/generator. The world's ocean may eventually provide us with energy to power our homes and businesses. Generating technologies for deriving electrical power from the ocean include tidal power, wave power, ocean thermal energy conversion, ocean currents, ocean winds and salinity gradients. Of these, the three most well-developed technologies are tidal power, wave power, and ocean thermal energy conversion.

7.2 WAVE ENERGY

Ocean wave energy is captured directly from surface waves or from pressure fluctuations below the surface. Waves are caused by the wind blowing over the surface of the ocean. In many areas of the world, the wind blows with

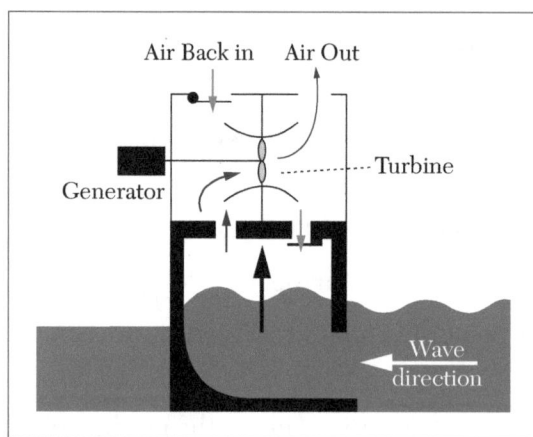

enough consistency and force to provide continuous waves along the shoreline. Ocean waves contain tremendous energy potential. Wave power devices extract energy from the surface motion of ocean waves or from pressure fluctuations below the surface. Kinetic energy (movement) exists in the moving waves of the ocean. That energy can be used to power a turbine. In Figure 7.2, the wave rises into a chamber. The rising water forces the air out of the chamber. The moving

FIGURE 7.2 Wave energy.

air spins a turbine which can turn a generator. When the wave goes down, air flows through the turbine and back into the chamber through doors that are normally closed. This is only one type of wave energy system. Others use the up-and-down motion of the wave to power a piston that moves up and down inside a cylinder. That piston can also turn a generator. Most wave energy systems are very small. But, they can be used to power a warning buoy or a small light house.

Wave energy conversion takes advantage of the ocean waves caused primarily by interaction of winds with the ocean surface. Wave energy is an irregular and oscillating low frequency energy source that must be converted to a 50 Hertz frequency before it can be added to the electric utility grid. Some systems extract energy from surface waves. Others extract energy from pressure fluctuations below the water surface or from the full wave. Some systems are fixed in position and let waves pass by them, while others follow the waves and move with them. Some systems concentrate and focus waves, which increases their height and their potential for conversion to electrical energy. A wave energy converter may be placed in the ocean in various possible situations and locations. It may be floating or submerged completely in the sea offshore or it may be located on the shore or on the sea bed in relatively shallow water. A converter on the sea bed may be completely submerged, it may extend above the sea surface, or it may be a converter system placed on an offshore platform. Apart from wave-powered navigation buoys, however, most of the prototypes have been placed at or near the shore. The visual impact of a wave energy conversion facility depends on the type of device as well as its distance from shore. In general, a floating buoy system or an offshore platform placed many kilometers from land is not likely to have much visual impact (nor will a submerged system). Onshore facilities and offshore platforms in shallow water could, however, change the visual landscape from one of natural scenery to industry. Many research and development goals remain to be accomplished, including cost reduction, efficiency and reliability improvements, identification of suitable sites, interconnection with the utility grid, and better understanding of the impacts of the technology on marine life and the shoreline. Also essential is a demonstration of the ability of the equipment to survive the salinity and pressure environments of the ocean as well as weather effects over the life of the facility.

Some of the issues that may be associated with permitting an ocean wave energy conversion facility include:

- Disturbance or destruction of marine life (including changes in the distribution and types of marine life near the shore)

- Possible threat to navigation from collisions due to the low profile of the wave energy devices above the water, making them undetectable either by direct sighting or by radar [Figure 7.3]. Also possible is the interference of mooring and anchorage lines with commercial and sport fishing.

- Degradation of scenic oceanfront views from wave energy devices located near or on the shore, and from onshore overhead electric transmission lines.

Ocean waves are caused by the wind as it blows across the water. Waves are a powerful source of energy. The problem is that it is not easy to harness this energy and convert it into electricity in large amounts. Thus, wave power stations are rare. Waves passing across the top of the unit make a piston move, which pumps sea water to drive generators on land. They are also involved with wind power and biofuel. Renewable energy resources are ones that won't run out. Wave power is renewable. Wave energy is generated by converting the energy of ocean waves (swells) into other forms

FIGURE 7.3 Ocean wave energy conversion unit.

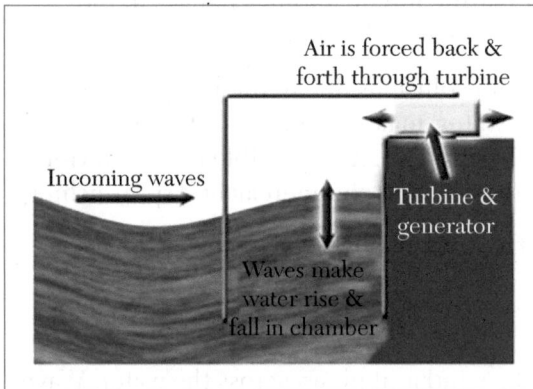

FIGURE 7.4 Waves to electricity conversion

of energy (currently only electricity). There are many different technologies that are being developed and tested to convert the energy in waves into electricity, as shown in Figure 7.4. Wave power varies considerably in different parts of the world, and whereas wind resource potential is typically given in gigawatts (GW), wave and tidal resource potential is typically given in terawatt-hours/year (TWh/yr).

Advantages:

- The energy is free—no fuel needed, no waste produced
- Inexpensive to operate and maintain
- Can produce a great deal of energy

Disadvantages:

- Depends on the waves—sometimes you'll get loads of energy, sometimes almost nothing

- Needs a suitable site, where waves are consistently strong

- Some designs are noisy, but then again, so are waves, so any noise is unlikely to be a problem

- Must be able to withstand very rough weather

7.2.1. Ocean Wave Energy Technologies

Wave technologies have been designed to be installed in nearshore, offshore, and far offshore locations. While wave energy technologies are intended to be installed at or near the water's surface, there can be major differences in their technical concept and design. For example, they may differ in their orientation to the waves or in the way they convert energy from the waves. Although wave power technologies are continuing to develop, there are four basic applications that may be suitable for deployment on the outer continental shelf (OCS): *point absorbers*, *attenuators*, *overtopping devices*, and *terminators*.

Terminators: Terminator devices extend perpendicular to the direction of the wave and capture or reflect the power of the wave. These devices are typically onshore or near shore, as seen in Figure 7.5; however, floating versions have been designed for offshore applications. The oscillating water column is a form of terminator in which water enters through a subsurface opening into a chamber, trapping air

FIGURE 7.5 Water column terminator and shore-based terminator.

above. The wave action causes the captured water column to move up and down like a piston, forcing the air though an opening connected to a turbine to generate power. These devices generally have power ratings of 500 kW to 2 MW, depending on the wave climate and the device dimensions.

Attenuators: Attenuators are long multi-segment floating structures oriented parallel to the direction of the waves, as seen in Figure 7.6. They ride the waves like a ship, extracting energy by using restraints at the bow of the device and along its length. The differing height of waves along the length of the device causes flexing where the segments connect. The segments are connected to hydraulic pumps or other converters to generate power as the waves move across. A transformer in the nose of the unit steps up the power-to-line voltage for transmission to shore. Power is fed down an umbilical cable to a junction box in the seabed, connecting it and other machines via a common subsea cable to shore.

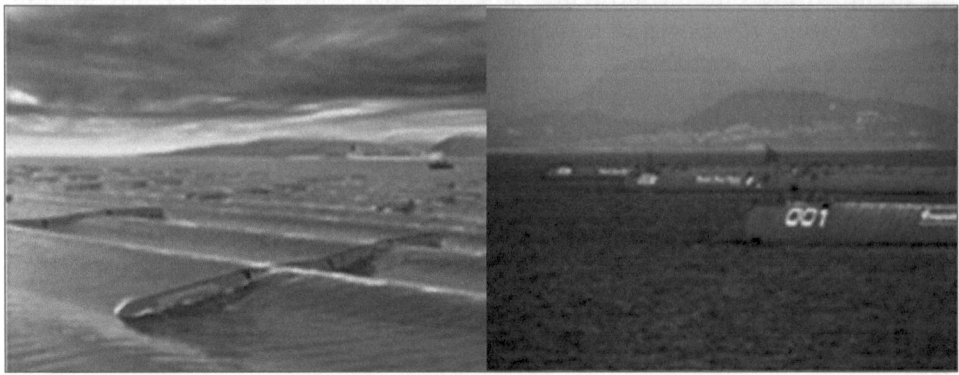

FIGURE 7.6 An array of attenuator wave energy devices.

Point absorber: A point absorber is a floating structure with components that move relative to each other due to wave action (e.g., a floating buoy inside a fixed cylinder) as in Figure 7.7. Point absorbers often look like floating oceanographic buoys. They utilize the rise and fall of the wave height at a single point for energy conversion. The relative up-and-down bobbing motion caused by passing waves is used to drive electromechanical or hydraulic energy converters to generate power.

FIGURE 7.7 Power buoy point absorber wave generation system.

Overtopping devices: Overtopping devices have reservoirs that are filled by incoming waves, causing a slight buildup of water pressure like a dam, as in Figure 7.8. The water is then released, and gravity causes it to flow back into the ocean. The energy of the falling water is used to turn hydroturbines to generate power. Specially built floating platforms can also create electricity by funneling waves through internal turbines and then back into the sea.

FIGURE 7.8 Wave dragon overtopping device.

7.3 TIDAL POWER

Another form of ocean energy is called *tidal energy*. When tides come into the shore, they can be trapped in reservoirs behind dams. Then when the tide drops, the water behind the dam can be let out just like in a regular hydroelectric power plant. Tidal energy has been used since about the 11th century, when small dams were built along ocean estuaries and small streams. The tidal water behind these dams was used to turn water wheels to mill grains. For tidal energy to work well, you need large increases in tides. An increase of at least 16 ft in height between low tide to high tide is needed. There are only a few places where this tide change occurs around the earth. Some power plants are already operating using this idea. Tidal power or current power systems capture the energy of ocean currents below the wave surface and convert them into electricity. Typically, these systems rely on underwater turbines, either horizontal or vertical, which rotate in either the ocean current or changing tide (either one way or bidirectionally), almost like an underwater windmill. These technologies can be sized or adapted for ocean or for use in lakes or nonimpounded river sites.

7.3.1 Tidal Barrage Operation

Tidal barrages work rather like hydroelectric schemes, except that the dam is much bigger. A huge dam called a *barrage* is built across a river estuary. When the tide goes in and out, the water flows through tunnels in the dam. The ebb and flow of the tides can be used to turn a turbine, or push air through a pipe, which then turns a turbine. Large lock gates, like the ones used on canals, allow ships to pass. A major drawback of tidal power stations is that they can only generate when the tide is flowing in or out—in other words, only for 10 hours each day. However, tides are totally predictable, so we can plan to have other power stations generating at those times when the tidal station is out of action. Tidal energy is generated from tidal movements. Tides contain both potential energy, related to the vertical fluctuations in sea level, and kinetic energy, related to the horizontal motion of the water. It can be harnessed using technologies using energy from the rise and fall of the tides or by technologies using energy from tidal or marine currents, as seen in Figure 7.9.

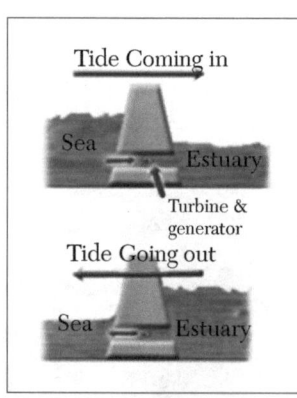

FIGURE 7.9 Tidal energy.

7.4 OCEAN THERMAL ENERGY CONVERSION (OTEC)

The temperature of water can be used to create energy. This ocean energy idea uses temperature differences in the ocean. It's warmer on the surface because sunlight warms the water. But below the surface, the ocean gets very cold. That's why scuba divers wear wet suits when they dive down deep. Their wet suits trap their body heat to keep them warm. Power plants can be built that use this difference in temperature to make energy. A difference of at least 38° Fahrenheit is needed between the warmer surface water and the colder deep ocean water. This type of energy source is called *ocean thermal energy conversion* (OTEC); its plants may be land-based, floating, or grazing. Renewable ocean energy can help diversify our energy portfolio and improve our environment. With the proper support, these resources will become a robust part of a reliable, affordable, clean electric supply infrastructure.

7.5 OCEAN TURBINE OPERATION

The turbine is composed of three sets of blades, as in Figure 7.10. The blades close when they are moving in the same direction as the flow of water, creating an obstacle that the water has to push out of its path of flow. When water pushes on the closed blades it causes the main shaft and generator to rotate. This rotational energy is collected in the form of electricity from the generator. When the blades are moving in the opposite direction as the flow of water, they open, creating minimal drag in the oncoming flow of water. This design creates very high surface area and drag on the power stroke side of the turbine while creating very low surface area and drag on the returning side. This design has proven to be the most efficient for collecting energy from ocean currents.

Benefits of ocean turbines:

- Energy output of 13.5 MW/hour
- 100% reliable energy—not dependent upon wind or sun, ocean currents are always flowing
- Doesn't add heat to environment
- No emissions
- Water safe materials and paints
- Antifouling, anticorrosive, won't rust, won't grow sea life

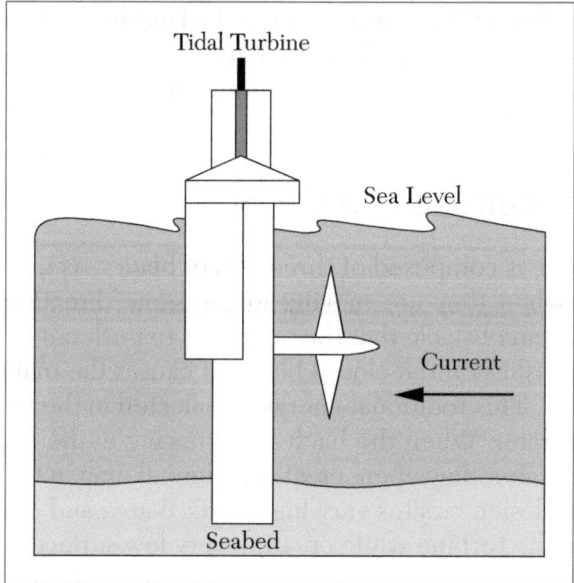

FIGURE 7.10 Ocean turbine operation.

- Can't see it, hear it, or smell it

- Huge survivability in storms (unlike other free energy devices)

- Can provide energy and clean water to hurricane/tsunami victims

- Components in the utility scale turbines are noncorrosive, nontoxic metals and high performance composite fiber

- Design of turbine has large acoustic signature so marine animals can echolocate it, moves slowly enough for them to move around it, will not harm marine life

- Less susceptible target for terrorism

- Doesn't take up high value land; places that need power have high population density, usually can't support large wind turbines or solar projects because the demand is too high for the available area for those projects

7.6 OCEAN CURRENT ENERGY

The relatively constant flow of ocean currents carries large amounts of water across the Earth's oceans. Technologies are being developed so that energy that can be extracted from ocean currents and converted to usable power. Ocean waters are constantly on the move. Ocean currents flow in complex patterns affected by wind, water salinity, temperature, topography of the ocean floor, and the Earth's rotation, as shown in Figure 7.11. Most ocean currents are driven by wind and the solar heating of surface waters near the equator, while some currents result from density and salinity variations of the water column. Ocean currents are relatively constant and flow in one direction, in contrast to tidal currents along the shore. While ocean currents move slowly relative to typical wind speeds, they carry a great deal of energy because of the density of water. Water is more than 800 times denser than air, so for the same surface area, water moving 12 mph exerts the same amount of force as a constant 110 mph wind. Because of this physical property, ocean currents contain an enormous amount of energy that can be captured and converted to a usable form.

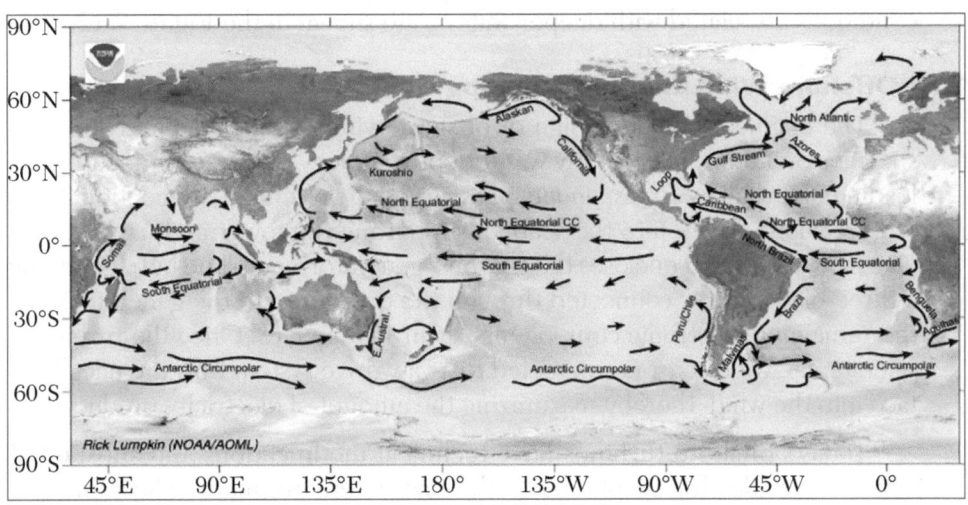

FIGURE 7.11 Ocean current surface.

7.7 OFFSHORE WIND ENERGY

Wind energy has been utilized by humans for more than 2,000 years. For example, windmills were often used by farmers and ranchers for pumping water or grinding grain. In modern times, wind energy is mainly used to generate electricity, primarily through wind turbines. All wind turbines operate in the same basic manner. As the wind blows, it flows over the airfoil-shaped turbine blades, causing them to spin. The blades are connected to a drive shaft that turns an electric generator to produce electricity. The newest wind turbines are highly technologically advanced and include many engineering and mechanical innovations to help maximize efficiency and increase the production of electricity. Offshore winds tend to blow harder and more uniformly than on land. The potential energy produced from wind is directly proportional to the cube of the wind speed. Thus, increased wind speeds of only a few miles per hour can produce a significantly larger amount of electricity. For instance, a turbine at a site with an average wind speed of 16 mph would produce 50% more electricity than at a site with the same turbine and average wind speeds of 14 mph. This is one reason that developers are interested in pursuing offshore wind energy resources. Commercial-scale offshore wind facilities are similar to onshore wind facilities. The wind turbine generators used in offshore environments include modifications to prevent corrosion, and their foundations must be designed to withstand the harsh environment of the ocean, including storm waves, hurricane-force winds, and even ice flows. New technologies, such as innovative foundations and floating wind turbines that will transition wind power development into the harsher conditions associated with deeper waters, are shown in the Figure 7.12.

7.7.1 Offshore Wind Energy Technology

The engineering and design of offshore wind facilities depends on site-specific conditions, particularly water depth, geology of the seabed, and wave loading. In shallow areas, *mono piles* are the preferable foundation type. A steel pile is driven into the seabed, supporting the tower and nacelle. The nacelle is a shell that encloses the gearbox, generator, and blade hub (generally a three-bladed rotor connected through the drive train to the generator) and the remaining electronic components, as in Figure 7.13. Once the turbine is operational, wind sensors connected to a yaw drive system turn the nacelle to face into the wind, thereby maximizing the amount of electricity produced.

Today's offshore turbines have technical modifications and substantial system upgrades for adaptation to the marine environment. These modifications include strengthening the tower to cope with loading forces from

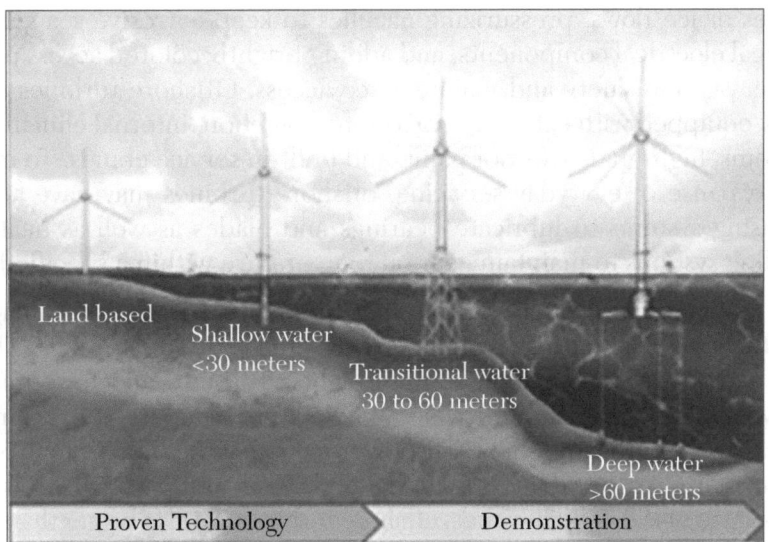

FIGURE 7.12 Progression of expected wind turbine evolution to deeper water.

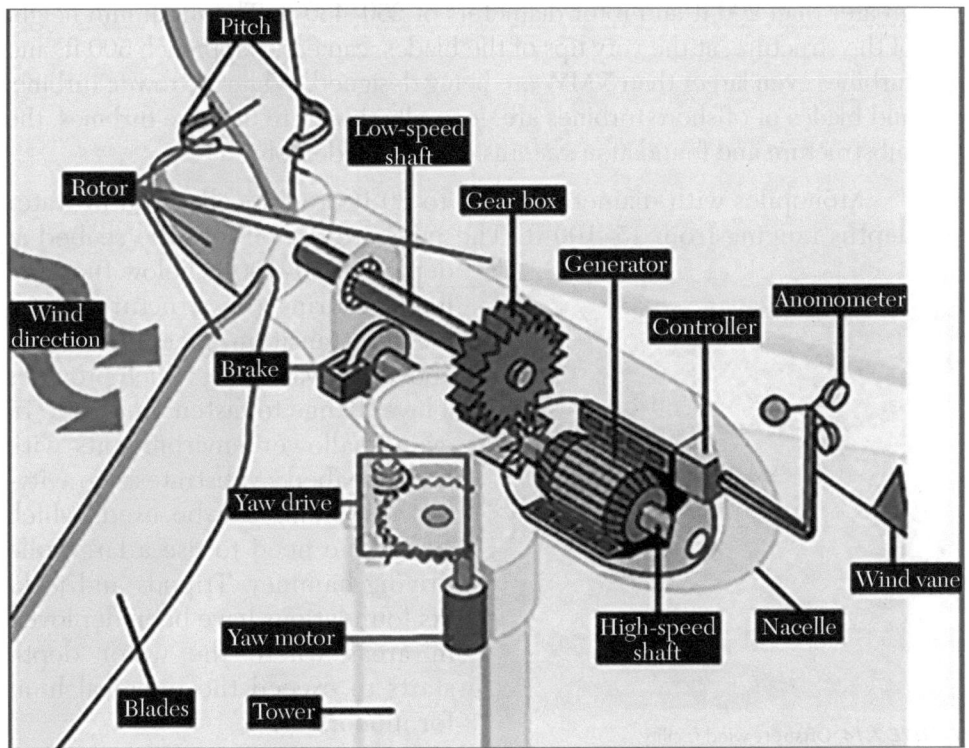

FIGURE 7.13 Schematic of wind turbine nacelle.

waves or ice flows, pressurizing nacelles to keep corrosive sea spray from critical electrical components, and adding brightly colored access platforms for navigation safety and maintenance access. Offshore turbines are typically equipped with extensive corrosion protection, internal climate control systems, high-grade exterior paint, and built-in service cranes. To minimize the expense of everyday servicing, offshore turbines may have automatic greasing systems to lubricate bearings and blades as well as heating and cooling systems to maintain gear oil temperature within a specified range.

Lightning protection systems help minimize the risk of damage from lightning strikes that occur frequently in some offshore locations. There are also navigation and aviation warning lights. Turbines and towers are typically painted light grey or off-white to help them blend into the sky, reducing visual impacts from the shore. The lower section of the support towers may be painted bright colors to increase navigational safety for passing vessels. To take advantage of the steadier winds, offshore turbines are also bigger than onshore turbines and have an increased generation capacity. Offshore turbines generally have nameplate capacities between 2 MW and 5 MW, with tower heights greater than 200 ft and rotor diameters of 250–430 ft. The maximum height of the structure, at the very tips of the blades, can easily approach 500 ft, and turbines even larger than 5 MW are being designed. While the tower, turbine, and blades of offshore turbines are generally similar to onshore turbines, the substructure and foundation systems differ considerably.

Monopiles with diameters of up to 20 ft are typically used in water depths ranging from 15–100 ft. The piles are driven into the seabed at depths of 80–100 ft below the mud line, ensuring the structure is stable. A transition piece protrudes above the waterline, which provides a level flange to fasten the tower. In even shallower environments with firm seabed substrates, gravity-based systems can be used, which avoids the need to use a large pile driving hammer. Tripods and jackets foundations have been deployed in areas where the water depth starts to exceed the practical limit for monopiles.

FIGURE 7.14 Offshore wind facility.

7.7.2 Semi-Submersible Offshore Wind Turbine

Offshore wind turbines have a considerably larger power output than their land-based counterparts. Previously, offshore turbines were restricted to shallow waters; however, they are now mounted on semi-submersible structures, allowing access to even better wind patterns further from the shore. Offshore wind turbines as shown in Figure 7.14 have a much higher capacity than the conventional turbines employed in land wind farms, mainly due to the enhanced wind patterns experienced offshore. Offshore turbines previously were mounted on structures fixed to the seabed, which restricted them to relatively shallow water at a maximum depth of 150 ft. However, offshore wind turbines can be mounted on floating structures, allowing them to operate in deeper waters further out to sea. This considers renewable energy from offshore wind turbines and from turbines mounted on a floating structure.

7.7.3 Offshore Floating Structures for Mounting Wind Turbines

Over the last 40 years or so, the offshore oil and gas industry has designed and developed various floating structures as in Figure 7.15 to support hydrocarbon production platforms. Most of these designs are

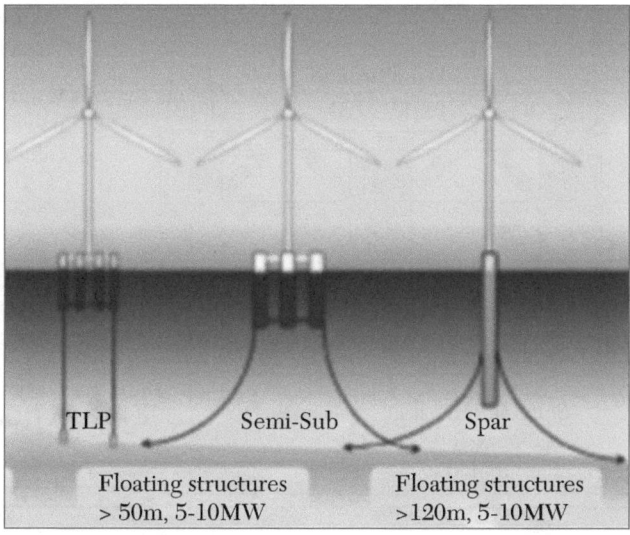

FIGURE 7.15 Floating structures.

suitable as support structures for offshore wind turbines; the more popular types are:

Semi-submersibles: These structures comprise hulls fabricated from large horizontal pontoons onto which vertical steel columns are welded, as in Figure 7.16. The columns and horizontal pontoons are interconnected and braced by a lattice of tubular steel supports. The structures

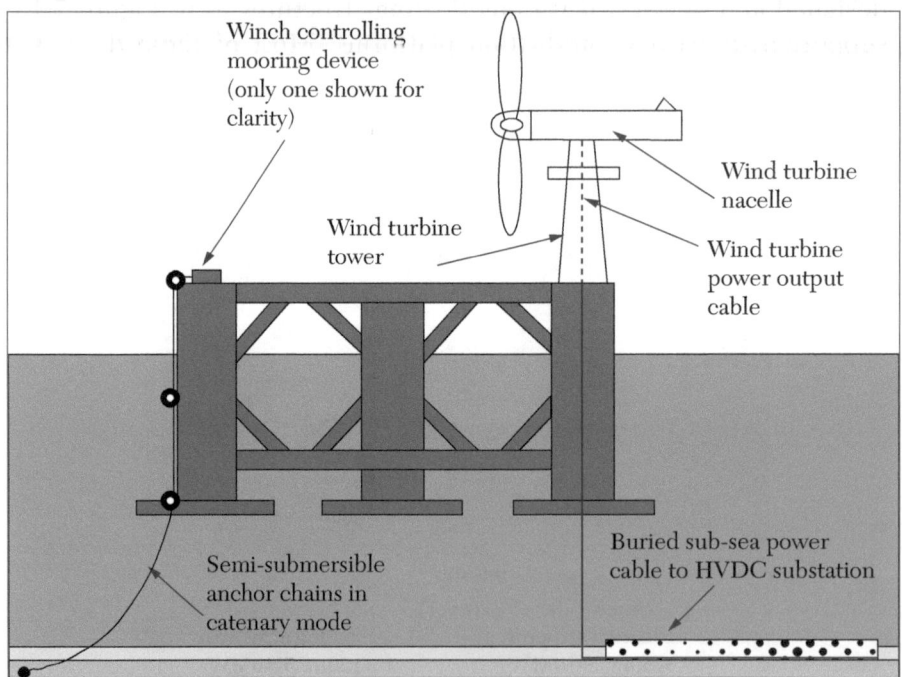

FIGURE 7.16 Semi-submersible.

are held to the seabed by anchors, whose chains are maintained in a catenary mooring mode by winches situated on the main deck. Various structures have been examined for potential use as floating supports for offshore wind turbines, including a multi-wind turbine support. However, this type of semi-submersible would require automatic weathervaning, controlled by satellite navigation that would turn the structure into the wind without interfering with the mooring system. Semi-submersibles can be self-propelled by their own marine diesel engines; the nonpropulsion type is towed to the required location. The offshore oil and gas industry has used semi-submersible floating structures for many years both as drilling rigs and production platforms, operating in depths of water of up to 6,000 ft.

Tension leg platform (TLP): This is a floating platform fixed by high-tensile solid steel round tie rods, attached to a specially designed template piled and grouted onto the subsea bedrock. The tie rods are kept in tension by overhead winches located under the main deck. TLPs can operate in depths of 6,000 ft.

Spar floating structure: The spar is just a large floating round hull that can be fabricated from steel or concrete. It has several sections built inside the hull which serve as ballast and oil storage tanks, being tethered to the seabed with conventional chains and winches. Spars can also operate in water depths of 6,000 ft. The Petronius Spar hydrocarbons production platform, weighing 45,000 tons, is currently operating at a depth of 1,750 ft in the Gulf of Mexico.

Offshore semi-submersible wind turbines in production: Hywind Statoil was the first company to produce electricity from a wind turbine mounted on a floating structure. In this case the structure was a steel fabricated spar type which was fabricated and towed in the horizontal position to an inshore assembly fiord. Here it was ballasted with sea water and brought to the vertical position, and further permanently ballasted with concrete. The tower was fitted onto the top of the spar, followed by the nacelle and rotor, and then the structure was towed to the offshore location where it was further ballasted and moored to the seabed. The wind turbine has a capacity of 2.6 MW, and there are plans to add more floating turbines, eventually constructing an offshore floating wind farm.

7.7.4 Transmitting Power Ashore through Subsea Cables

It is proposed to transmit power ashore through conventional subsea cabling, but also utilizing an innovative process known as *high voltage direct current* (HVDC). The units will be mounted on conventional bottom fixed structures such as steel jackets or monopods. The AC current produced by the wind turbine is converted to DC current before being transmitted ashore. This has several advantages over transmitting high voltage AC current:

- Less power loss giving greater efficiency

- Better control of distribution, less effects of land network faults

- Instant determination of power being generated and distributed ashore

Transport of wind-generated energy: All the power generated by wind turbines needs to be transmitted to shore and connected to the power grid. Each turbine is connected to an *electric service platform* (ESP) by a power cable. The ESP is typically located somewhere within the turbine array, and it serves as a common electrical collection point for all the wind turbines and as a substation. In addition, ESPs can be outfitted to function as a central service facility and may include a helicopter landing pad, communications station, crew quarters, and emergency backup equipment. After collecting the power from the wind turbines, high voltage cables running from the ESP transmit the power to an onshore substation, where the power is integrated into the grid. The cables used for these projects are typically buried beneath the seabed, where they are safe from damage caused by anchors or fishing gear and to reduce their exposure to the marine environment. These types of cables are expensive and are a major capital cost to the developer. The amount of cable used depends on many factors, including how far offshore the project is located, the spacing between turbines, the presence of obstacles that require cables to be routed in certain directions, and other considerations.

7.8 OFFSHORE SOLAR ENERGY

Solar energy technologies [Figure 7.17] potentially suitable for use in ocean environments include concentrating solar power technology and photonic technology. Every minute the Sun bathes the Earth

FIGURE 7.17 Solar energy.

in as much energy as the world consumes in an entire year. Since oceans cover more than 70% of the Earth's surface, they receive an enormous amount of solar energy. Deep ocean currents, waves, and winds all are a result of the Sun's radiant energy and differential heating of the Earth's surface and oceans.

7.8.1 Solar Energy Technologies

Solar radiation can be converted directly to usable energy through a variety of technologies. While there are no commercial solar energy facilities operating offshore at this time, solar energy technologies potentially suitable for use in offshore ocean environments include *concentrating solar power* (CSP) technology and *photonic technology*. CSP is a thermal solar technology that concentrates the Sun's rays to heat fluids or solids, and the heat is used to drive steam turbines or other devices to generate power. Photonic technologies convert the Sun's radiant energy directly to electricity or other useful forms of energy. Selection of the appropriate solar technology for a given situation depends in part on the intended use of the energy to be generated. CSP technologies might be more appropriate for generating and delivering electricity to shore, while photonic technology might be more appropriate for generation of electricity to be used on-site (such as on offshore platforms) and for supplying energy for activities such as hydrogen production or desalinization.

Concentrating solar power (CSP) technology: CSP plants generate electric power by using mirrors to concentrate (focus) the Sun's energy and convert it into high-temperature heat. That heat is then channeled through a conventional generator. The plants consist of two parts: one that collects solar energy and converts it to heat, and another that converts the heat energy to electricity. This approach requires large areas for solar radiation collection to produce electricity at commercial scale. CSP utilizes three technological approaches: trough systems, power tower systems, and dish/engine systems.

Trough systems: Trough systems use large, U-shaped (parabolic) reflectors (focusing mirrors) that have oil-filled pipes running along their center, or

FIGURE 7.18 Trough system.

focal point, as shown in Figure 7.18. The mirrored reflectors focus sunlight onto the pipes and heat the oil inside to temperatures as high as 750°F. The hot oil is used to boil water, which produces steam to run conventional steam turbines and generators.

Power tower systems: Power tower systems, also called central receivers, use many large, flat *heliostats* (mirrors) to track the Sun and focus its rays onto a receiver. As shown in Figure 7.19, the receiver sits on top of a tall tower in which concentrated sunlight heats a fluid, such as molten salt, to temperatures as high as 1,050°F. The hot fluid can be used to boil water, which produces steam to run conventional steam turbines and generators. Or the thermal energy can be effectively stored for hours, if desired, to allow for electricity production during periods of peak demand, even when the Sun is not shining.

FIGURE 7.19 Solar power tower—mirrors are concentrating sunlight at the top of tower.

Dish/engine systems: Dish/engine systems use mirrored dishes (about 10 times larger than a backyard satellite dish) to focus and concentrate sunlight onto a receiver, as shown in Figure 7.20. The receiver is mounted at the focal point of the dish. To capture the maximum amount of solar energy, the dish assembly tracks the sun across the sky. The receiver is integrated into a high efficiency "external" combustion engine. The engine has thin tubes containing hydrogen or helium gas that run along the outside of the engine's four piston cylinders and open into the cylinders. As concentrated sunlight falls on the receiver, it heats the gas in the tubes to very high temperatures, which causes hot gas to expand inside the cylinders. The expanding gas drives the pistons. The pistons turn a crankshaft, which drives an electric generator. The receiver, engine, and generator comprise a single, integrated assembly mounted at the focus of the mirrored dish.

Solar photonic technology: Solar photonic technology absorbs solar photons (particles of light that act as individual units of energy), and converts the energy to electricity (as in a photovoltaic [PV] cell) or stores part of the energy in a chemical reaction (as in the conversion of water to hydrogen and oxygen). PV technology converts sunlight directly to electricity. Concentrated PV (CPV) systems, which must track the Sun to keep the light focused on the PV cells, use various methods to concentrate sunlight such as mirrors or lenses. The primary advantages of CPV systems are high efficiency, low system cost, and low capital investment to facilitate rapid scale-up; reliability, however, is an important technical challenge for this emerging technological approach.

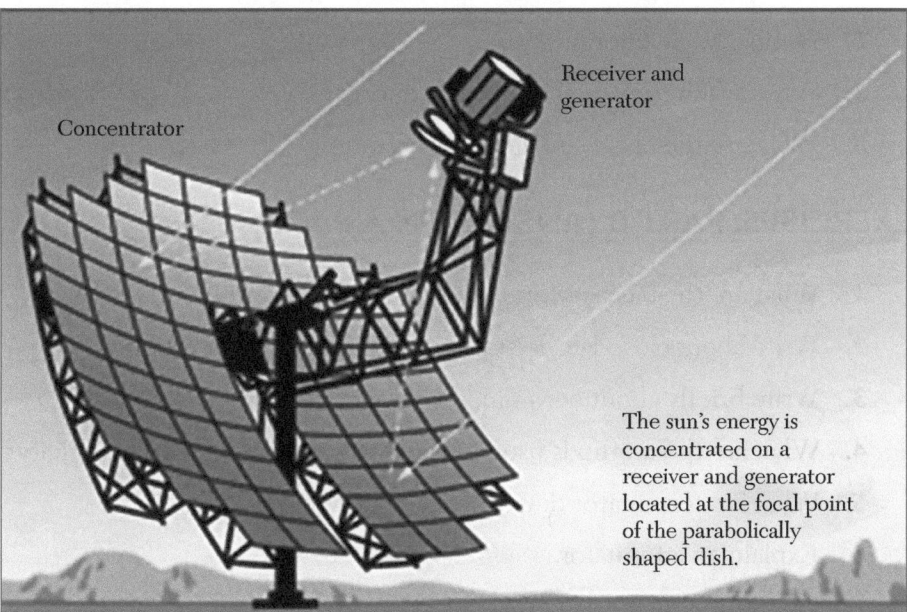

Concentrator

Receiver and generator

The sun's energy is concentrated on a receiver and generator located at the focal point of the parabolically shaped dish.

FIGURE 7.20 Solar dish/engine system.

7.8.2 Challenges in Ocean Power Technologies

Currently, most ocean power technologies are not economically competitive with conventional fossil fuel power. They tend to have low operating costs but high construction costs with a long payback period. Careful site selection is also extremely important to keep the environmental impacts of ocean power technologies to a minimum. Protecting shorefronts, keeping sea life migration patterns and habitat intact, and preventing alterations in ocean temperature or sedimentation processes must all be considered. In addition, water and geographic conditions must be right for the technologies to work.

EXERCISES: PART A (ANSWER IN A WORD OR A SENTENCE)

1. Define "ocean energy."

2. What are the two different types of energy produced from the ocean?

3. What are the three types of electricity conversion for ocean thermal energy?

4. What is desalinization?

5. How is ocean wave energy captured?

6. Define "wave energy."

7. What are the advantages of wave energy?

8. What are the disadvantages of wave energy?

EXERCISES: PART B (ANSWER IN A PAGE)

1. What are the basic criteria for a renewable energy source?

2. Write about closed cycle systems of electricity generation.

3. Write briefly about ocean mechanical energy.

4. What are the methods used to convert wave energy into electricity?

5. What is a terminator device?

6. Explain an attenuator.

7. What is a point absorber?

8. What are overtopping devices?

9. Explain briefly tidal energy.

10. Write about ocean thermal energy conversion.

11. Explain the operation of ocean turbines.

12. What are the benefits of ocean turbines?

13. Write in short about ocean current energy.

14. Write in short about offshore wind turbines.

EXERCISES: PART C (ANSWER WITHIN 2 OR 3 PAGES)

1. Write in detail about wave energy to electricity conversion.

2. Explain in detail the different wave energy technologies.

3. Explain offshore wind energy and its technologies.

4. Explain offshore solar energy and its technologies.

WEB LINKS

http://www.energy-without-carbon.org/OceanThermal

http://www.darvill.clara.net/altenerg/wave.htm

http://www.boem.gov/

CHAPTER 8

MARINE ELECTRONICS

This chapter discusses the role of electronics in the marine generator set, marine instruments, wireless control stations, navigation equipment, autopilot system, satellite phones, firefighting equipment, bubbler gauges, navigation instruments, AIS operation, electrical propulsion, and gas indicators.

8.1 THE ROLE OF ELECTRONICS IN THE MARINE GENERATOR SET

Top of form

The engine of a ship/boat/vessel has an electronic control system. The control system consists of the following components:

- Electronic control module (ECM)

- Software (flash file)

- Wiring

- Sensors

- Actuators

The electronic system of the vessel consists of the following components: *electronic control module* (ECM), *hydraulic electronic unit injectors* (HEUI), *injection actuation pressure control valve* (IAPCV), *wiring harness*, switches, and sensors. The ECM receives information from the sensors and the switches on the engine. The ECM processes the information that is collected to make decisions on control of the engine. By altering the fuel delivery of the injectors, the ECM controls the speed and power produced by the engine. The ECM consists of two main components: the *control computer* (hardware) and *personality module* (software). The control computer comprises the microprocessor and the electronic circuitry. The personality module contains the software for the control computer. The software contains operating maps that define the engine's horsepower, torque curves, and engine speed. The electronic control circuit diagram for the engine with a control panel is shown in Figure 8.1.

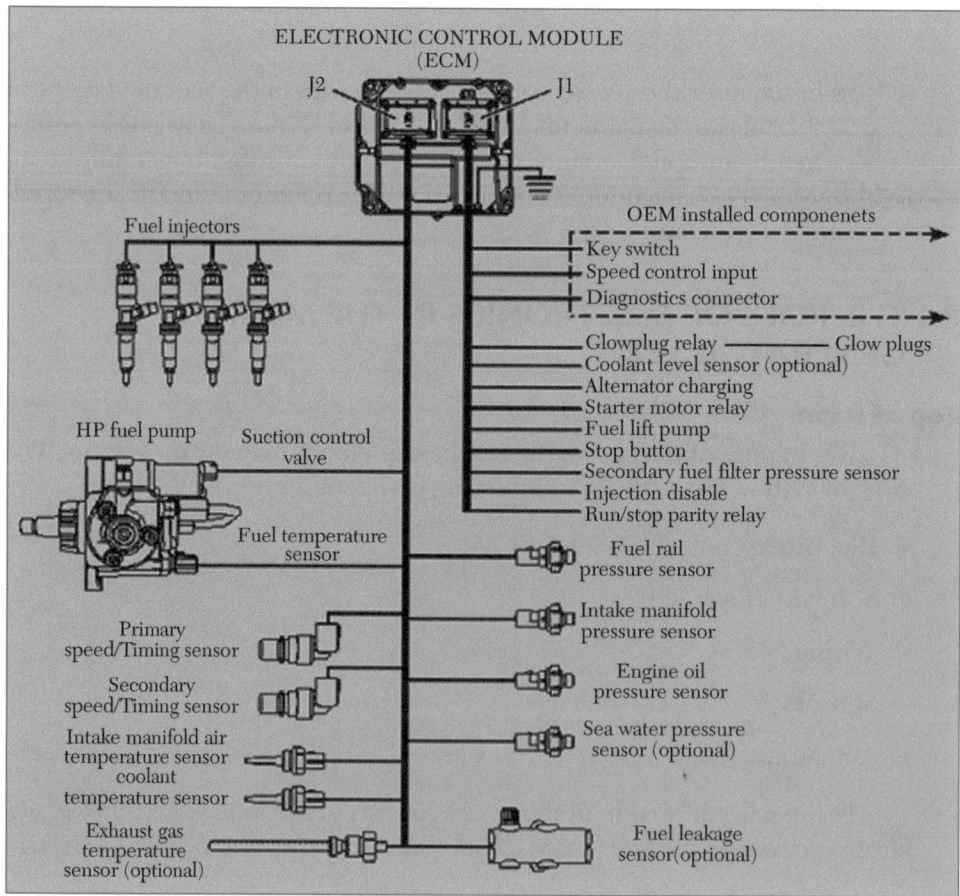

FIGURE 8.1 Electronic control system for engine.

The engine consists of the following parts:

1. air cleaner,

2. air inlet temperature sensor,

3. exhaust gas temperature sensor,

4. turbocharger,

5. air-to-air after-cooler,

6. engine,

7. coolant temperature sensor,

8. primary speed/timing sensor,

9. fuel injectors,

10. return fuel cooler,

11. seawater pressure sensor,

12. return fuel pressure relief valve,

13. secondary speed/timing sensor,

14. high pressure fuel pump/transfer pump/fuel temperature sensor,

15. fuel rail pressure sensor,

16. prefilter oil pressure sensor,

17. postfilter oil pressure sensor,

18. coolant pressure sensor,

19. oil temperature sensor,

20. fuel leakage detection sensor,

21. ECM,

22. electric fuel lift pump,

23. postprimary fuel filter pressure sensor,

24. preprimary fuel filter pressure sensor,

25. primary fuel filter,

26. intake manifold pressure sensor,

27. intake manifold air temperature sensor,

28. transfer pump inlet regulator,

29. secondary fuel filter,

30. fuel tank,

31. postsecondary fuel filter pressure sensor,

32. presecondary fuel filter pressure sensor,

33. GPS (global positioning system).

8.1.1 Engine Governor

The ECM governs the engine. The ECM determines the timing, the injection pressure, and the amount of fuel that is delivered to each cylinder. These factors are based on the actual conditions and on the desired conditions at any given time during starting and operation. The governor uses the throttle signal to determine the desired engine speed. The governor compares the desired engine speed to the actual engine speed. The actual engine speed is determined through interpretation of the signals that are received by the ECM from the engine speed/timing sensors. If the desired engine speed is greater than the actual engine speed, the governor injects more fuel to increase engine speed, as illustrated in Figure 8.2.

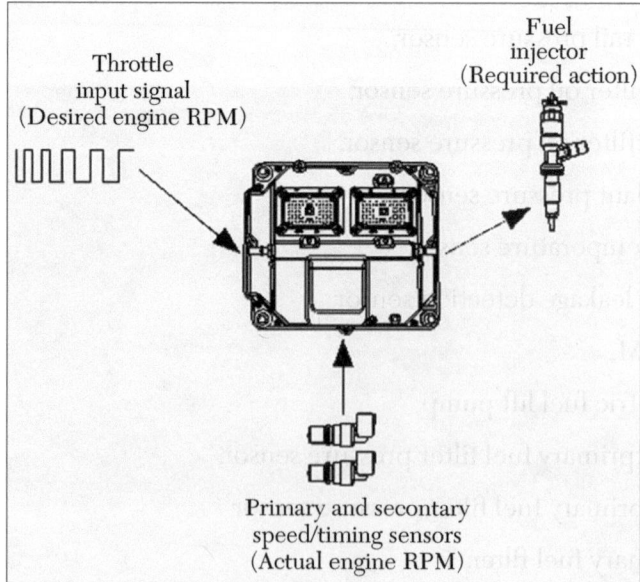

FIGURE 8.2 Engine control.

8.1.2 Timing Considerations

Once the governor has determined the amount of fuel that is required, it must determine the timing of the fuel injection based on input from the following components:

- Coolant temperature sensor

- Intake manifold air temperature sensor

- Intake manifold pressure sensor

The ECM adjusts timing for optimum engine performance and for economic fuel usage. Actual timing and desired timing cannot be viewed with the electronic service tool. The ECM determines the location of the top center of the number one cylinder from signals that are provided by the engine speed/timing sensors, determines when injection should occur relative to this position, and then provides the signal to the injector at the desired time. The ECM sends a high-voltage signal to the injector solenoids to energize them. By controlling the timing and the duration of the high-voltage signal, the ECM can control the following aspects of injection:

- Injection timing

- Fuel delivery

The flash file inside the ECM establishes certain limits on the amount of fuel that can be injected. The *fuel limit* is a limit that is based on the intake manifold pressure and is used to control the air/fuel ratio for control of emissions. When the ECM senses a higher intake manifold pressure, the ECM increases the fuel limit. A higher intake manifold pressure indicates that there is more air in the cylinder. When the ECM increases the fuel limit, the ECM allows more fuel into the cylinder. The *rated fuel limit* is a limit based on the power rating of the engine and on the engine rpm. The rated fuel limit is like the rack stops and the torque spring on a mechanically governed engine. It provides the power curves and torque curves for a specific engine family and a specific engine rating. All these limits are determined at the factory and cannot be changed.

ECM lifetime totals: The ECM maintains total data of the engine for the parameters such as: total operating hours, engine lifetime hours, total idle time, total idle fuel, total fuel, total max fuel, and engine starts.

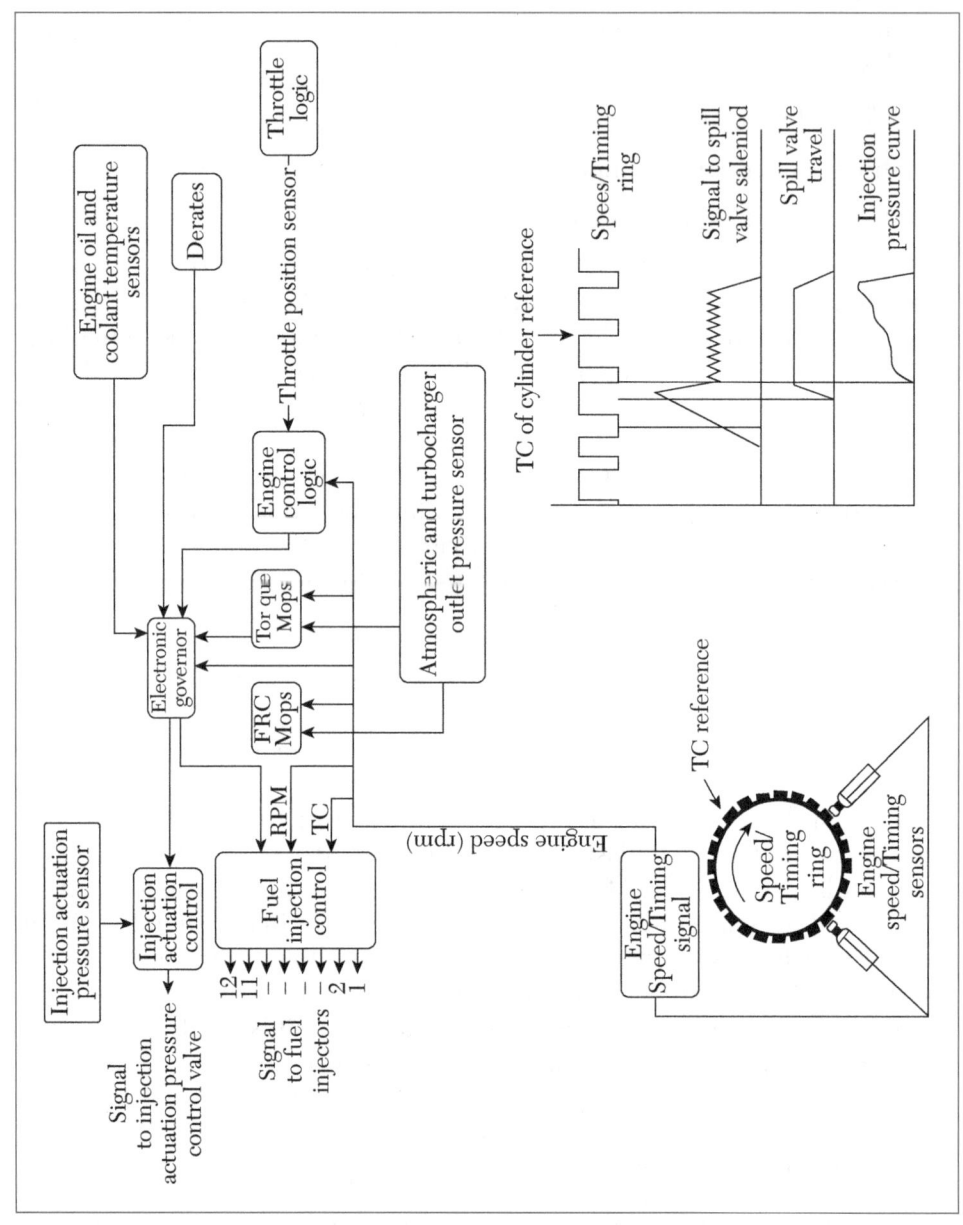

FIGURE 8.3 Diagram of electronic governor.

The *total operating hours* are the operating hours of the engine. The operating hours do not include the time when the ECM is powered but the engine is not running. *Engine lifetime hours* is the number of hours when electrical power has been applied to the engine. These hours will include the time when the ECM is powered but the engine is not running. *Total idle time* and *total idle fuel* can include operating time when the engine is not operating under a load. Fuel information can be displayed in liters. *Total fuel* is the total amount of fuel that is consumed by the engine during operation. *Total max fuel* is the maximum amount of fuel that could have been consumed by the engine during operation. *Engine starts* is the total number of times the engine has been started. Figure 8.4 is a visual representation of typical electronic engine system components, which vary by engine. The ECM governs the engine speed by controlling the amount of fuel that is delivered by the injectors. The desired engine speed is determined by input from the throttle switch. Actual engine speed is measured by the engine speed/timing sensors. The ECM changes the amount of fuel that is injected until the actual engine speed matches the desired engine speed as in Figure 8.3. Bottom of Form

8.1.3 Fuel Injection

Top of Form

The ECM controls the timing, duration, and pressure of the fuel that is injected. The block diagram of the fuel system is shown in Figure 8.5. The ECM controls the timing and the duration by varying the signals to the injectors. The injectors will inject fuel only if the injector solenoid is energized by a 105-volt signal from the ECM. By controlling the timing and the duration of the 105-volt signal, the ECM can control the timing of the injection and control the amount of fuel that is injected. The ECM modulates the injection pressure by varying the signal to the injection actuation pressure control valve (IAPCV). The IAPCV controls the pressure of the high-pressure oil that pressurizes the fuel in the injectors. By controlling the signal to the IAPCV, the ECM controls the pressure of the fuel that is injected into the engine.

The ECM limits engine power and modifies injection pressure and injection timing during *cold mode* operation. Cold mode operation

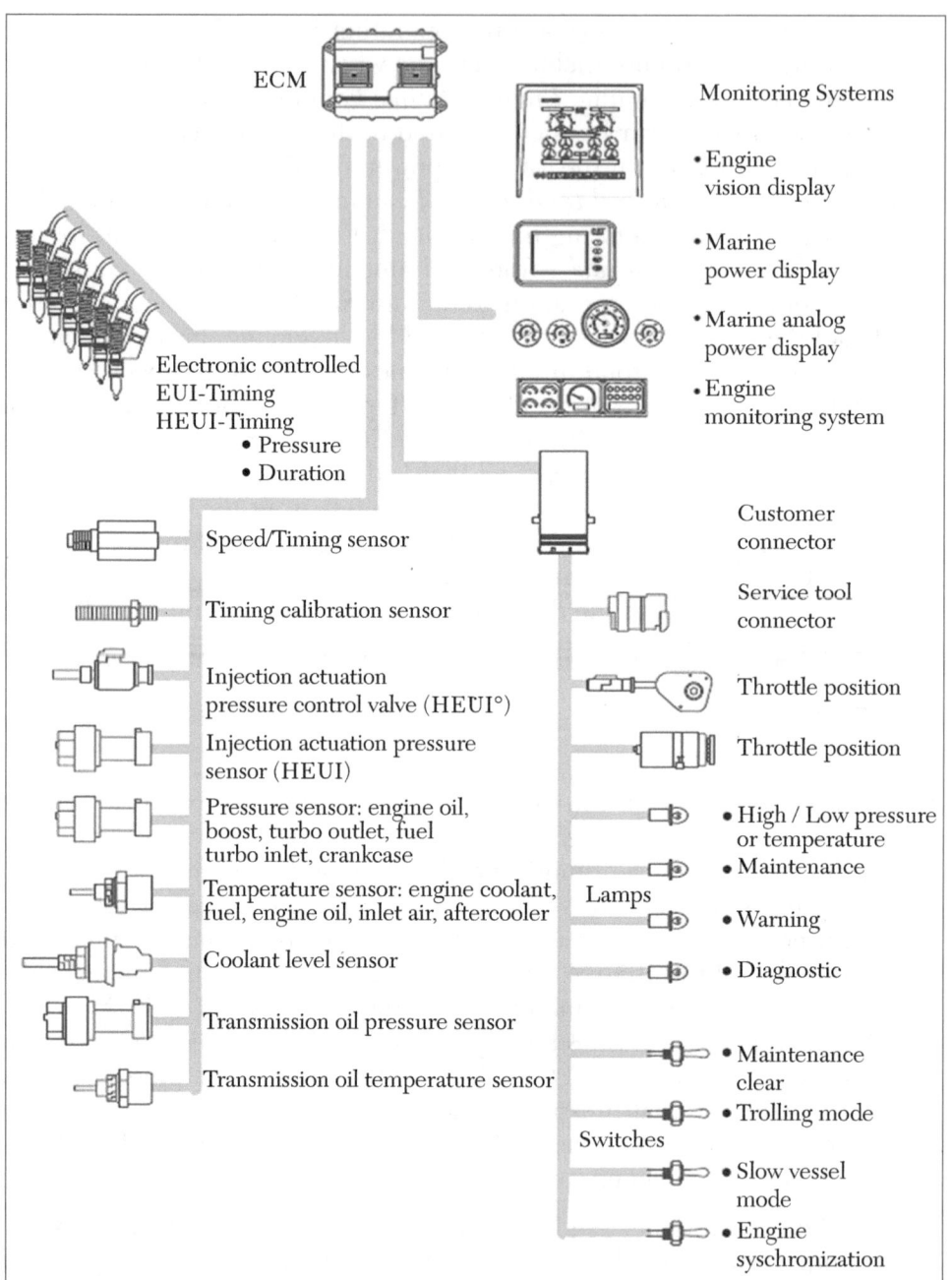

FIGURE 8.4 Visual representation of electronic engine system components.

has the following benefits: increased startability, reduced warm-up period, and reduced white smoke. Cold mode is active if the engine oil temperature falls below a predetermined value and other conditions are met and remains active until the engine has warmed or until a time limit has been exceeded. The personality module inside the ECM sets certain limits on the amount of fuel that can be injected. The *FRC limit* is a limit that is based on the boost pressure, which is calculated as the difference in pressure between atmospheric pressure and turbocharger outlet pressure. The FRC limit controls the air/fuel ratio for control of emissions. When the ECM senses a higher boost pressure, the ECM increases the FRC limit. A higher boost pressure indicates that there is more air in the cylinder. When the ECM increases the FRC limit, the ECM allows more fuel into the cylinder. The *rated fuel position* is a limit that is based on the power rating of the engine, similar to the rack stops and the torque spring on a mechanically governed engine. It determines maximum power and torque values for a specific engine family and a specific rating and is programmed in the personality module at the factory.

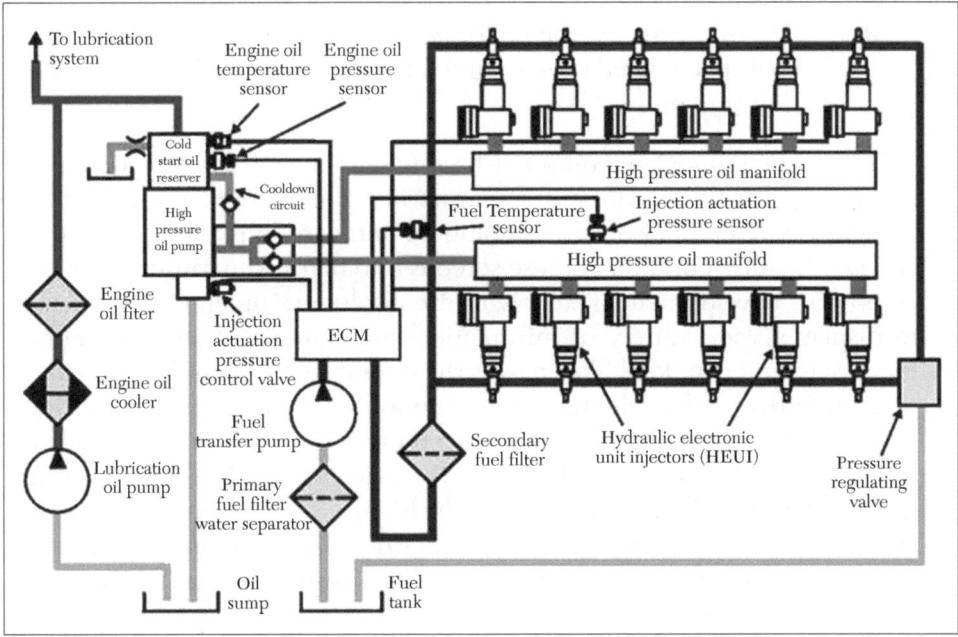

FIGURE 8.5 Block diagram of fuel system.

8.2 MARINE INSTRUMENTS

The following are some of the marine instruments that operate with electronics.

- *Engine monitoring tachometers* provide a warning light display notifying the operator of a potential engine error.

- *Depth sounders* are available with dual air and water temperature displays along with transducer configurations.

- *Digital gateway systems* monitor engines. With push of a button the operator can tell the status of the health of the engine including diagnostic messages, fault alerts and parameter information.

- *GPS speedometer* gathers GPS information from an internal GPS antenna. No external antenna required. Course over ground (COG) and actual heading (compass heading over ground) are displayed on the digital LCD. Speed data is shown by an analog pointer. This pointer is driven by a digital stepper motor for increased accuracy and minimized pointer bounce during vessel operation.

- *Marine electronic navigation instruments* provide functions such as depth, boat speed, wind speed, wind direction, true wind, motor, engine monitoring, and more.

The development of computer-based technologies has provided a fertile ground for the expansion of marine instrumentation. With the technological development of high resolution, antiglare, liquid crystal displays (LCD), the incorporation of these screens into marine instrumentation was an obvious development. The new user consoles are much larger and able to display a wide variety of information simultaneously and incorporate touch screen technology. This makes the operation of the onboard systems more intuitive and facilitates such common computer features as drag-and-drop. The chart plotter in Figure 8.6 provides considerable detail and can be zoomed in or out. Note the GPS, navigation, and weather information displayed at the top. The display in Figure 8.7 integrates twin engine and fuel statistics with navigation, weather details. The graph on the right-hand side of Figure 8.7 provides a graphical representation of the water depth and temperature.

One of the very latest developments in touch screen technology utilizes invisible infrared beams, making it unnecessary to even touch the screen

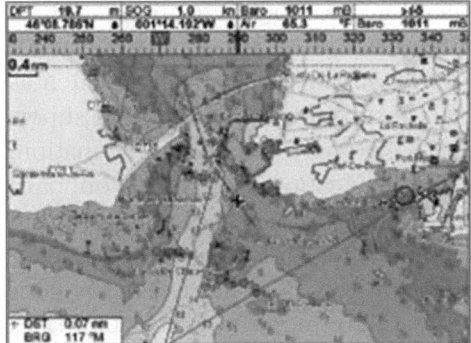

FIGURE 8.6 Chartplotter.

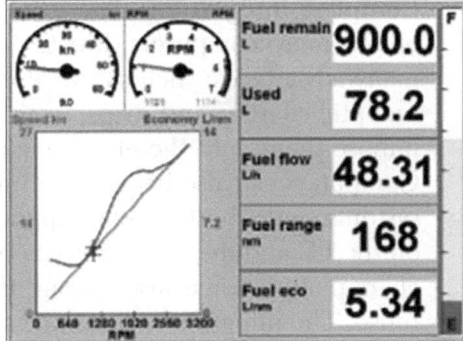

FIGURE 8.7 Display board.

to operate it. The operator's finger need only come into close proximity to the screen surface to activate the required function. The user can integrate data from many sources and display it all on one screen. For example, instrumentation read-outs, such as wind data and vessel speed, navigation information including electronic charts, GPS, radar, plus video input from different cameras around the vessel, access to onboard computers, Internet browsing, and DVD output can all be integrated into a single display. From a navigation point of view, some exciting developments have been made possible with the integration of large volumes of data and the use of multi-input, large-format LCD screens. It is now common for these screens to overlay data from several sources onto the one image to provide a more complete picture. Some of these include:

- Radar imagery superimposed onto the electronic chart.

- The integration of electronic charts with GPS, as well as data from instruments such as wind, depth sounder and the ship's log (vessel speed).

- Photographic images of land topography and buildings, etc., superimposed onto the electronic chart to give a better representation of what the captain is seeing immediately in front of him. This is particularly useful in conditions of limited visibility or when entering a port for the first time. The system can even pan around as well as look well ahead to tell the captain what is coming.

- Topographical images of the sea floor immediately in front of the vessel, again with the ability to pan around and look well ahead. This could be particularly useful when navigating in dangerous waters, such as around coral reefs.

Complete integration: The interconnection and complete integration of one or more computers directly into the instrumentation system provides many advantages, such as the ability to download weather charts (such as synoptic charts, wind maps, or wave charts) from the Internet and super-impose them onto the electronic chart to give more detailed information. Computers can also be used for improving navigation, monitoring onboard systems, and a host of other applications. Navigation programs can be run on the PC and the resultant data (waypoints, etc.) uploaded directly to the chart plotter, while also storing the data on the PC's hard disk for later reuse. The integration of marine radios with computer advances increases both knowledge and safety which is shown in Figure 8.8.

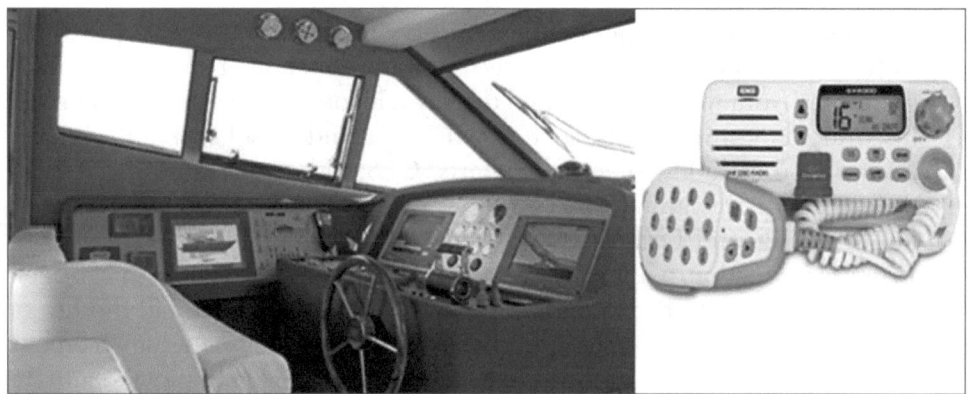

FIGURE 8.8 Integration of marine radios with computer.

Autopilots have also profited significantly from the development of multi-input data streams. The integration of a rate gyro with a flux gate compass has provided an autopilot with superior behavior characteristics resulting in performance similar to that of a human helmsman. The rate gyro measures and adjusts for the yaw of the vessel in a heavy seaway. This makes the vessel more comfortable for the occupants as well as reducing the wear and tear on the steering systems. Integration of marine radios with the GPS and other instruments provides position and barometric data and time, all of which is very useful during an emergency. The next and perhaps the last area to be integrated into the operator's console is engine manage-ment data. These systems provide a wide range of data on the operation of the engine(s) and fuel system. With the integration of systems, it is now pos-sible to input data from one user console and then move to another console (e.g., from the main helm station to the fly bridge) and continue to operate the vessel whilst being able to view the same data. This feature can be very useful when operating the vessel in confined spaces or when entering an unfamiliar port.

8.2.1 Wireless Control Stations

The wireless environment is one of the newest technologies to find immediate application afloat. Wireless hand controls for the autopilot and engine controls allow the skipper to stand on the deck and bring the vessel alongside the pier; a useful feature especially when docking a vessel in a confined area or in trying weather conditions. It is now possible to have integrated data at various control stations wirelessly, as in Figure 8.9. GPS technology has recently seen further improvements. There are three new features being utilized in the more sophisticated GPS systems: *wide area augmentation system* (WAAS); the *European geostationary navigation overlay service* (EGNOS) and *multifunction transport satellite* (MTSAT). WAAS covers the Americas; EGNOS covers Europe, and MTSAT covers the Asia Pacific region. They all provide increased positional accuracy and a much wider area of coverage compared to the existing land-based DGPS transmitters. There is also less interference due to weather and onboard electrical activity. The differential signals are sent on the same frequency as the standard GPS signal, thus negating the need for a separate receiver. *Automated identification system* (AIS) is a system required by all vessels over 300 gross tones. However, AIS receivers are finding their way aboard small craft to warn of an approaching ship. AIS transceivers transmit data such as the boat's unique ID number, position, course, and speed to all vessels nearby as well as to the VTS stations. Principally, the system integrates a VHF transceiver to send and receive data and the GPS to provide the vessel's data.

Man-overboard alert systems are being integrated into vessel electronics, too. Each member of the crew wears a small personal transmitter that maintains a constant link to the vessel. Should a person fall overboard, the system detects the event

FIGURE 8.9 Wireless control stations.

and sounds an alert, also logging the GPS position on a monitor or chart plotter. The logged position is displayed with the track back to the person in the water. Some systems have incorporated the ability for the electronic system to self-diagnose problems, alert the user to the problem, and provide some possible solutions. Some systems also incorporated Internet links into their engine electronics to facilitate remote diagnosis.

8.2.2 Salt Assault

The biggest enemy of all marine electronics is moisture, especially salt-laden moisture. Very few instruments can resist moisture ingress when the temperature varies greatly over an extended period, as happens on deck throughout the year in the temperate latitudes. During the summer months, the atmosphere inside the instrument is heated by the blazing sun and thus expands. This increases the atmospheric pressure inside the instrument and eventually forces out some of the enclosed gases. When the temperature drops during the winter months or while sailing in cold conditions, the temperature of the atmosphere inside the instrument drops. This causes a corresponding drop in pressure. After this has occurred a few times, the relative pressure inside the instrument is less than the outside ambient pressure. The resultant pressure differential causes a partial vacuum that draws in air from the surrounding atmosphere. The salt-laden moisture in the air can then corrode the internal workings of the instrument.

It is difficult to build instruments that are completely resistant to these harsh conditions, although some instruments are completely sealed. Instruments such as speed logs and depth sounders can often be calibrated. Instruments such as radar and autopilots require adjustment by a technician. A nautical chart is one of the most fundamentals tools available to the mariner. It is a map that depicts the configuration of the shoreline and seafloor. It provides water depths, locations of dangers to navigation, location and characteristics of aids to navigation, anchorages, and other features. Today's massive ships push the depth limit of many ports and harbors. Tools such as nautical charts, accurate positioning services, and ocean and weather observations play a key role in ensuring that shipments move swiftly and safely along our marine highways. The large-format monitor uses the latest infrared touch screen technology, as shown in Figure 8.10. It is not necessary to touch the screen; thus keeping the screen surface clean.

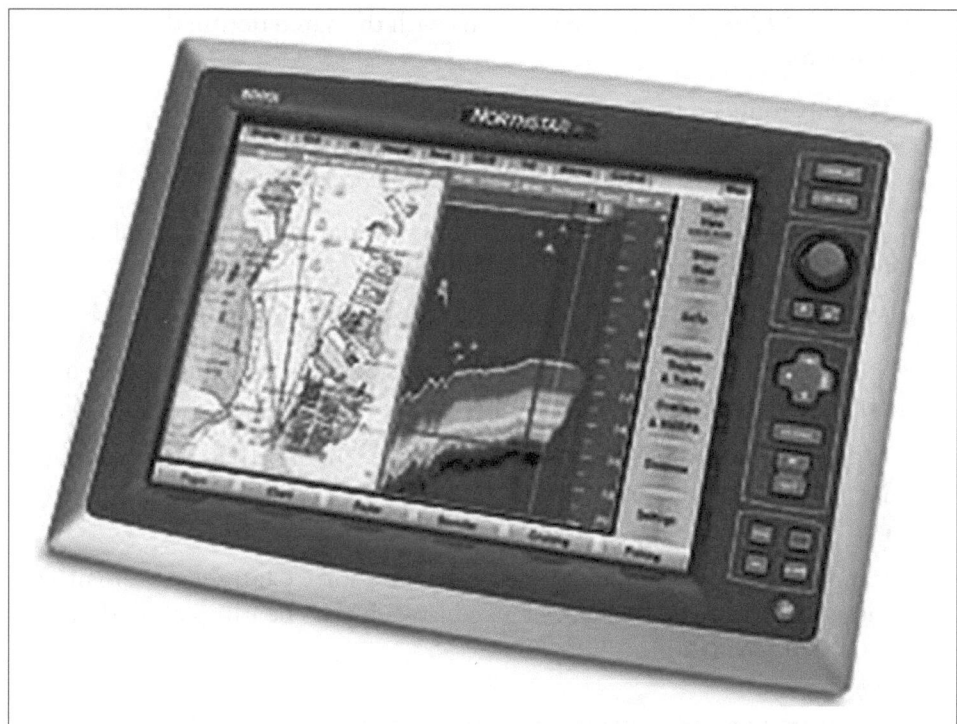

FIGURE 8.10 Infrared touchscreen.

8.2.3 Producing Nautical Charts

Mariners have special maps called as nautical charts. Much like road maps, nautical charts provide basic navigation information, such as water depths and the locations of hazards. These tools are used by mariners plan efficient routes and avoid dangerous or ecologically sensitive areas.

Information in the nautical chart:

- Depths reduced to chart datum: A sounding like 3_5 indicates 3½ meters of water under *lowest astronomical tide* (when the chart datum is "L.A.T."). An underlined sounding like 0_4 indicates a height of 40 cm above L.A.T. Depths are given from 0.1 to 20.9 in meters and decimeters, and from 21 to 31 in meters and half meters. Greater depths are rounded down to the nearest safest meter (for example, 32.7 meters is rounded down to 32 meters). The geographical position of a sounding is the center of the depth figure.

- Isobaths: Lines connecting positions with the same depth: depth contours.

- Heights reduced to chart datum: Heights of, for instance, lighthouses, mountains, and cliffs are more often reduced to another datum such as *mean high water* (M.H.W.) or *mean high water spring*.

- Tidal information: Details of both the vertical and the horizontal movement of the water are often included in the chart.

- Lighthouses, buoys, and marks: Lights, lateral, and cardinal marks.

- Seabed qualities: Pebbles, seaweed, rocks, wrecks, pipelines, sand, and other seabed characteristics for anchoring.

- Magnetic variation: The angle between the true North and the magnetic North varies in place and time. The local variation is indicated in the compass card.

- Landmarks: Conspicuous positions on the shore: Churches, radio masts, mountain tops, etc. that can be used for compass bearings and other means of navigation.

8.3 MARINE NAVIGATION EQUIPMENT

Navigation equipment such as GPS, speed log, echo-sounder, AIS, autopilot, gyrocompass, radar, ECDIS, and more are shown in Figure 8.11. Navigation software provides the visual information.

Hydrographical charts are used for navigation and position fixing. The charts should be kept up to date as per notice to mariner, issued at regular intervals. *Gnomonic charts* are used for Great Circle sailing. An *electronic chart display and information system* (ECDIS) meets the SOLAS requirements [Figure 8.12]. Other systems may be used as supplementary aids, along with paper charts.

8.3.1 GPS-Based Instrument Recovery Stray Line Buoys

Stray line instrument recovery buoys contain a GPS receiver and a radio modem that allows the buoy to report its position when it reaches the surface. The stray line buoy is based on a 10-in glass instrument housing with a depth rating of 6,700 mts. Contained within this are a GPS receiver, radio modem, PIC microprocessor, LED strobe light, solar panels, batteries, and the supporting electronics. Upon rising to the surface, the buoy assumes

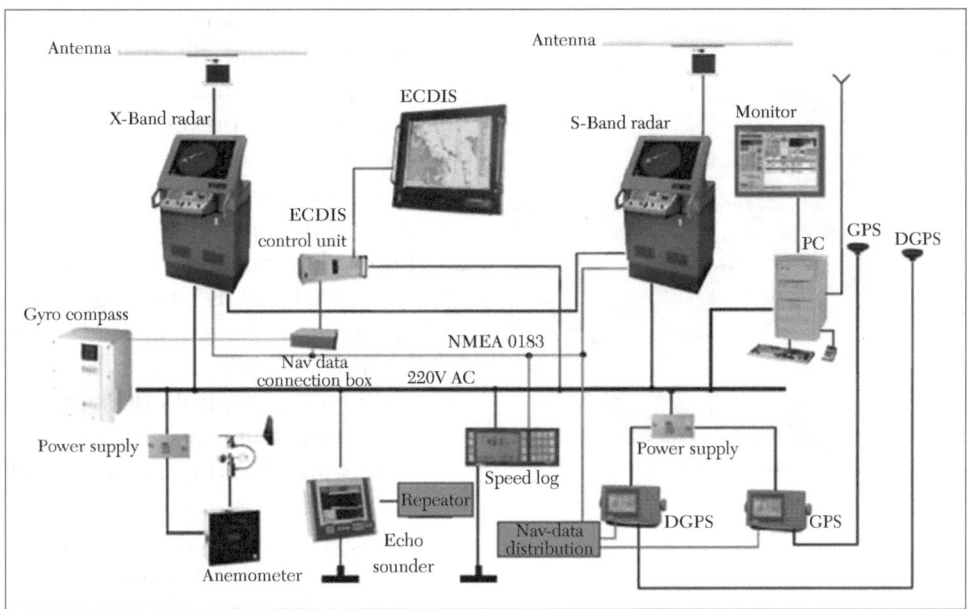

FIGURE 8.11 Marine navigation equipment.

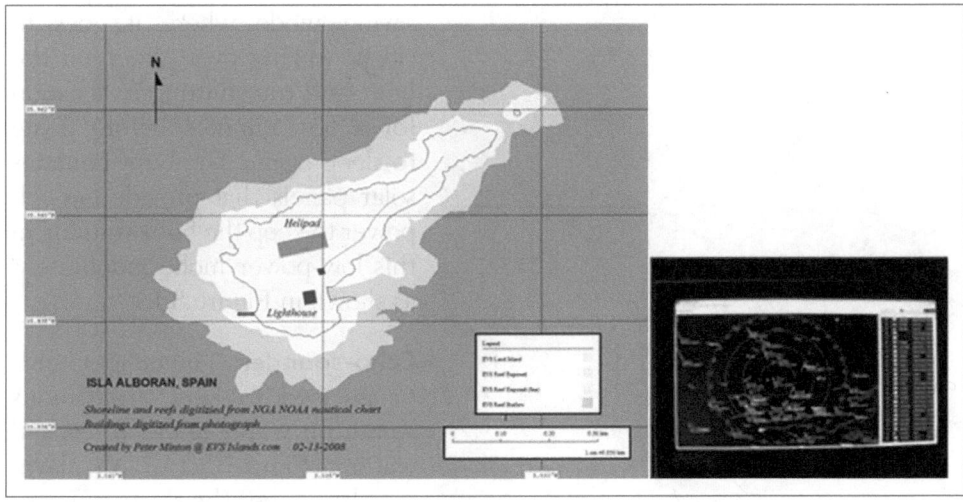

FIGURE 8.12 Navigation instruments.

FIGURE 8.13 GPS stray line buoys.

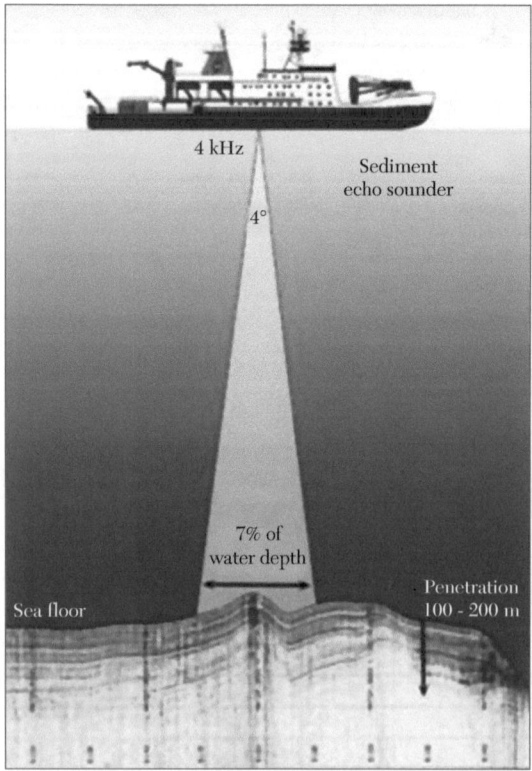

FIGURE 8.14 Echo-sounder.

its normal noninverted position and a mercury tilt switch powers up the system. A delay circuit is included to prevent the system turning off again on rough seas. The microprocessor then initializes both radios, transmits the buoy identification code plus an "ON SURFACE" message, and attempts to obtain a valid GPS location. At periodic intervals (user configurable, typically 30 seconds) the buoy transmits its identification and location or, if a valid GPS location is not available, the message "WAITING FOR GPS". After a defined amount of time (user defined typically 3 hours) it may be assumed that the instrument has not been recovered. At this time the buoy enters a power-saving mode where it goes to sleep, waking every hour on the hour and transmitting its location for a few minutes before going to sleep again. The buoy contains solar panels that provide enough power to keep the buoy running in this low-power mode indefinitely, as shown in Figure 8.13.

Echo-sounder: The echo-sounder is used to keep continuous watch on water depth in coastal waters [Figure 8.14]. An alarm may be set per the ship's requirements to warn of shallow waters. Depths recorded should be compared with those shown on the charts, giving allowance for the ship's draft and tidal effects. A hand lead is provided on the ship for taking soundings manually in shallow waters.

8.3.2 Weather Monitoring Systems

The world's oceans are divided in sixteen areas (NAVAREA) for dissemination of navigational and metrological warnings. Warnings are broadcast on radio as per list of radio signals. Weather facsimile recorders receive data from the weather satellite and display the latest weather maps, ice charts, and other forecasts. This helps in modifying the ship's routes as the voyage proceeds to minimize the effects of bad weather, as shown in Figure 8.15.

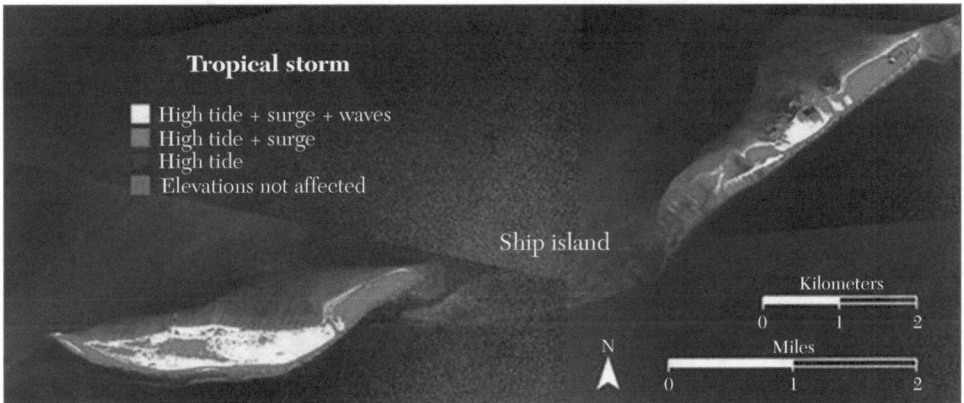

FIGURE 8.15 Weather monitoring systems.

8.3.3 Routing and Reporting

In coastal waters, ships must use government-approved routing systems. Governments have established *vessel traffic systems* (VTS) designed to contribute to safety of life, efficiency of navigation, and protection of the marine environment and adjacent areas. VTS is mandatory within territorial waters only. Automatic identification systems (AIS) transmit/receive information on ships, names, positions, courses, speeds, destinations, cargos, etc. by digital radio technology for the ships in objective area. They should be in operation at all times. They are useful in collision prevention, search and rescue, or operation of VTS. This equipment is normally placed on the bridge, and there are some further controls provided in UMS (*unmanned machinery space*) ships, as the machinery and boiler controls can also be controlled from the bridge.

8.4 INTEGRATED SAILBOAT INSTRUMENTS

A typical basic integrated sailboat instrument system for offshore cruising is shown in Figure 8.16, the primary components being:

- Depth instrument, receiving its input from a thru-hull transducer

- Speed instrument, receiving its input from a thru-hull paddlewheel log unit

- Wind instrument, receiving its input from a masthead anemometer

- GPS unit, or a Chart plotter

- Heading sensor, essentially a gyro-controlled fluxgate compass which may or may not have a display unit

- Autopilot

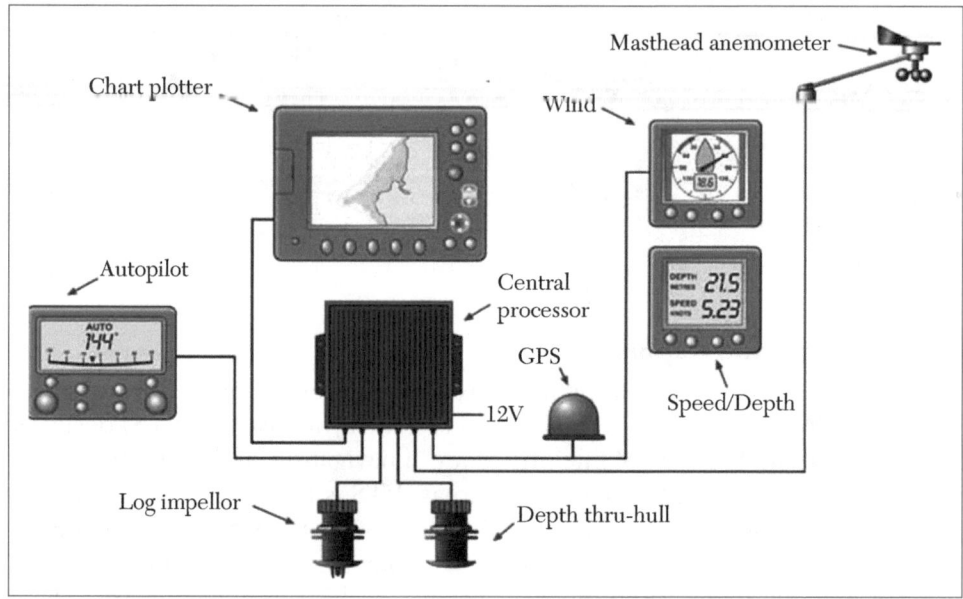

FIGURE 8.16 Sailboat instrument system.

The magnetic compass remains an independent cockpit instrument, having been made redundant in the integrated system by data provided by the heading sensor. For a number of years, all marine VHF radios and top-end single side band (SSB) radios have had *digital selective calling* (DSC) functionality built in, but the latest innovation is to incorporate automatic identification system (AIS) technology too. By connecting an AIS-enabled VHF to the chart plotter, full AIS data can be displayed on screen. DSC is just one of the subsystems of the *global marine distress and safety system* (GMDSS), others of which include satellite communication, the *maritime safety information* (MSI) system, the *electronic position indicating radio beacon* (EPIRB) system, and the *search and rescue transponder* (SART) system.

8.5 AUTOPILOT SYSTEM

Figure 8.17 is a block diagram of the major components of an autopilot system. The compass indicates the direction in which the boat is pointed, often referred to as the *actual heading*. Depending on the type of boat and installed equipment, the compass may be a magnetic compass, an electronic fluxgate compass, a gyroscopic compass, or a GPS compass. The actual heading is fed electronically from the compass to the processor (SPU), which is the "brains" of the autopilot. The SPU contains the microcontroller(s) and other electronic hardware and the sophisticated control software necessary to steer the boat on any desired heading. The *control head*, normally located in the wheelhouse, is the interface between

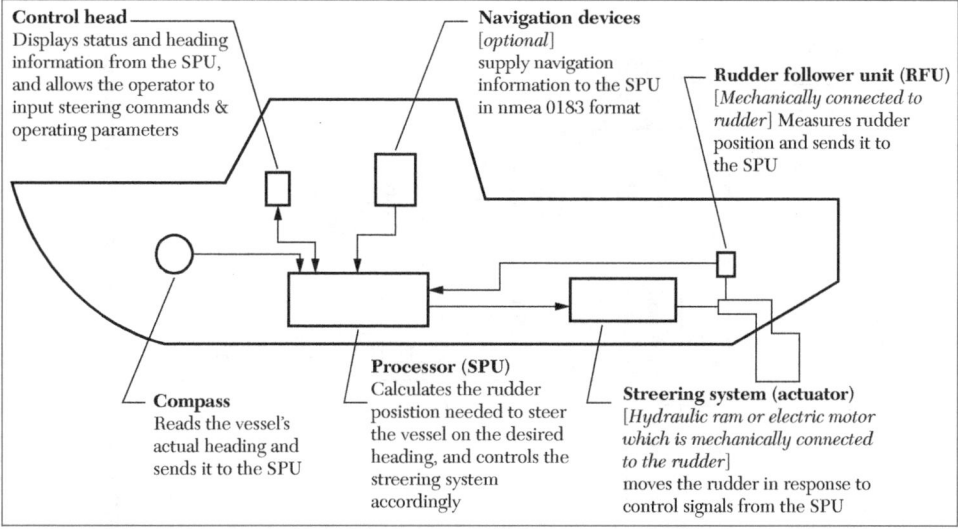

Control head
Displays status and heading information from the SPU, and allows the operator to input steering commands & operating parameters

Navigation devices
[*optional*]
supply navigation information to the SPU in nmea 0183 format

Rudder follower unit (RFU)
[*Mechanically connected to rudder*] Measures rudder position and sends it to the SPU

Compass
Reads the vessel's actual heading and sends it to the SPU

Processor (SPU)
Calculates the rudder position needed to steer the vessel on the desired heading, and controls the steering system accordingly

Streering system (actuator)
[*Hydraulic ram or electric motor which is mechanically connected to the rudder*]
moves the rudder in response to control signals from the SPU

FIGURE 8.17 Basic autopilot system.

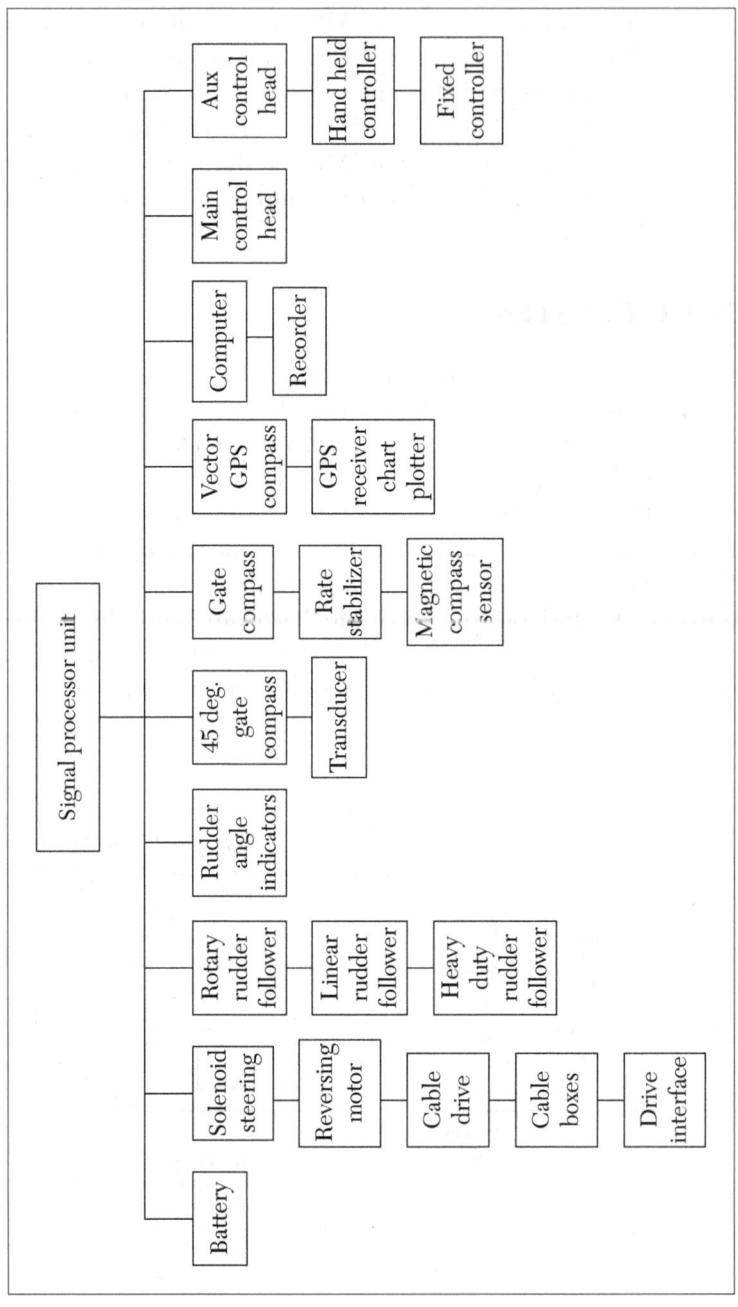

FIGURE 8.18 System block diagram of advanced autopilot system.

the boat's operator and the autopilot. The control head displays information about what the autopilot is doing, and it has various controls (buttons and/ or knobs) that allow commands to be sent to the autopilot. The final part of the picture is the *steering system*. For the autopilot to steer the boat, there must be a steering system capable of moving the rudder independently of the ship's helm. This might be a hydraulic ram that is connected to the rudder post or tiller quadrant, or an electric motor connected to the steering cables. But no matter what type of steering system the boat uses, electronic signals from the SPU tell the system to move the rudder, how far and in which direction. The autopilot may also have a *rudder follower unit* (also called a *rudder feedback unit*; RFU), a device that tells the SPU what position the rudder is in at any given time. Advanced autopilot systems, their functions, and their relationships to each other are shown in Figure 8.18, showing the interconnections between the elements of the system: the signal processor unit (SPU), the main control head, a compass or other heading sensor, solenoid(s), reversing motor or cable drive (which moves the rudder), the rudder follower unit, optional auxiliary control head(s) and remote controls, various optional accessories, external equipment, and other navigation equipment.

8.6 SATELLITE PHONES AND THEIR USE ON SHIP

Continuous contact with the shore is important to prevent accidents and emergency situations. Satellite phones are the only mode of communication. Communicating with people when on land is not a difficult task. The service is easy, fast, and reliable. Cell phones have stipulated towers to send and receive signals up to an approximate range of 1–5 miles. Satellite phone is the answer for a "No tower" communication system, directly connecting to the satellite without the use of any mediator. A satellite phone, like any other cell phone, uses radio waves for the transfer of signals. There are many satellite phone service providers but the quality of service depends on the type of satellite they use. The two main types of satellites are:

- Low earth orbit satellites (LEO)

- Geosynchronous satellites

The height of an LEO satellite orbiting the Eearth varies greatly. Some satellites are polar, which means they reach both the poles of the Earth while orbiting. Satellites that are not polar do not reach the Antarctic and the

Arctic regions. The quality of service a satellite phone provides depends on the height of its orbit: satellites at lower heights have lesser coverage area as compared to the higher ones. LEO satellite phones rarely face delay in connection but have lesser mobility due to smaller coverage area [Figure 8.19].

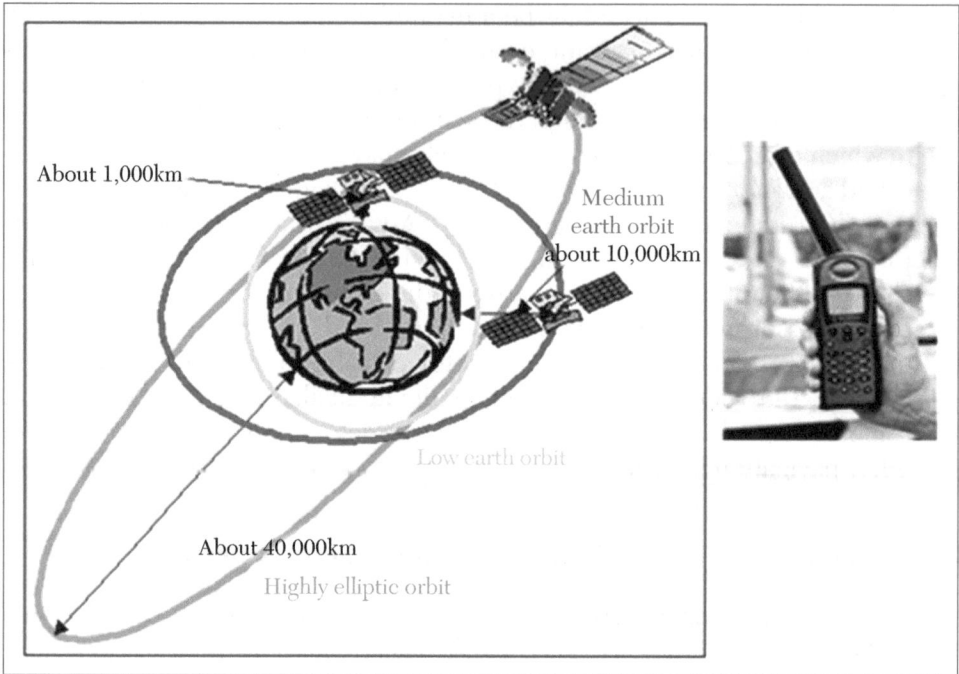

FIGURE 8.19 Satellite orbit and satellite phone.

Geosynchronous satellites are located at higher orbits, at an approximate height of 22,300 miles above the Earth's surface. These satellites move at the same rate as that of the Earth. As they are located at higher orbits, they have a very large coverage area. One satellite can cover around one-third of the Earth's surface, as shown in Figure 8.20. Phones using these satellites generally have long antennas facing the direction of the satellite and often face delay in connection. A clear and unobstructed sky is a must. There are also Iridium phones that use nondirectional antennas. This means that the phones antenna needn't be pointing in a specific direction. When a crew member makes a call from a satellite phone, the signal goes directly to the satellite of that service provider. The call is processed by the satellite and is then sent to the Earth through a gateway. The gateway directs the call through the regular landline or the local cell phone service providers. If a person from the ship wants to make a call to another satellite phone, the signal is sent to the satellite and is relayed to the receiver phone without the help of any gateway. Thus there is no intermediary in a satellite to satellite phone. The only issue regarding the usage of a satellite phone is that they have long antennas. The antennas should face the satellite whenever the call is being made. A satellite call will face interruption and delay if the call is made from inside of a building or the sky is not unobstructed. As satellite phones do not use any land infrastructure and provide uninterrupted service, the call charges are very high compared to that of a normal cell phone. The handsets also are large and heavy.

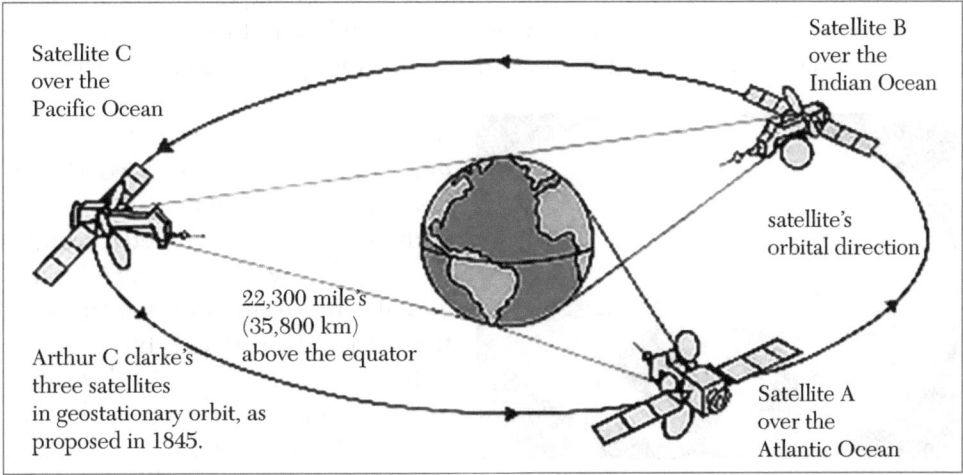

FIGURE 8.20 GEO to cover Earth.

Applications of satellite phones: A satellite phone is a boon for a ship and its crew members. It's the only and the most effective source of communication on ship. Satellite phones have made a lifesaving difference during emergencies. Weather conditions are often unpredictable at sea and satellite phones are used to caution a ship in case of storm, rogue waves, or any sudden change in the weather condition. Satellite phones are used for business purposes to remain in contact with clients at the next port of call and provide a data line to send and receive business information and transactional documents. In natural disasters, such as hurricanes, floods, or earthquakes, land-based cellular towers are often destroyed, cutting off communication with the affected areas. It is during such conditions that satellite telephony becomes the only mode of communication for calling help.

8.7 FIREFIGHTING EQUIPMENT

The engine room employs fire detection systems which have a master control panel on the bridge, with auxiliary panels in the engine control room and the fire control station. The system consists of different types of fire detectors [Figure 8.21], located at various places according to the risk of the type of fire. There are three phenomena associated with a fire: flames, smoke, and heat, which are detected by infrared flame detectors, smoke detectors, and heat detectors in the engine room. In workshops, where welding is always going on and smoke and naked flame are always present, there would be only a heat detector, or none at all as it is a certified hot work area. In the engine control room smoke detectors are used. Near boilers and incinerators where abnormal conditions can produce a naked flame, an infrared flame and an ionization type smoke detector are used. The flame detectors are used near fuel handling units like purifiers, hot filters, refiners, conditioners. Upon the detection of a fire, an audible alarm is sounded throughout the ship, with the control panel and alarm systems showing the location of the fire. If two types of fire detectors, i.e., flame and smoke, are simultaneously triggered, they would activate the *hyper mist system* in that zone automatically.

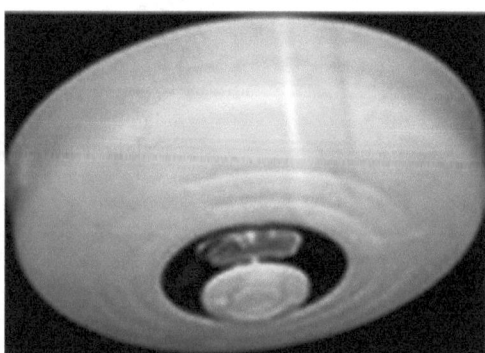

FIGURE 8.21 Fire detector.

8.8 SHIPBOARD LEVEL SENSORS

Ships have various tanks as a part of their structure, and measuring the contents of a tank is a quite a job. The usage may vary from ballast to sludge retention. All these tanks store certain liquids whose level (quantity) has to be measured and checked on a daily basis. Failure to do so will result in huge devastation, ranging from a pollution incident to stability problems.

Most tanks have high- and low-level float indicators provided on the tank walls. When the liquid level in the tank reaches maximum, the float is lifted by the rising liquid surface. The opposite end of the float device has a magnet which flips another magnet in the fixed body. The magnet's flipping action makes or breaks the circuit which causes an alarm. They also have a provision for testing the alarm manually.

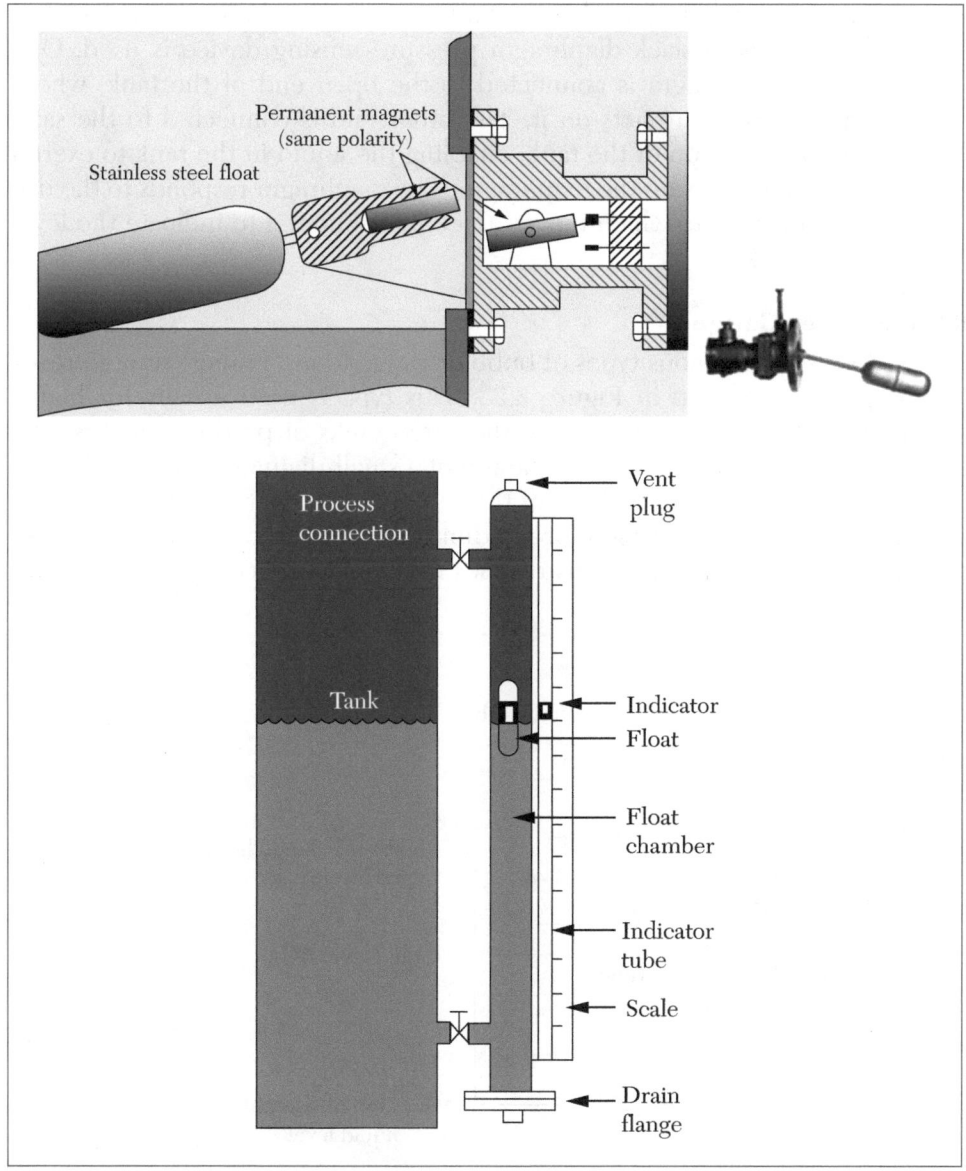

FIGURE 8.22 Floats to measure tank levels.

8.8.1 Pressure-Based Level Measurement

The head of available liquid in the tank is used and the head is converted into level measurement. As the level inside the tank varies, the head of liquid and thus the pressure varies accordingly, and this is used as reference for indicating the level in the tank. For sensing the pressure variation inside the tank, a slack diaphragm pressure-sensing device is used. One end of the diaphragm is connected to the open end of the tank, where atmospheric pressure acts on it. The other end is connected to the safe, bottom-most portion of the tank, enabling the liquid in the tank to exert a corresponding head on the diaphragm. The diaphragm responds to the difference in pressure, which moves a needle calibrated to indicate the level inside the tank.

8.8.2 Bubbler Gauge

One of the famous types of bubbler gauge found in most remote reading gauges is shown in Figure 8.23. This type is used usually for highly viscous oils in fuel oil tanks and the cargo tanks of product carriers, etc. Compressed air is admitted at the top of a small-diameter tube fixed vertically inside the tank. The base of the tube is left open to the tank's bottom surface. When compressed air is admitted from the tube top, it bubbles out from the bottom and thus reaches the liquid surface. The air pressure

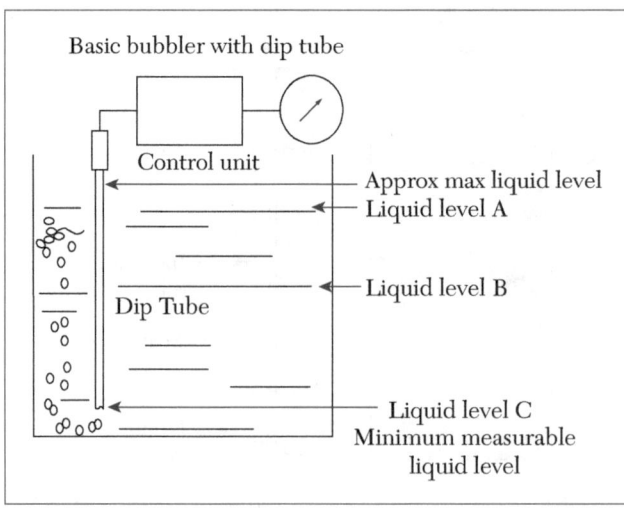

FIGURE 8.23 Bubbler gauge.

is dependent on the hydrostatic head of liquid present in the tank. This air pressure is measured and calibrated to indicate the tank level continuously. A three-way valve may also be provided to bypass the compressed air from the restrictor, thus allowing blowing off the tube inside the tank to avoid any deposits.

8.8.3 Ultrasonic/Microwave Level Sensor

These devices [Figure 8.24] use ultrasonic waves or microwaves to determine the ullage level of the liquid in the tank. The waves hit the surface of the liquid inside the tank and bounce back to the transmitter itself. The time taken to receive the signal back is calculated and the ullage of the tank is determined. The main advantage of this type is there are no moving parts; it is mainly used for LNG/LPG tankers where manual sounding is not possible.

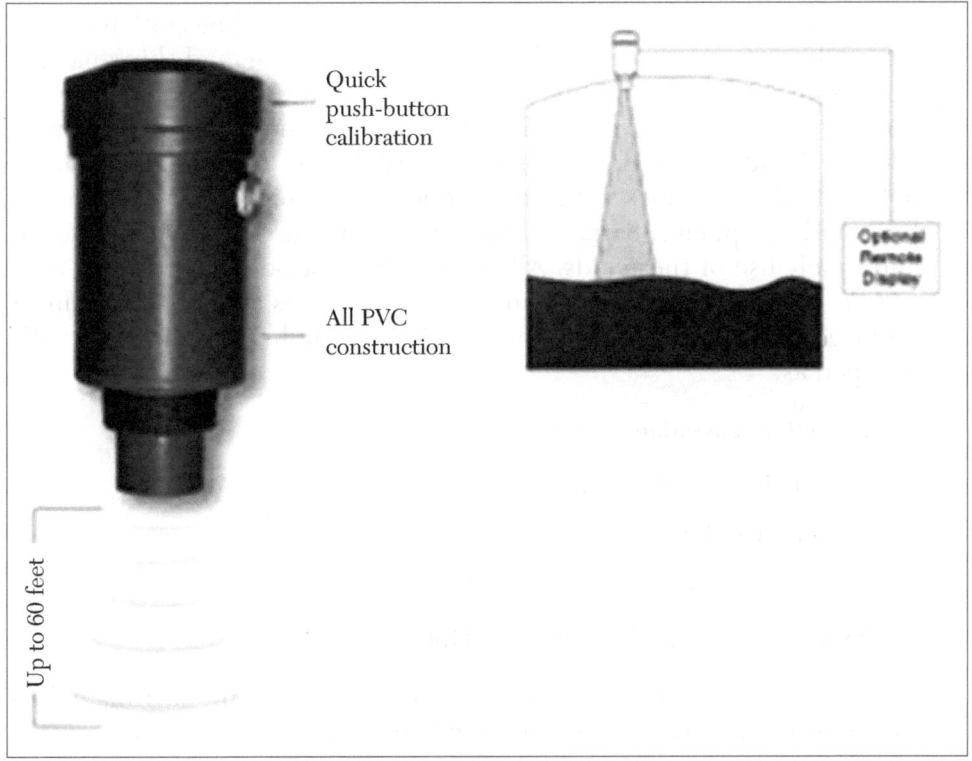

FIGURE 8.24 Ultrasonic level sensor.

8.8.4 Capacitive Level Sensor

A capacitor has two plates, separated by a dielectric medium. For measuring tank levels, the capacitance probe acts as a single plate and the tank side wall acts as the second plate. As the liquid level inside the tank varies, the potential between the tank wall and the probe varies accordingly. This causes a change in the current in the circuit and thus it is calibrated for indicating the tank level.

8.9 SHIP'S BRIDGE

A ship has to navigate around the clock through different waters, at times with restrictions, in changing weather and sea conditions. Communication must be maintained during routine voyages, in restricted waters, in emergencies, and for rescue operations. Various equipment, instruments, and appliances are provided for performing these functions. The ship's bridge serves as a controlling and commanding station for the entire ship. One can control all the machinery, boiler, and ship's navigation from the bridge [Figure 8.25]. This provides a common platform for the ship's alarming and controlling station for onboard machinery. Electrical and electronic equipment is installed so that the electromagnetic interference does not affect proper functioning of navigation systems and equipment. Safety of navigation depends on proper, efficient, and timely use of these aids. All these aids must be checked from time to time for their performance and accuracy. Errors and deviations must be logged. The common activities carried out on bridge can be broadly grouped as:

- Lookout and avoidance of collision

- Control of ship's speed and direction

- Navigation and position fixing

- Monitoring weather and sea condition

- Communication, both external and internal

Binoculars are used for long-distance viewing in daytime when the weather is clear. At night or when visibility is restricted, radar should be used for lookout. Range scale should be selected according to the traffic density, ship's speed, and distance from the coast. The heading marker should be checked against the compass heading and the ship's fore and aft line. Ships of 10,000 GT and above should have two radars, including one operating on X band, 9 GHz frequency. They should have *automatic radar plotting aid* (ARPA). Smaller ships are fitted with *automatic tracking aid* (ATA) or *electronics plotting aids* (EPA). Alarms should also be fitted for the failure of these lights on the bridge panel.

FIGURE 8.25 Ship's bridge.

8.9.1 Controlling Ship's Speed and Direction from Bridge

Ship's speed: Every ship should have an indicator for the propeller speed and direction of rotation on the bridge. If the propeller is of controllable pitch, there should be indication of pitch also. In the *integrated bridge system* (IBS), various operations such as passage execution, communications, machinery control, and safety and security are centrally monitored. Engines can be stopped in an emergency. Overriding provisions are made in the case of the main engine. The speed log shows the ship's speed in knots, and the distance indicator records distance covered in nautical miles. A calibration chart is provided; these readings should be verified by plotting the ship's position on the charts.

Direction controls: The ship's direction is set with the *standard magnetic compass*, *steering magnetic compass*, or *gyrocompass*. There should be communication between the standard magnetic compass and the steering position. An efficient periscope with sufficient magnification and adjustability is provided for comfortable viewing of the standard compass by the helmsman.

Maneuvering displays on bridge control screen: Maneuvering characteristics are displayed on the bridge. These include propeller RPM and ship's speed in knots corresponding to full, half, slow, and dead slow ahead positions on the telegraph. Astern power as percentage of ahead power, time change over from full head to full astern, distance to stop from full ahead, time to stop, turning circle at full speed, at maximum rudder angle, in loaded and ballast conditions time to steer from hard port to hard starboard and minimum speed to maintain course are also displayed. Both main and auxiliary steering systems must be tried out. Failure alarm and auto start systems should be located on the bridge. Charts showing procedure to change over to emergency steering should be displayed near the steering control.

Alarms fitted in bridge: Alarms are provided for failure of main propulsion, machinery, and steering systems. These and the alarms for navigation light failure, off course, radar warning, etc., must be acknowledged by the officer on watch within 30 seconds, or the alarm is sounded in cabins, office, and mess for back-up assistance.

8.10 TRACKING SHIPS USING AIS

Keeping track of ships sailing at sea has become an important aspect of maritime navigation. The automatic identification system (AIS) is a device which not only helps in tracking ships at sea, but also helps in avoiding accidents and traffic congestion. Increasing ship traffic at sea has led to several problems both near the ports and on high seas. Many accidents and collisions of ships in the past have been a result of lack of information of the nearby ship and erroneous instructions from port authorities. These incidents have been the main reason for the invention and usage of ship tracking devices. Moreover, many ship owners also need to keep a track of the cargo they are transporting. Ship tracking devices help in serving these needs also. The automatic identification system is one such versatile ship tracking device. An automatic identification system, as the name suggests, automatically helps procure detailed information about any ship at sea. A fully automatic system, AIS provides all the information regarding a particular ship to nearby ships and also to the coastal authorities.

8.10.1 AIS Operation

The automatic identification system (AIS) runs on the basic principle of transferring data electronically over a radiowave frequency. An AIS device consists of very high frequency (VHF) transmitters and receivers. The transmitter and receiver are attached to the ship's display and sensors systems through a communication link. To receive exact information about other ships and to send its own, the AIS also has a global positioning system (GPS) connected to a satellite. The GPS can be an internally attached device or a separately fitted system. AIS is also connected to all the other systems of the ship, and that is how it receives the ship's details and sends them across to other ships. AIS is also known as the most important and safest navigation system onboard a ship because it runs automatically and continuously, sending and receiving information regardless of the ship's position with respect to the shore. Moreover, though only one channel is required to transfer the details, AIS has a secondary channel to prevent any kind of interference or loss of information.

As each ship has its own AIS system, there are high chances of an increase in traffic and congestion in the channels; however, this never happens. AIS has an automatic system that resolves the contention between itself and other stations despite rise in load. This is possible because each station has its own transmission slot. AIS is so designed that transmission slots are automatically assigned to a particular station on the basis of the traffic history of the station. Each of the slots is of 26.6 milliseconds, which means that each station can transmit information during that much time before the chance goes to the next station. For this, there are in total 2,250 slots. The information sent from a station in a time slot serves as a reference for the next information, which gets stored in some other randomly organized slot. The information received and transmitted through these slots is immediately transferred to any vehicle that comes in the same radio range. Thus the 2,250 slots together serve as a common network for providing information of all the ships coming under a particular frequency channel.

8.10.2 Information Transfer

AIS transfers every detail regarding a particular ship. It procures details such as a ship's name, speed, position, direction, rate of turn, destination, etc. It also gets physical dimensions of the ship, such as length, breadth, tonnage, beam, and draft. All these items of information are directly sent to the display system of the ship, where every detail is continuously displayed in real time. All these details are extremely helpful in avoiding ship collisions, reducing and monitoring traffic, assisting in navigation, search and rescue operations, and in investigation and research. Thus, AIS helps not only in tracking the ships but also in avoiding many unfavorable situations, as shown in Figure 8.26.

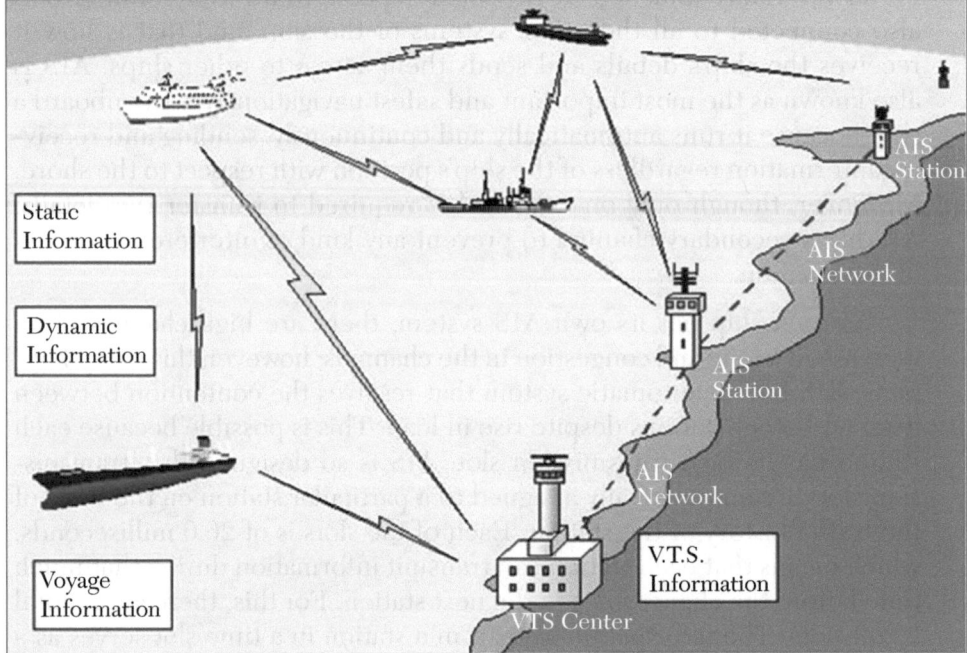

FIGURE 8.26 AIS information.

8.11 ELECTRICAL PROPULSION SYSTEM

Recent trends and developments in power electronics have paved way for more advanced and modern trends in electrical propulsion. LNG, with a boiling point of minus 161.5 degrees Celsius at atmospheric pressure, is highly volatile and thus demands a containment system that can maintain a temperature near or below its boiling point. However intact the containment system may be, a certain percentage of vapors (*boil-off*) accumulate, tending to increase the tank pressure. If this increase is not relieved, it will have a huge impact on insulation walls and membranes. To solve this problem, the boil-off is piped into the engine room. LNG carriers have a natural advantage in using boil-off gas for propulsion.

8.11.1 Boil-Off for Propulsion

The boil-off piped into engine room is used as fuel in the boilers, and the high-pressure steam produced is used for propulsion in steam turbines. More recent methods use boil-off in diesel engines.

- Using boil-off gas for power generation (generator engines) and having an electrical motor for propulsion in *dual fuel diesel electric propulsion* (DFDE) and *tri-fuel diesel electric propulsion* (TFDE) ships

- Using boil-off gas for the main propulsion engine itself; these are normal two-stroke, cross-head propulsion engines

It is widely known that steam turbines have very low thermal efficiency, and because of this the marine world has started to prefer diesel engines. With new technologies, it is currently possible to burn natural gas in the engines.

Figure 8.27 shows a typical platform layout of a TFDE-electrically propelled vessel. When the main propulsion diesel engine is replaced by two electrical motors, so much space is saved in the engine room that the entire bottom platform looks empty. The weight of the motor is less compared to the diesel engine, and thus more cargo can be carried. Since the main propulsion diesel engine requires more auxiliaries for its operation, the entire plant must be huge and complex. With electrical propulsion, the auxiliaries are just some thyristors and exciter control panels that occupy much less space and hardly require any maintenance.

8.11.2 TFDE Propulsion Layout

The name TFDE originates mainly due to the power generation engines being able to use three different types of fuel, thus the name "tri-fuel diesel electric propulsion." With respect to propulsion, the most modern trend is to have electrical motors instead of diesel engines. Power generated in the

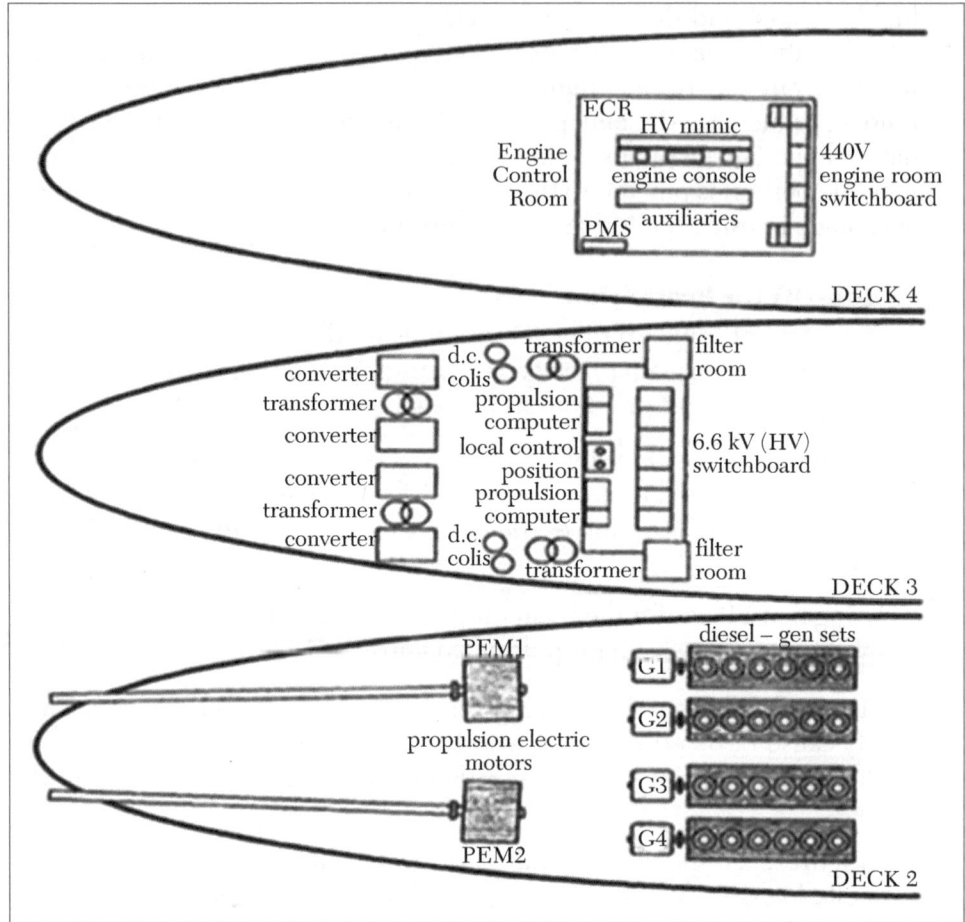

FIGURE 8.27 TFDE-electrically propelled vessel.

alternators is in the range of 10 to 20 megawatts, which then feeds the main propulsion motors. Figure 8.28 specifies various propulsion motor options with fixed and controllable pitch propellers.

Electric propulsion offers advantages like layout flexibility, low running cost and maintenance, better efficiency, and good maneuverability. A typical electrical propulsion system has four generators, each generating 8–12 mW of power at 6,600 or 11,000 Volts, 60 Hz. This is fed to the main bus bar, from where the propulsion motors are fed. There are two propulsion motors coupled with a reduction gear, driving the propeller as in Figure 8.29.

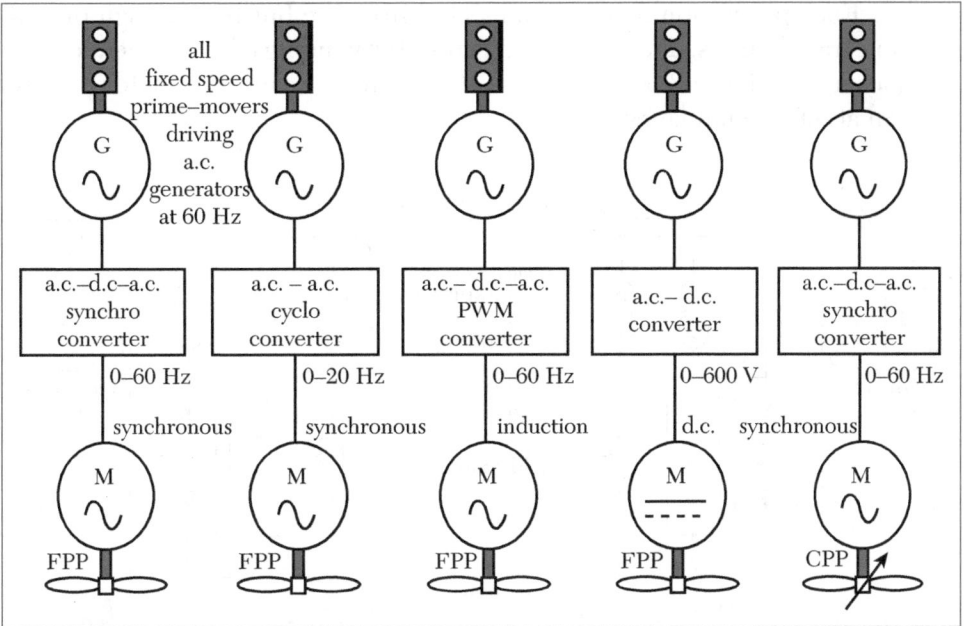

FIGURE 8.28 Electrical propulsion options.

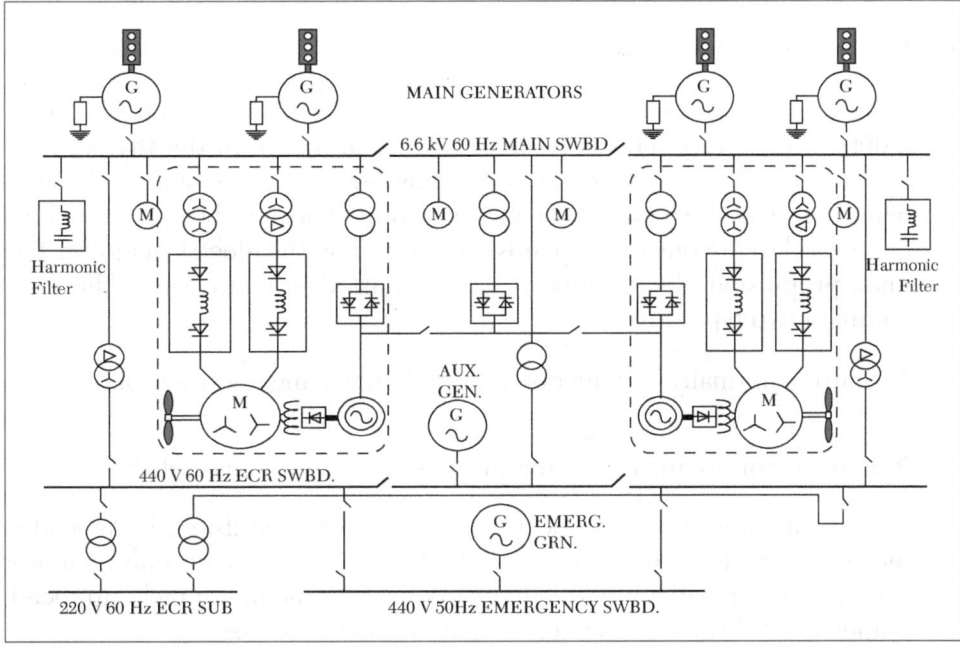

FIGURE 8.29 Power distribution.

Each propulsion motor is connected from the bus bar through propulsion transformers. The motor has two stator windings for redundancy reasons, as in Figure 8.30. When one stator winding fails, the motor can still run at 50% redundancy.

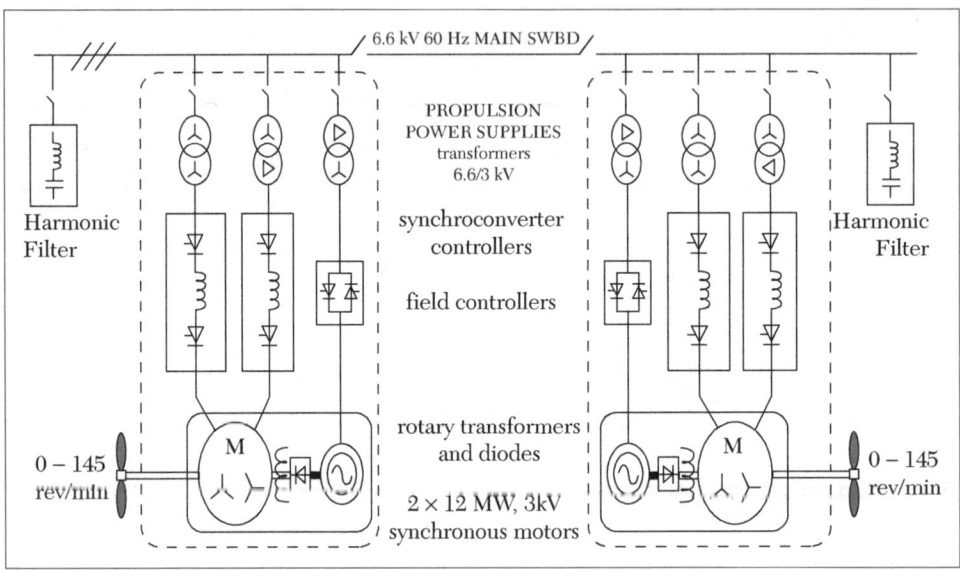

FIGURE 8.30 Propulsion motor supply.

Even though the generated voltage is 6.6 kV, the motors run on 3,000 Volts or less. This is because of the limitation in the thyristor firing circuits. The motors can even run as low as one or two revolutions, which offer better control for maneuvering. The number of generators connected to the bus bar depends on the load on the electrical motor. The main propulsion electric motors can be started in many ways. The most common two types are

1. Starting normally as induction motor, then making it a synchronous motor

2. Pony motor driving the motor, making it a synchronous motor

On all ships that have electrical motors for propulsion, kVAr load is more important that the conventional kW load. When the propulsion motor starts, the kVAr load is more than the kW load. As the motor picks up speed, gradually kW load increases as the kVAr starts to stabilize.

8.12 GAS DETECTION METERS FOR SHIPS

Various gas meters used onboard ships measure the hydrocarbon content, explosion hazard risk, and also the oxygen analyzers. Under the following circumstances the cargo tank or any enclosed space onboard the ship must be evaluated to ensure that the space is gas free and has ample oxygen for personnel to work there if required. Tank evaluation is done to ensure that the atmosphere inside the tank is safe enough for personnel to make an entry. There is different equipment available on board for the evaluation of tank atmosphere. Some of them are: combustible gas indicators or explosimeters, tank scope or noncombustible gas indicators, multi-gas analyzers, and oxygen analyzers.

8.12.1 Combustible Gas Indicators or Explosimeters

An *explosimeter* is a device used to detect the amount of combustible gases present in a sample of a given atmosphere. This gives a reading in terms of percentage of the LFL (*lower flammable limit*). "Resistance proportional to heat" is the working principle. The equipment consists of a Wheatstone bridge in which one of the resistances is variable. The circuit is shown in Figure 8.31.

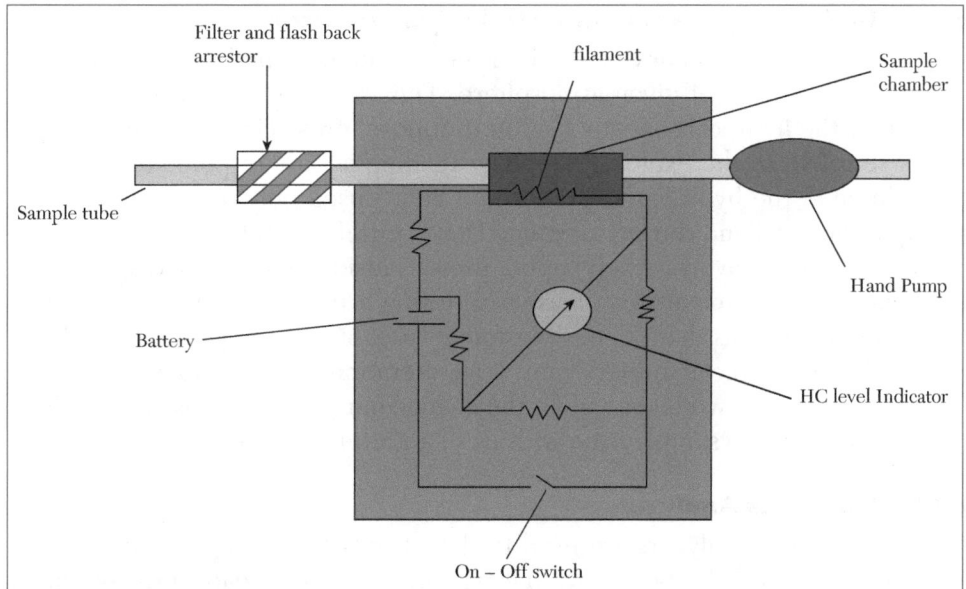

FIGURE 8.31 Combustible gas indicator.

It consists of four resistances in which one varies according to the amount of the gas present. A hand pump is used to draw the gas or the atmosphere containing the gas inside the device. A filter and flash back arrestor is used to filter the gas and acts as a flame arrestor. The device is switched on. As the hand pump is operated to suck a sample of gas from the cargo tank, simultaneously the filament gets heated. Any combustibles in the sample will land on the filament in the sample chamber. The combustibles will burn as the filament is already hot, causing an increase in resistance which disturbs the Wheatstone bridge. The reading can be read from the indicator. The instrument gives the reading in percentage of the lower flammable limit or lower explosive limit, which is 1%. This type of gas meter can only be used if the gas content is very low—i.e., this instrument should not be used if the atmosphere contains:

- H/C + inert gas—then the gas will not burn as there is no oxygen

- H/C + oxy-acetylene—then the burning will be too violent

- H/C + oxy-hydrogen—same as above

- Lead petroleum vapors—lead oxide deposits on the filament cause a reduction in sensitivity

8.12.2 Tankscope or Noncombustible Gas Indicator

A *tankscope* is a device used for measurement of hydrocarbon gas content in a sample of given atmosphere. This instrument is meant for measuring the hydrocarbon vapor in inert atmospheres. This instrument is not as sensitive as the explosimeter. The reading is only in percentage of the volume of the hydrocarbon vapor and hence used only during the gassing up operations and during inerting. This is purely meant for measuring the volume of the hydrocarbon vapors present inside any enclosed space, and hence it is not meant for measuring during a man entry. It works on the same principle as that of an explosimeter except that the gas does not burn inside the sample chamber; there is an alteration in the temperature of the heated filament which enhances the change in resistance. It is always advisable to flush the sample tube with fresh air after every use.

8.12.3 Multi-Gas Analyzers

Multi-gas analyzers are used to detect only targeted gases and vapors [Figure 8.32]. It is specific to that type of gas only, so care must be taken to ensure that correct tubes are used for the particular type of gas. The multi-gas analyzer consists of a portable bellows pump and detector tubes. The detector tube is like a vial filled with reagent that will react with the specific chemical. Both the ends of the tube are closed. To use it, we have to break the two ends of the tube and insert it into the pump according to the directions on the tube. Now start pumping three to four times (or as specified by the manufacturer) to suck in the gas from the atmosphere. If the atmosphere contains that gas or vapor, the color of the tube changes. The length of the color change can be read from the tube and compared to obtain the level of that gas or vapor. Some of the gases include carbon monoxide, chlorine, hydrogen sulphide, organic arsenic compounds, arsine, and phosphoric acid esters. An extension hose is provided to measure the concentration of vapor present at a different height. The oxygen analyzer is a device used to measure the concentration of oxygen in a given atmosphere. This device plays a vital role.

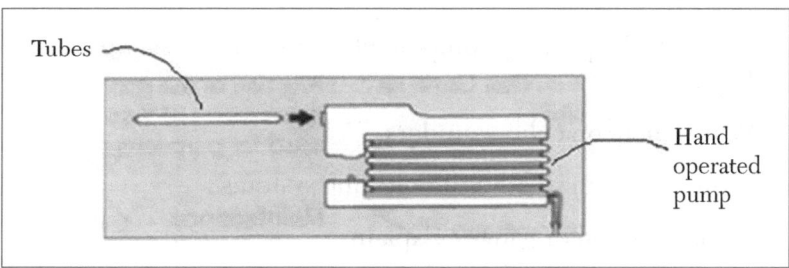

FIGURE 8.32 Multi-gas analyzers.

EXERCISES: PART A (ANSWER IN A WORD OR A SENTENCE)

1. Define "nautical chart."

2. A _____ phone is an answer for a no-tower communication system.

3. Geosynchronous satellites are located at orbits of an approximate height of _____ miles above the Earth's surface.

4. _____ phones use nondirectional antennas.

5. How many GEO satellites are needed to cover Earth?

6. The echo-sounder is used for _____.

7. What is the use of AIS?

8. What information is transferred by AIS?

9. What is the use of a tank scope device?

10. What is an oxygen analyzer?

EXERCISES: PART B (ANSWER IN A PAGE)

1. What are the roles of ECM in engine?

2. Write about fuel injection to the engine.

3. In short, explain wireless control stations.

4. List the information provided in a nautical chart.

5. Write about the marine navigation system.

6. What are the primary components of integrated sailboat instruments? Explain.

7. Write a note on echo-sounders.

8. Write a note on weather monitoring systems.

9. What is a bubbler gauge? Explain.

10. Explain the function of ultrasonic-microwave level sensor.

11. How do you measure a liquid level using the capacitive level sensor?

12. Write about the noncombustible gas indicator.

EXERCISES: PART C (ANSWER WITHIN 2 OR 3 PAGES)

1. Explain the role of electronics in a marine generator set.

2. Explain marine instruments in detail.

3. Write in detail about marine navigation equipment and instruments.

4. What are the components of an autopilot system? Explain the components' operation.

5. In detail, explain the operation of satellite phones.

6. Explain shipboard level sensors.

7. Write about the ship's bridge in detail.

8. Write in depth about the operation of AIS and its information.

9. Explain the modern electrical propulsion system for LNG tankers in detail.

10. Explain gas detection meters for ships.

WEB LINKS

http://www.nautilo.fr/help

http://www.sailboat-cruising.com/sailboat-instruments.html

CHAPTER 9

OCEANOGRAPHIC INSTRUMENTS

This chapter deals with instruments and their measured parameters, oceanographic instrumentation, Argo robots, measurements of hydrographic properties, measurement of dynamic properties, BIOMAPER, and many others.

9.1 INSTRUMENTS AND THEIR MEASURED PARAMETERS

Oceanographic instruments measure or sample physical, chemical, and biological quantities in the water column. There is often more than one way to measure a quantity (i.e., temperature, currents) and there are many quantities to measure, hence there are many instruments. The following is the list of quantities commonly measured these include: depth (meters), temperature (degree Celsius), salinity (practical salinity units of chlorine, sodium, sulfate, magnesium and potassium) oxygen, phosphate, nitrate, silicate, pH, density (kilos per cubic meter), water clarity (Forel scale to measure color), sound or ambient noise (Hertz), Sound Speed, bioluminescence, seabed sediment, current, and waves. Some physical properties of seawater are conservative (i.e., temperature and salinity); this means that away from the surface the only way they can change is by mixing. All the above quantities are normally accompanied by a measurement of date, time, and position, i.e., latitude and longitude. Measurement of time and position are made with a global positioning system (GPS) and are relative to GMT time. Accuracy is of the order of +/− 10^{-15}m.

9.1.1 Depth

In the measurement of depth in relation to the instrument or the sea-floor, it is fundamental to reference the location of the measurement. The simplest way to measure the depth of the measurement is to mark the wire or rope to

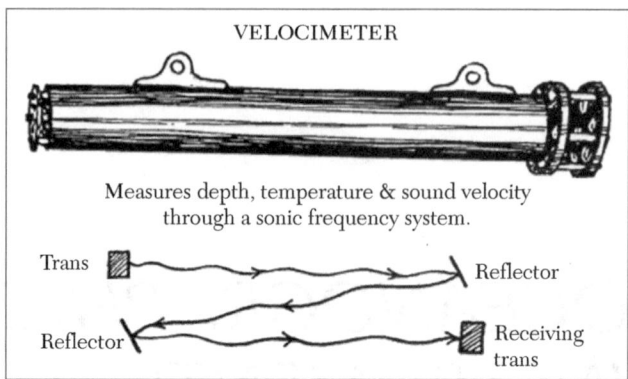

FIGURE 9.1 Velocimeter, sonic frequency system

which the instrument is attached and measure the length used. The more accurate way is to measure the water pressure at the instrument level and convert this to a depth. The simplest way to measure sea bed depth is to lower a weight with a wire or rope and measure the length used till the weight has reached the bottom. Yet another is to use an *echo-sounder* and time how long takes for the sound pulse to echo back, as in Figure 9.1. *Sonar* (sound navigation and ranging) and SOFAR (sound fixing and ranging) use sound for measurement. *Conductivity, temperature, and depth* (CTD) instruments use a strain gauge pressure transducer to sense depth.

9.1.2 Temperature

There are several ways of measuring temperature, the simplest being a mercury in glass or stem thermometer. A stem thermometer is commonly used to measure sea surface temperature by placing it in a bucket of seawater. A *protected reversing thermometer* is special kind of stem thermometer which measures subsurface temperature. "Protected" means that the thermometer is isolated from water pressure. The reversing thermometer is attached to a water sampling bottle. When the sampling bottle is closed, the thermometer is inverted and, as a result of its construction, the mercury "breaks" at a particular point and runs down the other end of the capillary to record the temperature at the depth of the reversal. After the thermometer has been reversed, it becomes almost insensitive to subsequent changes of temperature and it is read when it is bought back on deck. After corrections for scale errors and for the small change in reading due to any difference between the in-situ temperature and that on the deck, the reversing thermometer is accurate to about +/− 0.02 C in routine use.

An *unprotected reversing thermometer* is exposed to water pressure. The water pressure compresses the glass in the bulb and causes the thermometer to indicate a higher temperature than the protected thermometer. The difference between the two thermometers is a measure of the compression of the glass, which depends on a known compressibility and upon pressure, i.e., depth. A protected/unprotected reversing thermometer pair is used to measure both temperature and depth. The depth is then used to reference the depth at which the water bottle sample was taken. Depth measured in this way is accurate to +/− 0.5% of depth or +/− 5 meters, whichever is greater.

Another widely used temperature sensor is the *thermistor*. The thermistor is usually a small piece of electrically conductive material such as a resistor. The thermistor relies upon measuring electrical resistance, which is directly or inversely proportional to temperature. Thermistors are used on CTD (*conductivity, temperature and depth*) and XBT (*expendable bathythermograph*) instruments. The CTD instrument is composed of a rosette with weights, 12–24 Nisking water bottles [see Figure 9.13], reversing protected and unprotected thermometers, and other instruments such as a fluorimeter and a transmissometer. The water samples are used for collection and for calibration and chemical/nutrient analysis. The instrument provides real-time sections or profiles of temperature, salinity, density, geostrophic currents, and dissolved oxygen. The temperature/salinity diagrams are used for water mass identification. The XBT is composed of a torpedo-shaped probe that contains a thermistor and a very fine coiled copper wire which is unreeled from both the probe and the canister, from which it is deployed as shown in Figure 9.2. It is launched from the ship, being either hand-held, deck-mounted or hull-mounted. The depth is calculated from the elapsed time and the expected fall rate (~6.5 m/s) and comes in four types (T4 ~460m, T7 ~760 m, T10 ~200m, TDeep ~760). This instrument provides a large body of data (global dataset) over many seasons/years for statistical evaluation of climatic

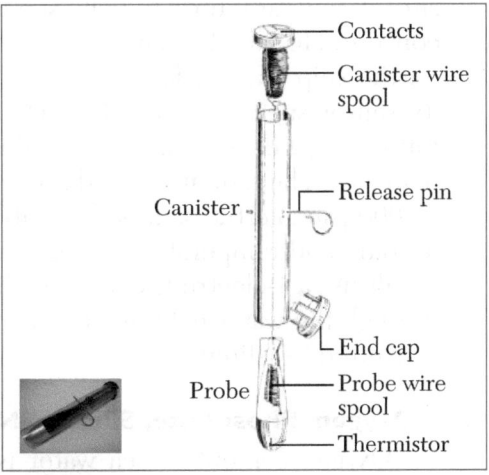

FIGURE 9.2 Expendable bathythermograph (XBT)

changes. *Thermistor chains*, consisting of a cable with a number of thermistor elements at intervals, are sometimes moored along with current meters to record temperature at a number of depths. A *data logger* samples each thermistor at regular intervals and records temperatures as a function of time.

A *seasor* is an instrument that looks like a small airplane, which is lowered in the water and moored by the ship at a known depth or at the surface. With the aid of a conducting cable, the data is real-time monitored and produces typical sections such as virtual CTD profiles. Sea surface temperature may be inferred by measuring the infrared radiation emitted by the sea surface. This method is used by airborne platforms such as satellites. Use of satellites or other remote-sensing methods permits rapid measurement over a wide area.

9.1.3 Salinity

Salinity is a measure of the quantity of salt in a volume of sea water in *practical salinity units* (psu). In very approximate terms, a salinity of 35 psu means there are 35 grams of salt in a kilogram of water. Salinity is an important quantity to measure as salinity in conjunction with temperature and depth/pressure enables calculation of the density of the seawater and sound speed. Salinity in combination with temperature also allows the oceanographer to label water masses in the ocean, study their movement, and hence infer ocean currents. Salinity is obtained by first measuring the electrical conductivity of the seawater at a known temperature and pressure. Using an internationally standard formula, the conductivity is then converted to salinity. Salinity may be measured from a water bottle sample on the ship using a laboratory *salinometer* or in situ using a conductivity sensor such as on a CTD. The salinometer measures a conductivity ratio using a conductance cell and conductance bridge which is balanced manually. The ratio measured is between a standard sample of sea water at 35.000 psu and the sample. Salinity is then obtained from the conductivity ratio and temperature using standard tables or formulas. CTDs commonly use a 4-electrode cell through which water flows as the cell is moved through the water column. The cell measures conductivity which is then converted to salinity.

9.1.4 Oxygen, Phosphate, Silicate, Nitrate, pH

Oxygen content of sea water is commonly measured in conjunction with CTD measurements or water bottle sampling. CTD units sometimes have an oxygen sensor attached enabling an in-situ measurement of oxygen, but these measurements produce noisy results which need to be filtered. Oxygen content from water samples is determined by chemical titration. The titration method is often more reliable than using an in-situ sensor, which may suffer from drift or calibration problems. Phosphate, silicate, and nitrate are determined by chemical means from water samples. Other chemical parameters such as pH may be determined from water samples.

9.1.5 Water Clarity

Water clarity may be taken as a measure of the degree of transmission of visible light through the sea so they are taken whenever possible during daylight hours. The more turbid the water, the less light is transmitted. Water clarity has always been measured to estimate the silt run-off from rivers, monitoring pollution streams as monitoring algae growth. The simplest device for measuring water clarity is the *Secchi disc,* which consists of a white place 30 cm in diameter which is lowered into the water (a lead weight is suspended under the disc to ensure that it will sink rapidly and vertically) and the depth at which it is lost to sight is noted. The greater the depth at which the disc is no longer visible, the clearer the water. During the procedure, the color of the disc is classified with the *Florel Ule scale.* The Secchi disc is only a semiquantitative device, but being simple and low cost, it is often used.

A more quantitative device is the *transmissometer,* which measures the attenuation of a beam of light of known wavelength over a fixed path length. The transmissometer uses a light source and a photoelectric cell and measures the beam attenuation coefficient C over a direct path from the source to the photocell. The beam attenuation coefficient is a function of the shapes and amount of particulate material in the water.

Another instrument is the *nephelometer,* which measures light scattered through an angle rather than from a direct path from the light source to the photocell. Nephelometer measurements have a more direct relationship with the quantity of suspended solids in the water than a transmissometer. It is thus used when one wants determine sediment concentrations in the water, i.e., in grams/volume units, The nephelometer is useful in muddy, highly turbid water such as near coastal estuaries, rivers, and near sea bottom (nepheloid layer).

9.1.6 Sound

Sound is measured in the sea to support many applications, most notably naval sonar operations, geophysical studies, and studying sounds emitted by marine life. The basic instrument used to measure sound is the *hydrophone.* A hydrophone is a transducer which converts sound energy (pressure) into electrical energy (current). Some substances such as quartz or certain ceramics, when placed under pressure, acquire an electrical charge or voltage across the crystal surface. This behavior is called *piezo-electricity*. Thus, the hydrophone consists of sensor made of a crystalline or ceramic substance. The basic unit of sound measurement is the *decibel*, which is measure of the pressure exerted by the sound wave or measure of the sound intensity. The other important quantity is the frequency of the wave measured in Hertz.

Sound in the sea may be thought of as consisting of many superimposed waves at varying levels of intensity and frequency resulting in a complex wave arriving at the sensor. Using mathematical techniques, the complex wave is decomposed into waves of discrete frequency and intensity. The result of this analysis is a graph of intensity (decibels) versus frequency (Hertz) called a *spectrum*. The sound or background noise is frequently referred to as *ambient noise*. Many hydrophones consist of more than one sensor element, or an *array*. The advantages of an array over a single hydrophone are several. First the array is more sensitive, since lots of elements will generate more voltage (if connected in series) or more current (if connected in parallel) than a single element exposed to the same sound field. Second, the array possesses directional properties that enable it to discriminate from sounds arriving from different directions. Third, the array has an improved signal to noise ratio since it discriminates against isotropic noise in favor of a signal arriving in the direction that an element of the array is pointing. Sonar domes fixed to the hull of naval vessels have a cylindrical array. A recent development is the *towed array*, a flexible line of hydrophone elements towed from a ship. Another type of hydrophone is used on a *sonobuoy*, which is dropped from an airborne platform to record ambient noise and locate ships or submarines. Sonobuoy is compact expendable device containing a small radio transmitter for relaying signals picked up by the hydrophone.

Sound speed: Sound speed is a measure of how quickly sound propagates through the ocean. Sound speed is needed for measuring the water depth with echo-sounders since the water depth is calculated by multiplying the mean sound speed in the water column by half the time that it takes for the sound to echo back from the sea bed. The basic means of measuring sound speed is to measure the time it takes for a pulse of sound to travel a known distance. The sensor consists of a sound source, a metal reflector, and a receiver. The sensor may be attached to a probe which is lowered into the water or on an expendable instrument such as the *expendable sound velocity* (XSV) probe. The XSV works on simple principle to the XBT. Sound speed may be calculated from a well-known relation among sound speed, temperature, salinity, and depth. In the absence of direct sound speed measurements, sound speed can be estimated quite accurately provided temperature, salinity, and depth are known.

9.1.7 Currents

Measurement of ocean current is fundamental to a general understanding of the ocean. Ocean currents transport heat and effect climate and weather. Currents transport marine life and sediments. Currents also effect the passage of ships and the oceanographic instruments themselves. Current is a vector quantity having both a direction and a speed. Therefore, a current instrument must measure both direction and speed.

There are different basic ways of measuring a current:

1. The oldest method of tracing a current is a *drift bottle* or *card*. These floating items are designed so that their movement is determined only by surface currents and contain a request to the finder to notify the time and place of recovery.

2. A *moored current meter* in the sea produces measurements of speed and direction at fixed time intervals.

3. A current meter attached to a moving platform such as a ship measures the relative current. The absolute current is then determined by such vector subtracting the motion of the ships over the ground, determined by navigation. This method is used with the *acoustic Doppler current profiler* (ADCP). An ADCP may also be moored at a fixed location on the sea-bed.

4. A *drifting buoy* measures current through observation of the motion, i.e., the change of position at fixed time intervals, of an object flowing or floating with the current.

5. A *dynamic method* recognizes that the surface of the sea is not level but has hills and valleys. Therefore, contour maps may be drawn of its surface showing these elevations and depressions. Given that in the Southern Hemisphere ocean currents flow in an anticlockwise direction around areas of high sea level, a map of sea flow of an area can be drawn

from a knowledge of the topography of the surface. These contours are not visible by the naked eye but are correlated to the density of the water, the water below a hill being of lower density than that below a valley. The density of the water, instead of being directly measured, is computed from the temperature and the salinity of the ocean water.

Owing to turbulence, current is a quantity that often fluctuates over short periods of time, and averaging or smoothing is usually required to present the data in useful form. This process of averaging is termed *vector averaging*. Current meters which use vector averaging sample the current at short intervals, say once every 2 seconds, and then compute the average over a longer interval, say 10 minutes. The resulting output of the meter is a time series of current at the averaging interval. There are four common techniques of current measurement:

1. *Propellor-type meters* count (meter) the rotation of a rotor or propeller placed in the current. Direction is sensed by a vane which is orientated parallel to the current. The orientation of the vane is then referenced against a magnetic compass. An example of such a current meter is the Aanderaa model, which incorporates other in-situ sensors such as water temperature and water pressure that allow recording of other data in conjunction with current.

2. *Electromagnetic current meters* use the principle that a voltage will be induced by a conductor which moves across a magnetic field. The conductor in this case is seawater, which readily conducts a current due to salt ions. Electromagnetic current meters consist of pairs of electrodes, an internally generated magnetic field, and a flux gate compass. Water flows through the magnetic field, thereby producing a voltage which is proportional to the current speed. The current amplitude is measured along the axis of each electrode as well as the compass output. Relative current direction, corrected with compass data and current amplitude, is computed internally. The results are then stored in two absolute current vector components (North and East). The components are then vector averaged over a user programmed averaging interval.

3. *Acoustic current meters* utilize the Doppler effect—the change in frequency of sound reflected by a stationary object relative to the frequency at which it is moving. The *acoustic Doppler current profiler* (ADCP) emits a beam of sound of known frequency which

reflects off small particles moving with the water. The beam reflected to the receiver will have a change in frequency proportional to the speed of the particles and thus the current speed. One sound beam will give the component of current in the direction of the beam. However, three orthogonal components are needed to get the true current vector so the ADCP utilizes more than one beam. Four beams are typically used to obtain a redundant velocity measurement for data checking and improved instrument reliability. ADCPs measure water speed at multiple water depths or *range cells* along the path of the acoustic beams. This is achieved by periodically transmitting short pulses (typically 1–50 milliseconds) of sound, then making multiple measurements of the frequency of the echoes at discrete time intervals after the initial sound pulse. A depth profile of the current is assembled after some averaging processes. On a moving platform such as a ship, the motion of the ship must be subtracted from the ADCP relative current to get the true current. Usually the motion of ship is determined by using GPS navigation, which offers the best means of accurately determining the ship's motion at frequent intervals. The result of ADCP data processing is a time series of current versus depth profile along the track of ship. In the case of moored ADCP, it is a time series at a fixed point.

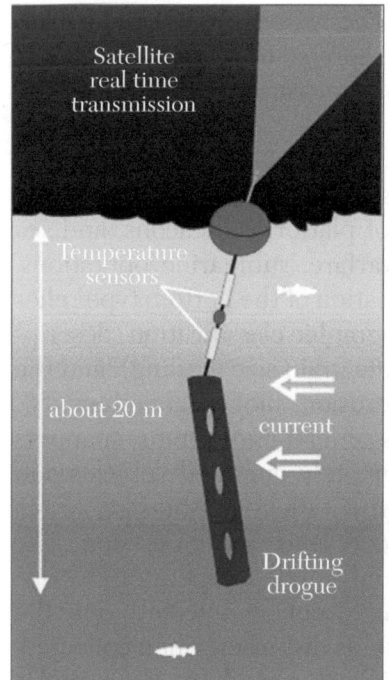

FIGURE 9.3 Drifting buoys.

4. *Drifting buoys* float on the ocean surface to follow the current. The buoy usually includes a *drogue*, which is device such as a parachute or sheet to drag the buoy with current. The drogue is attached to the buoy below the water, as shown in Figure 9.3. Some method is needed of tracking the buoy's position. Position tracking may be done using:

- Visual sighting with a bearing compass

- Attaching a metal target/reflector to the buoy so it can traced by radar

- Using a radio transmitter and tracking the buoy location with a radio receiver such as satellite

Drifting buoys may have other sensors attached, such as thermistors which measure sea surface temperature, a pressure gauge to measure air pressure, and an anemometer which measures wind speed and direction. Another type of Lagrangian device is a spherical free falling acoustic device with a skirt that reduces turbulence as it falls. This device, with the help of a triangular configuration of transponders on the seabed, can be located with high precision while sinking and drifting with the currents. This device produces a profile with real current components with depth and after each profile, so after it has reached the seabed, it reaches the surface again, waiting for collection. Other means of tracing 3-D movements of water are acquired with the aid of tracers such as dye and pollutants.

9.1.8 Waves and Tides

The most common instrument used for measuring waves is a *waverider buoy:* a buoy which follows the movements of the water surface. The wave height is measured with an *accelerometer*, which measures the vertical acceleration of the water. The acceleration is integrated to produce a time series of displacement of the sea surface, i.e., a time series of wave height. The buoy is typically attached to a flexible mooring line, which in turn is attached to a heavy weight on the sea bed. The wave height signal is usually transmitted by radio signal to a ship or shore station from an antenna attached to the buoy. Tides are most commonly measured with a *tide gauge*.

9.1.9 Seabed Sampling

Seabed information is very useful to estimate anchorage safety locations, offshore engineering (such as sitting of platforms, beacons, and sea walls), mineral exploration, fishing, mine warfare, submarine operations, and sonar acoustic performance. The classification of the bottom type relies on a geological/scientific classification, hydrographic classification (description of the most predominant components, the grain size grading), and the nature of the sea floor (washed materials, erosion, biological sediments). Seabed sampling can be obtained by means of lead lines, grabs, snappers and scoops, corers, dredges, divers, and remotely operated vehicles and submersibles. *Leadlines* are weights armed with a sticky substance to which particles adhere and on which heavier objects leave an impression. The advantage of the leadline is that it is cheap and simple to operate. The disadvantages are that the heavier materials might not be detected, only the very top surface layer is sampled and the latter is disturbed when collected. *Grabs, snappers, and scoops* are used for collecting medium-size samples from the surface and immediate subsurface layer of the sea floor. They usually comprise a bucket or scoop which is activated on hitting the sea floor. Some are spring loaded; others close when they are raised off the seabed. The disadvantages of the grabs are that they are not suitable for collecting soft or liquid mud, as the sample is often washed out of the bucket before it reaches the surface.

The *Shipek grab* consists of two concentric half cylinders that on striking the sea floor rotate through 180°. During this rotation, the bucket scoops a sample from the seabed and then remains closed while the grab is hauled to the surface. This type of grab is most effective on unconsolidated sediments, but the impact of the grab on the consolidated and compacted seabed can make it bounce and cause only a superficial grab. *Corers* are used to obtain an undisturbed vertical sample of the sea floor. They comprise a tube- or box-shaped cutting mechanism similar to an apple corer or pastry cutter. They are driven into the seafloor and when withdrawn retain the undisturbed sample of the sediment layers by the means of different methods such as vacuum suction and shutters. Corers can be drawn in the seafloor by means of their own weight, explosives, pneumatics, or mechanical vibration. Cores contain liners which allow the sample to be removed and stored with the minimum disturbance.

Dredges are designed to drag along the sea floor collecting loose materials and sediments. They often incorporate a filter which allows smaller sediments to pass through. Samples are obviously disturbed but do reflect the seabed materials over a reasonable large area. *Divers* inspection allows a positive identification of the shallow sea floor. Large as well as small features can be identified.

9.1.10 Bioluminescence

Bioluminescence is the emission of light by living organisms. The process is such that light is radiated but very little heat is emitted. The process is an animal production and approximately 240 groups of organisms have been identified as bioluminescent, such as dinoflagellates, jellyfishes, copepods, euphausiids, squids, and some fish. Bioluminescence is the result of a substance such as luciferin being oxidized in the presence of a catalytic enzyme, the luciferase. Bioluminescence is triggered by unexplained internal actions or external actions such as surface and internal waves; ship, fish, and whale movements; and upwelling. Bioluminescence can be measured and collected visually and recorded in coded form from standard tables. Data collected includes information about stimulus (causes such as light, wave action, rain or fish), color, kind (continuous, patches, bands, blobs, shapes), duration (seconds or continuous), and extent (size of patches or continuous) of the bioluminescence. Various *mesh nets* with preset depth catchers catch larger plankton, while the smallest plankton type must be centrifugated because it is too small for the smallest mesh.

9.2 OCEANOGRAPHIC INSTRUMENTATION

The following table summarizes the range of instrumentation used at sea and its content.

Range of Instrumentation Used at Sea.

Research need	Available equipment / instrumentation
Provision of observing platform	• Research vessels • Moorings • Satellites • Submersibles • Towed vehicles • Floats and drifter
Measurement of hydrographic properties (temperature, salinity, oxygen, nutrients, tracers)	• Reversing thermometers • Water bottles • CTDs • Multiple water sample devices • Thermosalinographs • Remote sensors
Measurement of dynamic properties (currents, waves, sea level, mixing processes)	• current meters • wave measurements • tide gauges • remote sensors • shear probes

Platforms: All measurements at sea require a reasonably stable platform to carry the necessary instrumentation. The platform can be at the sea surface, at the sea floor, in the ocean interior, or in space. The choice of platform depends on its capabilities to collect the required information in space and time.

9.2.1 Research Vessels

Research vessels must be seaworthy and capable of riding out bad weather. The weather conditions in the investigation area thus define the minimum size for the vessel. Additional requirements, such as the handling of heavy equipment at sea, can increase the minimum size. Typical ocean going research vessels are 50–80 meters long, have a total displacement of 1,000–2,000 tonnes, and provide accommodation for 10–20 scientists [Figure 9.4]. The minimum laboratory requirements consist of a wet laboratory for the handling of water samples, a computer laboratory for data processing, an electronics laboratory for the preparation of instruments, and a chemical laboratory for water sample analysis. Larger research vessels designed for multi-disciplinary research have additional biological, geophysical, and geological laboratories. Figure 9.4 shows a typical deck arrangement on a medium sized research vessel.

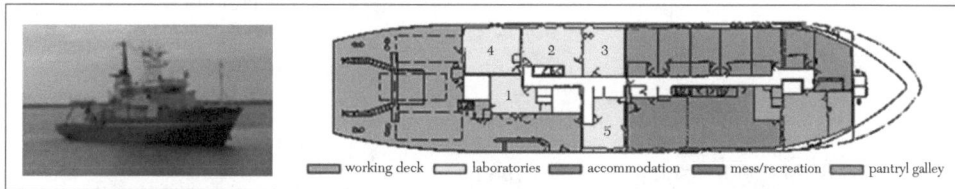

working deck   laboratories   accommodation   mess/recreation   pantryl galley

FIGURE 9.4 Research vessels and deck arrangement.

9.2.2 Moorings

Moorings are appropriate platforms wherever measurements are required at one location over an extended period. Their design depends on water depth and the type of instrumentation for which the mooring is deployed, but the basic elements of an oceanographic mooring are an anchor, a mooring line (wire or rope), and one or more buoyancy elements which hold the mooring upright and preferably as close to vertical as possible. Subsurface moorings are used in deep water in situations where information about the surface layer is not essential to the experiment. The main buoyancy at the top of the mooring line is placed some 20–50 meters below the ocean surface. This has the advantage that the mooring is not exposed to the action of surface waves and is not at risk of being damaged by ship traffic or stolen. Figure 9.5 shows a typical deep sea mooring. The main buoyancy is at the top of the mooring line. To protect the mooring against fish bites, wire is used for the upper 1,000 meters or so of the mooring line, while rope is used below. At the bottom of a deep-sea mooring, just above the anchor, is a remotely controllable release. It can be activated through a coded acoustic

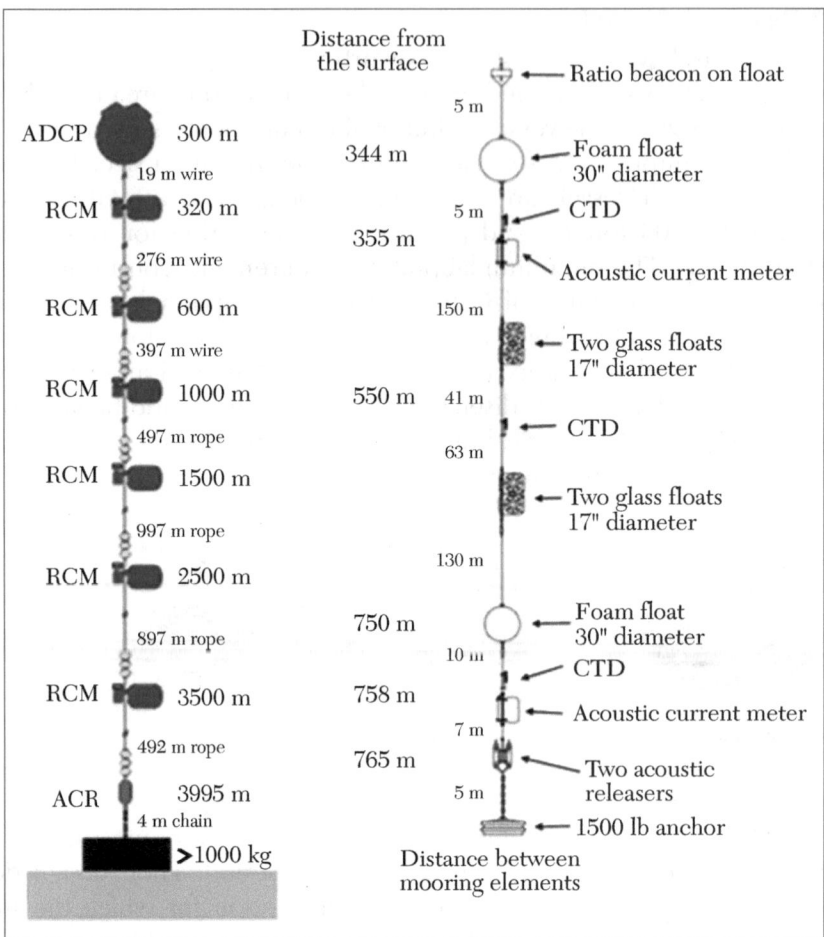

FIGURE 9.5 Moorings.

signal from the ship when it is time to recover the mooring. Triggering the release brings the mooring to the surface. The anchor, usually a concrete block or a clump of disused railway wheels, is left at the ocean floor.

An experiment that includes the surface layer or the collection of meteorological data requires a surface mooring. The main buoyancy for such a mooring takes the shape of a substantial buoy that floats at the surface and can carry meteorological instrumentation as in Figure 9.6. In the deep ocean, surface moorings are mostly *taut moorings*. They use only rope for the mooring line and make it a few percent shorter than the water depth.

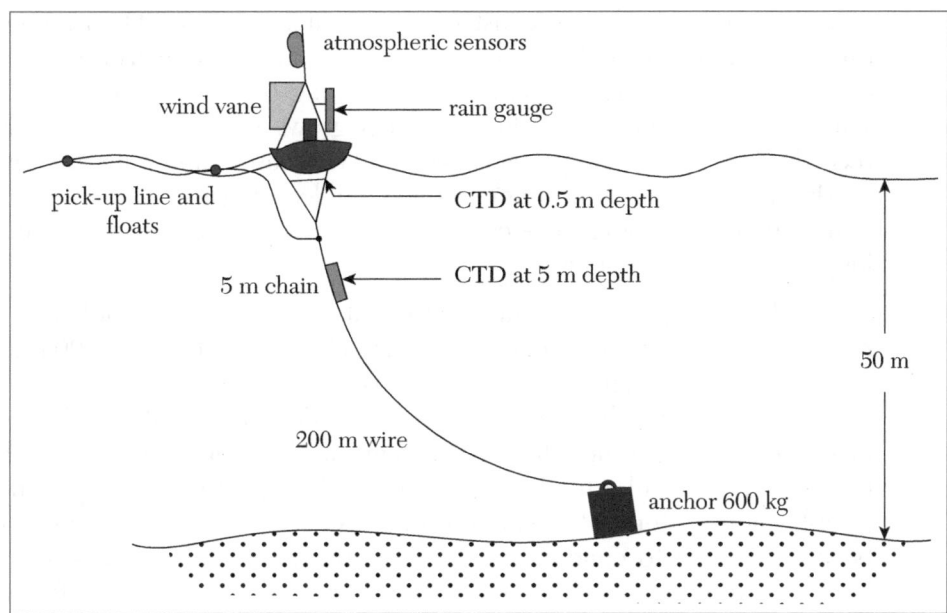

FIGURE 9.6 Surface mooring.

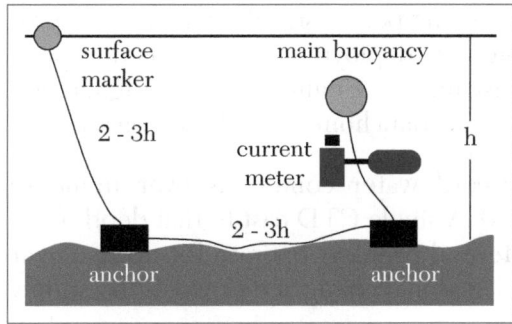

FIGURE 9.7 U-type mooring.

This stretches the rope and keeps it under tension to keep the mooring close to vertical.

Moorings on the Continental Shelf, where the water depth does not exceed 200 meters, do not require acoustic releases if a U-type mooring is used. A *U-type mooring* consists of a surface or subsurface mooring to carry the instrumentation, a ground line of roughly twice the water depth, and a second mooring with a small marker buoy, as in Figure 9.7. When the time comes to bring the mooring in, the marker buoy is recovered first, followed by the two anchors, and finally the mooring itself. U-type moorings are usually *slack moorings*; the mooring line is longer than the water depth, and the mooring swings with the current.

9.2.3 Moored Profiler

A *moored profiler* makes repeated measurements of ocean currents and water properties up and down through almost the entire water column, even in very deep water. The basic instruments it carries are a CTD for temperature and salinity and an ACM (*acoustic current meter*) to measure currents, but other instruments can be added, including bio- optical and chemical sensors. As interest in global climate change grows, so does our need for long records of ocean variability. Data collected over seasons and years are critical for developing ocean atmosphere models and learning about air sea interactions. Moored profilers were conceived as a cost-effective way to conduct long-term ocean sampling.

The moored profiler is attached to a subsurface mooring cable that can run from a depth of 50 meters (165 ft) down to the sea floor at 5,000 meters (3 miles) or more. The profiler uses a battery-powered traction motor to climb up and down the mooring. Its sensors document water and current properties as the profiler climbs or descends. The profiler has enough battery power to travel a total of about one million meters per deployment. (Deployments often last a year at a time.) An onboard microprocessor can be programmed to conduct complex sampling schedules. These include *burst sampling*, in which several profiles are collected in a day followed by several days of no sampling, with the whole pattern repeated through the deployment. Burst sampling minimizes «alias errors,» where unresolved high frequency motions contaminate long period signals. The moored profiler stores its data for the duration of its deployment. Scientists download the data when they recover the instrument. Engineers and designers are looking ahead to the prospect of sending data home in real time via satellite.

Advantages: Moored profilers record water conditions over immense depths, up to 5,000 meters (3 miles). A single CTD cast to that depth from a ship would take many hours. Moored profilers operate for up to a year at a time, so they record daily and seasonal changes that would be simply impossible to collect from a ship.

Limitations: Moored profilers climb up and down subsurface mooring cables, which usually are festooned with instruments measuring many aspects of the water. The profilers need a clear run of cable, so scientists can't place instruments in a profiler's path. To get around this problem, subsurface moorings are often set out in pairs, with a moored profiler on one and other instruments on the sister mooring. As with most subsurface instruments, moored profilers are limited by being isolated from the surface. They must carry their own batteries to power instruments and the motor, and the life of the batteries limits how long the profiler can stay out. Also, the profiler never comes to the sea surface, so it's cut off from radio contact with satellites. Researchers must wait until the deployment is over to see their data.

9.2.4 Satellites

The advent of satellite technology opened the possibility of measuring property fields and dynamic quantities from space. The advantage of this method is the nearly synoptic coverage of entire oceans and ease of access to remote ocean regions. Satellites have therefore become an indispensable tool for climate research. The major restriction of the method is that satellites can only see the surface of the ocean and therefore give only limited information about the ocean interior. Most satellites are named for the sensors they carry. Strictly speaking, the satellite and its sensors are two separate things; the satellite is a platform, the sensors are instruments.

As platforms, satellites fall into three groups. Most satellites follow *inclined orbits*: their elliptical orbits are inclined against the equator. The degree of inclination determines how far away from the equator the satellite can see the Earth. Typical inclinations are close to 60°, so the satellite covers the region from 60°N to 60°S. It covers this region frequently, completing one orbit around the Earth in about 50 minutes. Some satellites have an inclination of nearly or exactly 90° and can therefore see both poles; they fly on *polar orbits*. The typical height of satellites on polar and on inclined orbits is 800 km. The third and last group is the *geostationary* satellites. They orbit the Earth at the same speed the Earth rotates around its axis and are therefore stationary with respect to the Earth. This situation is only possible if the satellite is over the equator and orbits at a height of 35,800 km, much higher than all other satellites. Geostationary satellites therefore cannot see the poles. The selection of a satellite as a platform logically includes the selection of a sensor and a suitable orbit. An ice sensor to monitor the polar ice caps does not achieve much on a geostationary satellite; a cloud imager for weather forecasting is not placed in a polar orbit.

9.2.5 Submersibles

Submersibles are not a frequently used platform in physical oceanography. Three basic types can be distinguished, manned submersibles, remotely controlled submersibles, and autonomous submersibles. *Manned submersibles* are used in marine geology for the exploration of the sea floor and occasionally in marine biology to study sea floor ecosystems. They are not a tool for physical oceanography. *Remotely controlled submersibles* are commonly used in the offshore oil and gas industry and for retrieving flight recorders from aircraft that fell into the ocean. In science they find similar uses to manned submersibles but are again not a tool for physical oceanography. *Autonomous submersibles* are self-propelled vehicles that can be programmed to follow a predetermined diving path. Such vehicles have great potential for physical oceanography. Some major oceanographic research institutions are developing vehicles to carry instrumentation such as a CTD and survey an ocean area by regularly diving and surfacing along a track from one side of an ocean region to the other and transmitting the collected data via satellite when at the surface. It will be some time, however, before these vehicles will come into regular use. Eventually, autonomous submersibles will greatly reduce the need for research vessels for ocean monitoring.

9.2.6 Towed Vehicles

Towed vehicles are used from research vessels to study oceanic processes which require high spatial resolution, such as mixing in fronts and processes in the highly variable upper ocean. Most systems consist of a hydrodynamically shaped underwater body, an electromechanical (often multi-conductor) towing cable, and a winch. The underwater body is fitted with a pair of wing-shaped fins which control its flight through the water. In addition to the sensor package (usually a CTD, sometimes additional sensors for chemical measurements), it carries sensors for pressure, pitch, and roll to monitor its behavior and control its flight. The data are sent to the ship's computer system via the cable. The same cable is used to send commands to the underwater body to alter its wing angle.

Figure 9.8 shows a towed vehicle during deployment. A typical flight path for this vehicle covers a depth range of about 250–500 meters, which can be chosen to be anywhere between the surface and 800 meter depth. The vehicle is towed at about 6–0 knots (10–18 km/h) and traverses the 250 meter depth range about once every 5 minutes. When fitted with a CTD, this results in a vertical section of temperature and salinity with a horizontal resolution of about 1 km. An alternative towed system does not employ an underwater body to carry the sensor package but has sensors (for example, thermistors) built into the

FIGURE 9.8 Towed vehicle.

towed cable at fixed intervals. Because the distance between the sensors is fixed and the sensors remain at the same depth during the tow, these *thermistor chains* do not offer the same vertical data resolution as undulating towed systems and are only rarely used now.

9.2.7 Floats and Drifters

The main characteristic of floats and drifters is that they move freely with the ocean current, so their position at any given time can only be controlled in a very limited way. Until a decade ago, these platforms were mainly used in remote regions such as the Southern Ocean and in the central parts of the large ocean basins that are rarely reached by research vessels and where it is difficult and expensive to deploy a mooring. They have now become the backbone of a new observing system that covers the entire ocean. Strictly speaking, a *float* is a generic term for anything that does not sink to the ocean floor. A *drifter*, on the other hand, is a platform designed to move with the ocean current. To achieve this, it has to incorporate a floatation device or float, but it is usually more than that.

Two basic types can be distinguished. *Surface drifters* have a float at the surface and can therefore transmit data via satellite. If they are designed to collect information about the ocean surface, they carry meteorological instruments on top of the float and a temperature and occasionally a

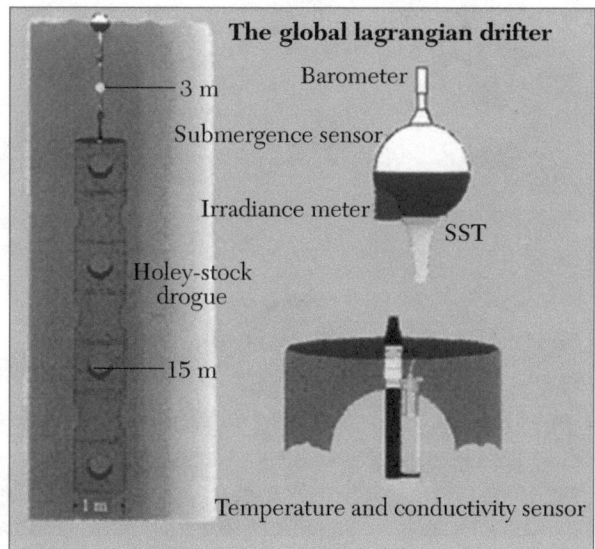

FIGURE 9.9 Drifter.

salinity sensor underneath the float. To prevent them from being blown out of the area of interest by strong winds, they are fitted with a *sea anchor* at some depth, as in Figure 9.9. If they are designed to give information on subsurface ocean properties, additional sensors are placed between the surface float and the sea anchor. The depth range of surface drifters is usually limited to less than 100 meters. Floats used for *subsurface drifters* are designed to be neutrally buoyant at a selected depth. These drifters have been used to follow ocean currents at various depths, from a few hundred meters to below 1,000 meters depth. The first such floats transmitted their data acoustically through the ocean to coastal receiving stations. Sound travels well at the depth of the sound velocity minimum (the SOFAR channel). These SOFAR floats can only be used at about 1,000 meters depth.

Modern subsurface floats remain at depth for a time, come to the surface briefly to transmit their data to a satellite, and return to their allocated depth. These floats can therefore be programmed for any depth and can also obtain temperature and salinity (CTD) data during their ascent. The most comprehensive array of such floats, known as Argo, began in the year 2000. *Argo floats* measure the temperature and salinity of the upper 2,000 meters of the ocean, as in Figure 9.10.

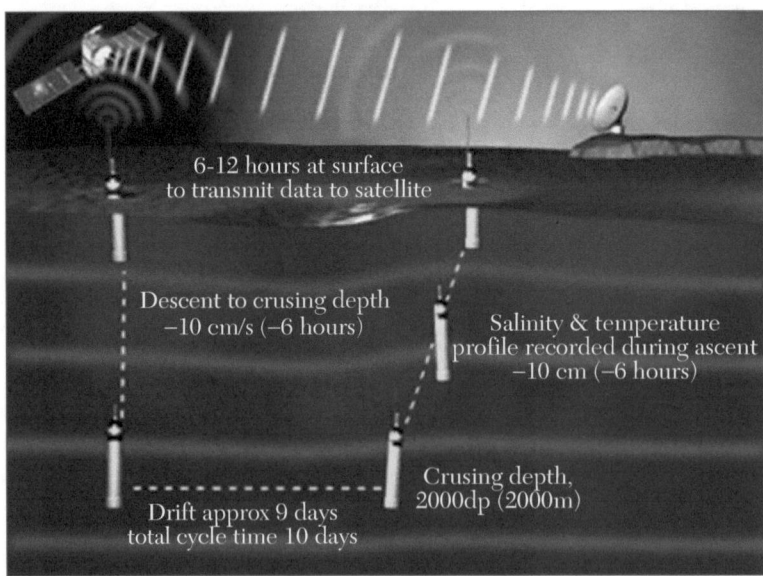

FIGURE 9.10 Operation of an Argo float in "park and profile" mode.

This will allow continuous monitoring of the climate state of the ocean, with all data being relayed and made publicly available within hours after collection.

Argo floats: Robots at sea: Due to the continuous increase of adverse effects of global warming, monitoring drastic variations in the parameters of the sea has become cumbersome. To overcome this difficulty, autonomous Argo float robots are used. Argo floats are a special kind of autonomous robots that remain at sea for a prolonged time, collecting and monitoring various parameters such as temperature, salinity, velocity, and similar properties of seawater. At present there are around 3,000 Argo robots floating in various oceans of the world. These robots make high-quality temperature and salinity profile on the basis of data procured from the upper 2,000-meter ice-free layer of the global ocean currents. The structure of Argo robot is divided into three main parts.

- Hydraulics

- Microprocessors

- Data transmission system

The hydraulics system controls an inflatable internal bladder that makes the robot float or dive. The mechanism consists of a hydraulic piston which operates with the help of a batteries and a small pump. The buoyancy is controlled by the piston moving downward and inflating the hydraulic bladder. The microprocessor controls and administers the functions of the robot. Microprocessors

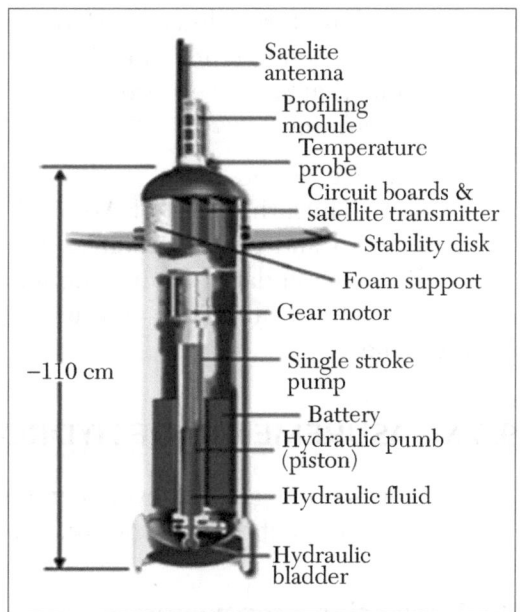

FIGURE 9.11 Schematic of ARGO floats.

are preprogrammed, which makes the system autonomous and self-reliant, though it can be controlled from a remote station. The data transmission system controls the communication functions with the satellite. The whole Argo robot is approximately 25 kg in weight and 110 cm in length, with maximum operating depth of 2,000 meters. Figure 9.11

shows the detailed cut-out diagram on a typical Argo float with the various parts marked clearly.

Based on the operating region and profile requirement, the float descends to the cruising depth and drifts at that level for several days. The buoyancy is maintained at the required depth by being neutrally buoyant, i.e., to adjust the density equal to the ambient pressure and compressibility less than the seawater. Once the data is collected, the robot ascends to the surface by pumping fluid into the external bladder, to make salinity and temperature profiles. This completes one cycle. Floats are designed to make 150 such cycles. All the parameters taken at various depths, such as drift depth, vertical sampling resolution, and timing, are precisely recorded. This data is sent to the satellite when the robot ascends to the surface. Ships sail at sea at the mercy of the elements and hence it is very important not only to know the immediate weather conditions but also to study long-term climate and parameters of the oceans in detail. Argo robots are used to study global changes in ocean parameters. This is how we know that sea levels are rising at a rate of 3 mm/year as a result of global warming.

The ice at the poles is also reducing, leading to extreme weather changes. Thus, the need of Argo robots becomes more valid in order to understand and predict changes in atmosphere and ocean parameters. Lack of such updated environmental data will make us incapable of predicting natural calamities and disasters, leading us to become victims of the same.

9.3 MEASUREMENTS OF HYDROGRAPHIC PROPERTIES

This section gives an overview of sensors and instrument packages for the measurement of temperature, salinity, oxygen, nutrients, and tracers.

9.3.1 Reversing Thermometers

The earliest temperature measurements at some depth below the surface were made by bringing water sample up to the deck of a ship in an insulated bucket and measuring the sample temperature with a mercury thermometer. Although these measurements were not accurate, they gave the

FIGURE 9.12 Reversing thermometers.

first evidence that below the top 1,000 meters, the ocean is cold even in the tropics. They also showed that highly accurate measurements are required to resolve the small temperature differences between different ocean regions at those depths. The first instrument that (through multiple sampling and averaging) achieved the required accuracy of 0.001°C was the reversing thermometer. It consists of a mercury-filled glass pipe with a 360° coil. The pipe is restricted to capillary width in the coil, where it has a capillary appendix [Figure 9.12]. The instrument is lowered to the desired depth. Mercury from a reservoir at the bottom rises in proportion to the outside temperature. When the desired depth is reached, the thermometer is turned upside down (reversed), but the flow of mercury is now interrupted at the capillary appendix, and only the mercury that was above the break point is collected in the lower part of the glass pipe. This part carries a calibrated gradation that allows the temperature to be read when the thermometer is returned to the surface. To eliminate the effect of pressure, which compresses the pipe and causes more mercury to rise above the break point during the lowering of the instrument, the thermometer is enclosed in a pressure-resistant glass housing. If such a *protected reversing thermometer* is used in conjunction with an *unprotected reversing thermometer* (a thermometer exposed to the effect of pressure), the difference between the two temperature readings can be used to determine the pressure and thus the depth at which the readings were taken. The reversing thermometer is thus also an instrument to measure depth. The measurement of salinity and oxygen, nutrients, and tracer concentrations requires the collection of water samples from various depths. This essential task is achieved with *water bottles*. When the bottle is lowered to the desired depth it is open at both ends, so the water flows through it freely. At the depth where the water sample is to be taken the upper end of the bottle disconnects from the wire and the bottle is turned upside down. This closes the end valves and traps the sample, which can then be brought to the surface. In an *oceanographic cast*, several bottles are attached at intervals on a thin wire and lowered into the sea. When the bottles have reached the desired depth, a metal weight (*messenger*) is dropped down the wire to trigger the turning mechanism of the uppermost bottle. The same mechanism releases a new messenger from the bottle; that messenger now travels down the wire to release the second bottle, and so on until the last bottle is reached.

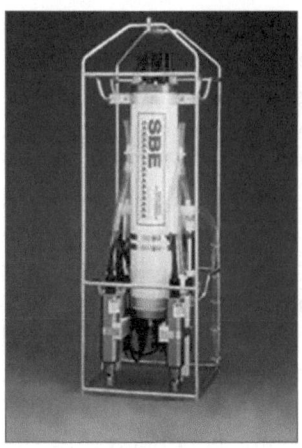

FIGURE 9.13 CTD

9.3.2 CTDs

Today's standard instrument for measuring temperature, salinity, and often also oxygen content is the CTD, which stands for *conductivity, temperature, depth* [Figure 9.13]. It employs the principle of electrical measurement. A platinum thermometer changes its electrical resistance with temperature. If it is incorporated in an electrical oscillator, a change in its resistance produces a change of the oscillator frequency, which can be measured. The conductivity of seawater can be measured in a similar way as a frequency change of a second oscillator, and a pressure change produces a frequency change in a third oscillator. The combined signal is sent up through the single conductor cable on which the CTD is lowered. This produces a continuous reading of temperature and conductivity as functions of depth at a rate of up to 30 samples per second. Electrical circuits allow measurements in quick succession but suffer from "instrumental drift", which means that their calibration changes with time. CTD systems therefore have to be calibrated by comparing their readings regularly against more stable instruments. They are therefore always used in conjunction with reversing thermometers and a multi-sample device.

9.3.3 Multiple Water Sample Devices

Multiple water sample devices allow the use of Niskin (name of scientist who designed bottle) water bottles on electrically conducting wire. They have different names, such as rosette or carousel consists of the Niskin water bottles are arranged on a circular frame as shown in Fig.9.14, with a CTD usually mounted underneath or in the centre. The advantage of multi sample devices over the use of a hydrographic wire with messengers is that the water bottles can be closed by remote control. This means that the sample depths do not have to be set before the bottles are lowered. As the device is lowered and data are received from the CTD, the operator can look for layers of particular interest and take water samples at the most interesting depth levels.

FIGURE 9.14 Multiple water sample device

9.3.4 Thermosalinographs

The introduction of the CTD opened the possibility of taking continuous readings of temperature and salinity at the surface. Water from the cooling water intake of the ship's engines is pumped through a tank in which a temperature and a conductivity sensor are installed. Such a system is called a thermosalinograph. The schematic is shown in Figure 9.15.

9.3.5 Remote Sensors

Most oceanographic measurements from space or aircraft are based on the use of *radiometers*, instruments that measure the electromagnetic energy radiating from a surface. This radiation occurs over a wide range

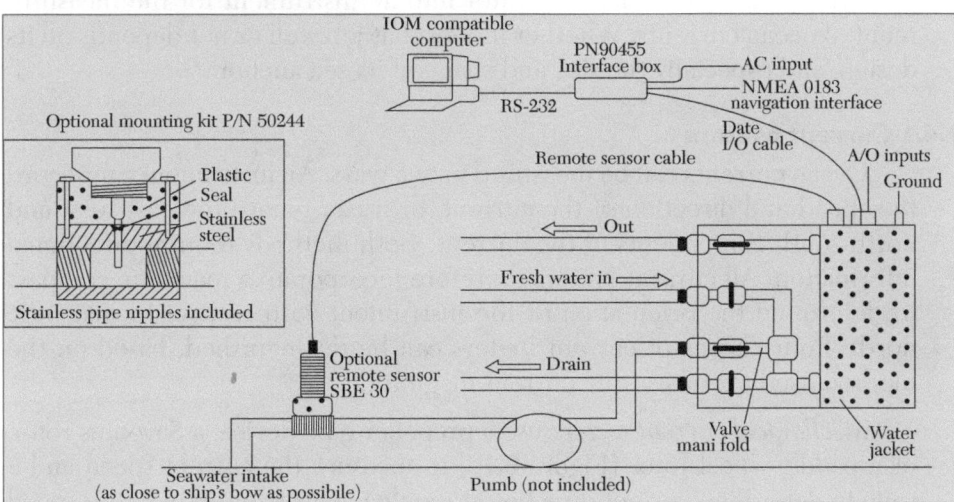

FIGURE 9.15 Schematic of thermosalinograph's operation.

of wavelengths, including the emission of light in the visible range, of heat in the infrared range, and at shorter wavelengths such as radar and X rays. Most oceanographic radiometers operate in several wavelength bands. Radiometers that operate in the infrared are used to measure sea surface temperature. Their resolution has steadily increased over the years; the AVHRR (*advanced very high resolution radiometer*) has a resolution that comes close to 0.2°C. Multi-spectral radiometers measure in several wavelength bands. By comparing the radiation signal received at different wavelengths it is possible to measure ice coverage and ice age, chlorophyll content, sediment load, particulate matter, and other quantities of interest to marine biology. Measurements at radar wavelengths are made by an instrument known as SAR (*synthetic aperture radar*). It can be used to detect surface expressions of internal waves, the effect of rainfall on surface waves, the effect of bottom topography on currents and waves, and a range of other phenomena. Many of these phenomena belong into the category of *dynamic properties*.

9.4 MEASUREMENTS OF DYNAMIC PROPERTIES

All the instruments summarized below are designed to measure movement in the ocean. An elementary way of observing oceanic movement is the use of drifters. Drifters are platforms designed to carry instruments. But all measurements obtained from drifters are of little use unless they can be related to positions in space. A geolocation (GPS) device which transmits the drifter location to a satellite link is therefore an essential instrument on any drifter, and this turns the drifter into an instrument for the measurement of ocean currents. Whether it does that job well or not depends on its design, and especially the size and shape of its sea anchor.

9.4.1 Current Meters

Ocean currents can be measured in two ways. An instrument can record the speed and direction of the current, or it can record the east-west and north-south components of the current. Both methods require directional information. All current meters therefore incorporate a magnetic compass to determine the orientation of the instrument with respect to magnetic north. Four classes of current meters can be distinguished, based on the method used for measuring current magnitude.

Mechanical current meters use a propeller type device, a Savonius rotor, or a paddle-wheel rotor [Figure 9.16] to measure the current speed and a vane to determine current direction. Propeller sensors often measure speed correctly only if they point into the current and have to be oriented to face the

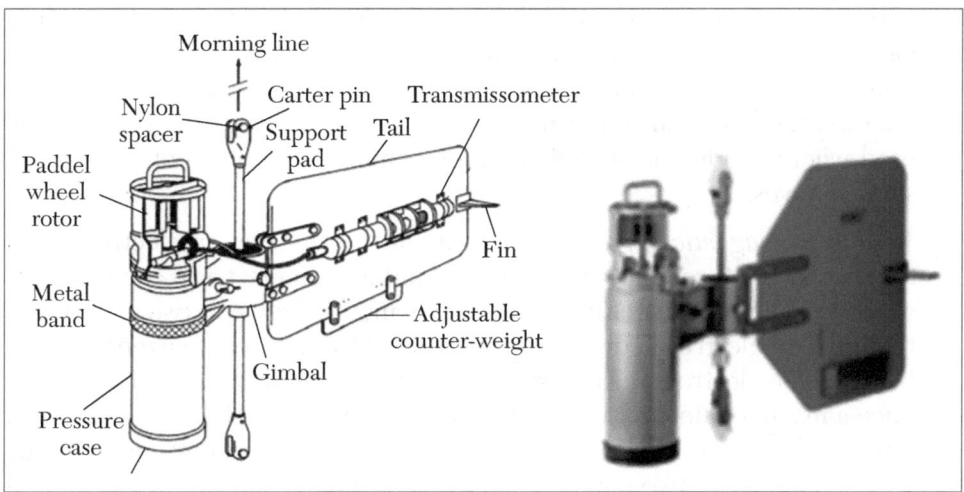

FIGURE 9.16 Mechanical current meters.

current all the time. Such instruments are therefore fitted with a large vane, which turns the entire instrument and with it the propeller into the current.

Propellers can be designed to have a cosine response with the angle of incidence of the flow. Two such propellers arranged at 90° will resolve current vectors and do not require an orienting vane. The advantage of the Savonius rotor is that its rotation rate is independent of the direction of exposure to the current. A Savonius rotor current meter therefore does not have to face the current in any particular way, and its vane can rotate independently and be quite small, just large enough to follow the current direction reliably. With the exception of the current meter that uses two propellers with cosine response set at 90° to each other, mechanical current meters measure current speed by counting propeller or rotor revolutions per unit time and current direction by determining the vane orientation at fixed intervals. In other words, these current meters combine a time integral or mean speed over a set time interval (the number of revolutions between recordings) with an instantaneous reading of current direction (the vane orientation at the time of recording). This gives a reliable recording of the ocean current only if the current changes slowly in time. Such mechanical current meters are therefore not suitable for current measurement in the oceanic surface layer, where most of the oceanic movement is due to waves. The Savonius rotor is particularly problematic in this regard. Suppose that the current meter is in a situation where the only water movement is from waves. The current then alternates back and forth, but the mean current is zero. A Savonius rotor will pick up the wave current irrespective of its direction, and the rotation count will give the impression of a strong mean current. The paddle wheel rotor is designed to rectify this; the paddle wheel rotates back and forth with the wave current, so that its count represents the true mean current [Figure 9.16]. Mechanical current meters are robust, reliable, and comparatively low in cost. They are therefore widely used where conditions are suitable, for example at depths out of reach of surface waves.

Electromagnetic current meters exploit the fact that an electrical conductor moving through a magnetic field induces an electrical current. Seawater is a very good conductor and if it is moved between two electrodes, the induced electrical current is proportional to the ocean current velocity between the electrodes. An electromagnetic current meter has a coil to produce a magnetic field and two sets of electrodes, set at right angles to each other, and determines the rate at which the water passes between both sets. By combining the two components the instrument determines speed and direction of the ocean current.

Acoustic current meters are based on the principle that sound is a compression wave that travels with the medium. Assume an arrangement with a sound transmitter between two receivers in an ocean current. Let receiver A be located upstream from the transmitter, and let receiver B located downstream. If a burst of sound is generated at the transmitter, it will arrive at receiver B earlier than at receiver A, having been carried by the ocean current. A typical acoustic current meter will have two orthogonal sound paths of approximately 100 mm length, with a receiver/transmitter at each end. A high-frequency sound pulse is transmitted simultaneously from each transducer and the difference in arrival time for the sound travelling in opposite directions gives the water velocity along the path. Electromagnetic and acoustic current meters have no moving parts and can therefore take measurements at a very high sampling rate (up to tens of readings per second). This makes them useful not only for the measurement of ocean currents but also for wave current and turbulence measurements.

Acoustic Doppler current profilers (ADCPs) operate on the same principle as acoustic current meters but have transmitter and receiver in one unit and use reflections of the sound wave from drifting particles for the measurement. Seawater always contains a multitude of small suspended particles and other solid matter that may not be visible to the naked eye but reflects sound. If sound is transmitted in four inclined beams at right angles to each other, the Doppler frequency shift of the reflected sound gives

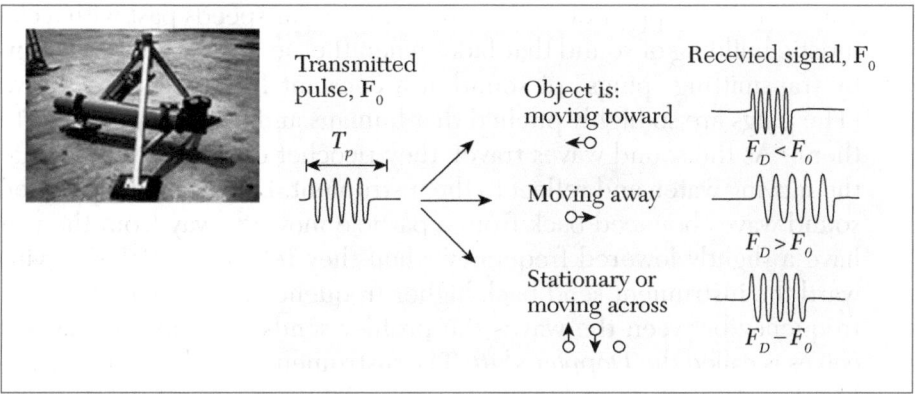

Transmitted
pulse, F_0

T_p

Object is:
moving toward

Recevied signal, F_0

$F_D < F_0$

Moving away

$F_D > F_0$

Stationary or
moving across

$F_D - F_0$

FIGURE 9.17 ADCP.

the reflecting particle velocity along the beam. With at least three beams inclined to the vertical, the three components of flow velocity can be determined. Different arrival times indicate sound reflected at different distances from the transducers, so an ADCP provides information on current speed and direction not just at one point in the ocean but for a certain depth range; in other words, an ADCP produces a current profile over depth.

Different ADCP designs serve different purposes, as shown in Figure 9.17. Deep ocean ADCPs have a vertical resolution of typically 8 meters (they produce one current measurement for every 8 meters of depth increase) and a typical range of up to 400 meters. ADCPs designed for measurements in shallow water typically have a resolution of 0.5 meters and a range of up to 30 meters. ADCPs can be placed in moorings, installed in ships for underway measurements, or lowered with a CTD and multi-sample device to give a current profile over a large depth range. An *acoustic Doppler current profiler*, or *acoustic Doppler profiler* (ADCP) is used to measure how fast water is moving across an entire water column. An ADCP anchored to the seafloor can measure current speed not just at the bottom, but also at equal intervals all the way up to the surface. The instrument can also be mounted horizontally on seawalls or bridge pilings in rivers and canals to measure the current profile from shore to shore, and to the bottoms of ships to take constant current measurements as the boats move. In very deep areas, they can be lowered on a cable from the surface.

Operation: The ADCP measures water currents with sound, using a principle of sound waves called the Doppler effect. A sound wave has a higher frequency, or pitch, when it moves to you than when it moves away.

You hear the Doppler effect in action when a car speeds past with a characteristic building of sound that fades when the car passes. The ADCP works by transmitting "pings" of sound at a constant frequency into the water. (The pings are so highly pitched that humans and even dolphins can't hear them.) As the sound waves travel, they ricochet off particles suspended in the moving water, and reflect to the instrument. Due to the Doppler effect, sound waves bounced back from a particle moving away from the profiler have a slightly lowered frequency when they return. Particles moving toward the instrument send back higher frequency waves. The difference in frequency between the waves the profiler sends out and the waves it receives is called the *Doppler shift*. The instrument uses this shift to calculate how fast the particle and the water around it are moving. Sound waves that hit particles far from the profiler take longer to come back than waves that strike close by. By measuring the time it takes for the waves to bounce back and the Doppler shift, the profiler can measure current speed at many different depths with each series of pings.

Platform requirements: Bottom-mounted ADCPs need an anchor to keep them on the bottom, batteries, and an internal data logger. Vessel-mounted instruments need a vessel with power, a shipboard computer to receive the data, and a GPS navigation system (so the ship's own movements can be subtracted from the current data). ADCPs have no external read-out, so the data must be stored and manipulated on a computer. Software programs designed to work with ADCP data are available.

Advantages:

1. In the past, measuring the current depth profile required the use of long strings of current meters. This is no longer needed.

2. Measures small scale currents.

3. Unlike previous technology, ADCPs measure the absolute speed of the water, not just how fast one water mass is moving in relation to another.

4. Measures a water column up to 1,000 meters long.

Disadvantages:

1. High frequency pings yield more precise data, but low frequency pings travel farther in the water, so scientists must make a compromise between the distance that the profiler can measure and the precision of the measurements.

2. ADCPs set to «ping» rapidly also run out of batteries rapidly.

3. If the water is very clear, as in the tropics, the pings may not hit enough particles to produce reliable data.

4. Bubbles in turbulent water or schools of swimming marine life can cause the instrument to miscalculate the current.

5. Users must take precautions to keep barnacles and algae from growing on the transducers.

9.4.2 Wave Measurements

The parameters of interest in the measurement of surface waves are wave height, wave period, and wave direction. At locations near the shore, wave height and wave period can be measured by using the principle of the stilling well described for tide gauges (below), with an opening wide enough to let pass surface waves unhindered. Wave measurements on the shelf but at some distance from the shore can be obtained from a pressure gauge. An instrument suitable for all locations including the open ocean is the *wave rider*, a small surface buoy on a mooring which follows the wave motion. A vertical accelerometer built into the wave rider measures the buoy's acceleration generated by the waves. The data are either stored internally for later retrieval or transmitted to shore. Wave riders provide information on wave height and wave period. If they are fitted with a set of three orthogonal accelerometers they also record wave direction.

9.4.3 Tide Gauges

Tides are waves of long wavelength and known period, so the major properties of interest for measurement are wave height, or tidal range, and wave-induced current. The latter is measured with current meters; any type is suitable. Two types of tide gauges are used to measure the tidal range. The *stilling well gauge* consists of a cylinder with a connection to the sea at the bottom. This connection acts as a low-pass filter: it is so restricted that the backward and forward motion of the water associated with wind waves and other waves of short period cannot pass through; only the slow change of water level associated with the tide can enter the well. This change of water level is picked up by a float and recorded as in Figure 9.18. Stilling

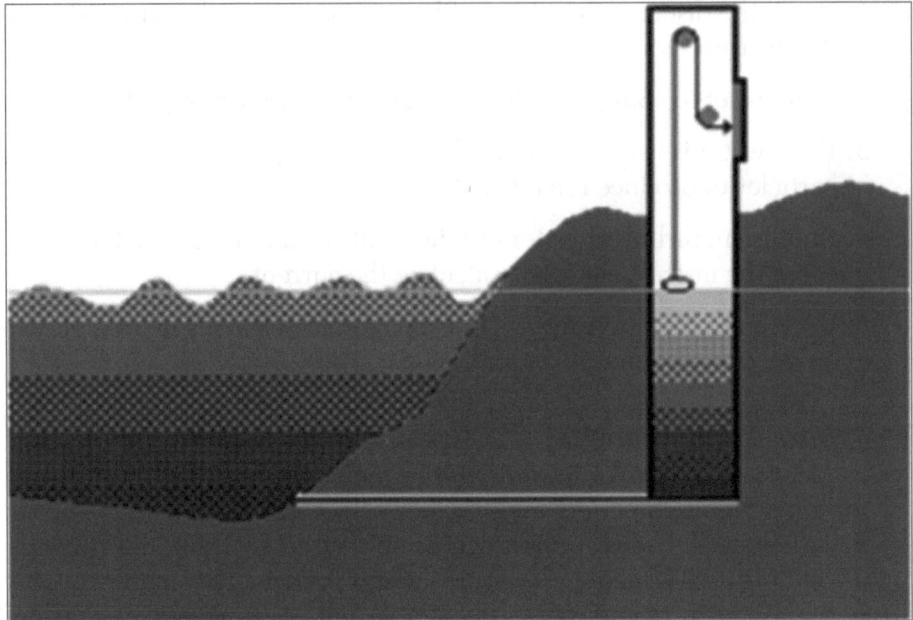

FIGURE 9.18 Tide gauges.

well gauges allow the direct reading of the water level at any time but require a somewhat laborious installation and are impracticable away from the shore. In offshore and remote locations, it is often easier to use a pressure gauge. Such an instrument is placed on the sea floor and measures the pressure of the water column above it, which is proportional to the height of water above it. The data are recorded internally and not accessible until the gauge is recovered.

Tide gauges are increasingly used to monitor possible long-term changes in sea level linked with climate variability and climate change. The expected rate of sea level change is only a few millimeters per year at most, so very high accuracy is required to verify such changes. Most tide gauges are not suitable for such a task, for a number of reasons. For example, a long-term trend in observed sea level can also be produced by a rise or fall of the land on which the tide gauge is built. This is known as *benchmark drift*. The wire in a stilling well gauge that connects the float with the recording unit stretches and shrinks as the air temperature rises and falls. Such effects are insignificant when the gauge is used to verify the depth of water for shipping purposes but not when it comes to assessing trends of millimeters per year. A new generation of tide gauges is being installed worldwide which gives water level recordings to absolute accuracies of a few millimeters with long-term benchmark stability. In these instruments the float and wire arrangement of the stilling well gauge is replaced by a laser distance measurement, and the data are transmitted via satellite to a world sea level center which monitors the performance of every gauge continuously.

9.4.4 Remote Sensors

Sea level can also be measured from satellites. An *altimeter* measures the distance between the satellite and the sea surface. If the satellite position is accurately known, this results in a sea level measurement. Modern altimeters have reached an accuracy of better than 5 cm. The global coverage provided by satellites allows the verification of global tide models. When the tides are subtracted, the measurements give information about the shape of the sea surface and, through application of the principle of *geostrophy*, the large-scale oceanic circulation.

9.4.5 Shear Probes

This extremely brief overview of oceanographic measurement techniques can only cover the essentials of the most important platforms and instruments. Special equipment exists, and new special equipment is being designed every day, to address specific problems. The *shear probe* may serve as an example. It is designed to give insight into oceanic turbulence at the centimeter scale. Turbulence is characterized by currents which vary over short distances and short time intervals, so an instrument designed to measure turbulence has to be able to resolve differences in current speed and direction over a vertical distance of not more than a meter or so. One such shear probe is a cylindrical instrument of less than 1 meter in length with two electromagnetic or acoustic current meters at its two ends. By measuring current speed and direction at two points less than 1 meter apart, it allows the determination of the current shear over that distance. To allow a reliable measurement not influenced by the heaving motion of the ship, the probe falls slowly and freely through the ocean. Its maximum diving depth is programmed before the experiment, and the probe returns to the surface when that depth is reached. It is then picked up by the ship, and the internally recorded data are retrieved. Another type of free fall instrument uses microstructure sensors that measure velocity fluctuations on a special scale of about 10 mm. It uses a piezo-electric beam that generates small voltages as the turbulent velocity varies the lift and thus the bending force on an aerofoil as it moves through the water.

9.4.6 Ferry Box

The *ferry box* is an automated measuring system used for the measurement of physical and biogeochemical parameters in surface waters. It is mounted on ships of opportunity, such as ferries or container ships, on their regular routes across the North Sea or on shore-based installations. Water is pumped from a subsurface inlet into the measuring circuit of multiple sensors. An important feature is the regular automated cleaning and antifouling procedure of the box. All processes can be controlled remotely via the telemetry line from land. Data are transmitted and made available after each transect via the Internet. Due to the automated biogeochemical instrumentation of the ferry boxes, the regular transects give detailed insights into the processes and are a main source for data assimilation into models. The detailed block diagram is shown in Figure 9.19.

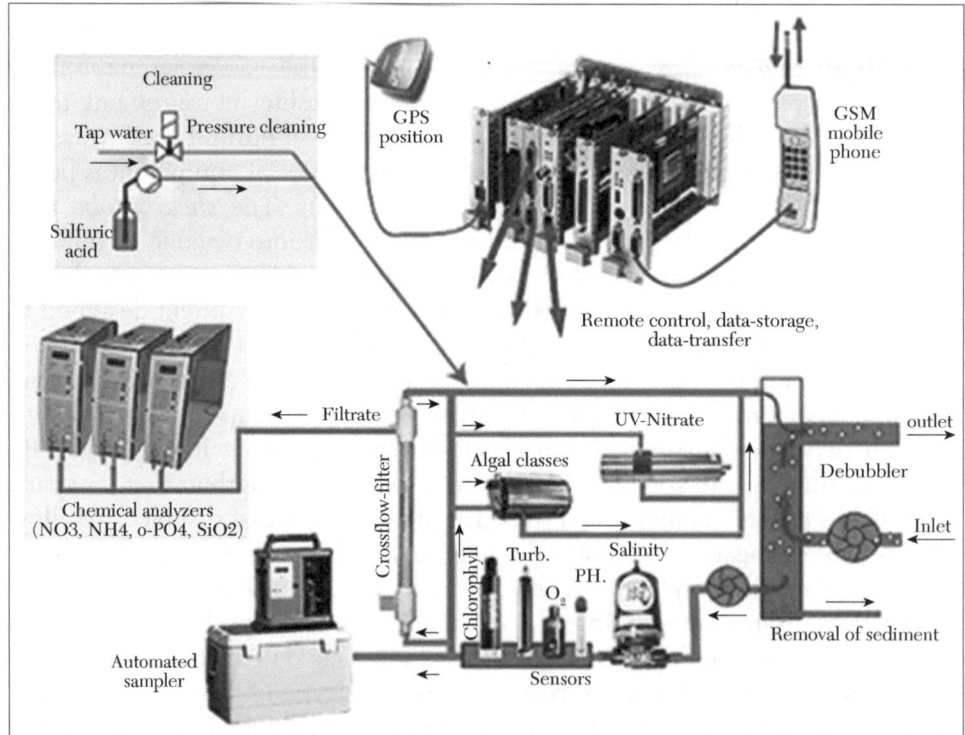

FIGURE 9.19 Ferry box.

9.4.7 Glider

The *underwater glider*, or glider in short, is an autonomous underwater vehicle. It is buoyancy-driven and extremely energy efficient. The glider is a relatively new measurement platform, originally developed as a low-cost, long-endurance device for observing the oceans. Being a host for a multitude of sensors, such as CTD, optical backscatter, and fluorescence sensors, gliders can be programmed to follow set way points. Depending on the sensor payload and battery type, glider missions may last from a couple of weeks to a couple of months. During the mission, the glider pilot can adjust sampling strategies, way points, etc. from behind his/her desk using a satellite link to communicate with the glider. Concurrently, while in operation, (a subset of the) data can be sent to a land-based server via a satellite link, providing scientists with near real-time data. In contrast to conventional underwater vehicles, the glider has no propeller. Instead, it has a buoyancy engine, which allows the glider to control its buoyancy by +/- 250 g. Consequently, the glider can attain a vertical motion up or down, depending on its buoyancy. A pair of wings converts a part of the vertical velocity into forward motion, so that the glider profiles the water column in a sawtooth manner. Its low cruising speed of about 40 cm/s makes the glider extremely efficient and it operates at about 1–2 Watts, giving the glider the unique endurance.

9.5 RADAR DOPPLER CURRENT PROFILER (RDCP)

Two radar systems operating in the high frequency (HF band) and microwaves (X band) regime are used to probe the sea surface. The X band combines local spatial coverage with high resolution, whereas the HF band provides broad coverage at the expense of resolution. Ground-based radars are used to obtain synoptic maps of hydrographic parameters, particularly waves, currents, and local water depth (bathymetry). Maps of surface currents yield ground truth for satellite remote sensing and calibration data for hydrodynamic models. Radar remote sensing of the sea surface provides a broad variety of observations. While single radar images give a kind of instantaneously «frozen» surface, the surface dynamics can be observed by tracking features in subsequent radar measurements. In contrast to the satellite systems, the incident angle of ground-based radars is only a few degrees, and near to zero (grazing) at far ranges. In this case, the working range of the radar strongly depends on the operating frequency. Due to their continuous measurement capabilities, the information provided by ground-based instruments installed at the coast or on offshore platforms does not suffer from the episodic character of the satellite or airborne radar observations.

9.5.1 X Band Radar

Microwave (X band) radars are limited by line of sight propagation to the horizon. Due to the large available bandwidth in the microwave frequency range, the spatial resolution can be as fine as 5 meters at a 5-km working range. By installing microwave radar on board a ship, the observation area can be extended along the ship track to a regional scale. X band wave radar ($\approx$ ship radar) can be used with special electronics and sophisticated evaluation algorithms for the measurement of waves, currents, and bathymetry. It uses an A/D converter, a PC, and processing software connected to marine X band radar of 9.41 GHz. The system is applied from shore and from ship stations. The bathymetric maps in Figure 9.20 show the seafloor relief, calculated from X band radar signals, before and after a severe storm passed. Within five days a volume of 50.000 m³sand (error about 25%) was transported into the observation area.

9.5.2 HF Radar

Decameter wave radars, also known as high frequency (HF) radars, make use of ground wave propagation far beyond the horizon. HF radars provide about 1.5 km up to 200 km working range. Surface currents are hard to measure by conventional means. The unique advantage of the HF radar

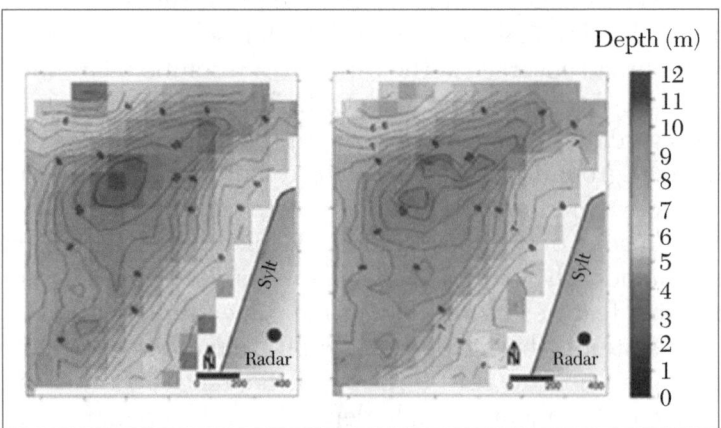

FIGURE 9.20 X band radar observations of bathymetry.

is the ability to map the horizontal variability of currents which is needed for several applications. Eddy dynamics, such as propagation and decay, can be studied, as well as the spatial variability of tidal currents. Maps of surface currents yield ground truth for satellite remote sensing and calibration data for hydrodynamic models.

9.6 SATELLITE REMOTE SENSING

Remote sensing is a unique technique used to observe large areas of ocean and land surface simultaneously. It is possible to measure concentrations of chlorophyll, suspended matter, and yellow substance (also referred to as CDOM) in the visible light spectrum. The algorithms for the open, blue ocean are well established, whereas for coastal regions with highly variable waters they are still the subject of research. In addition, *sea surface temperature* (SST) data from other satellites are used. Figure 9.21 shows the chlorophyll concentrations in sea measured by a satellite.

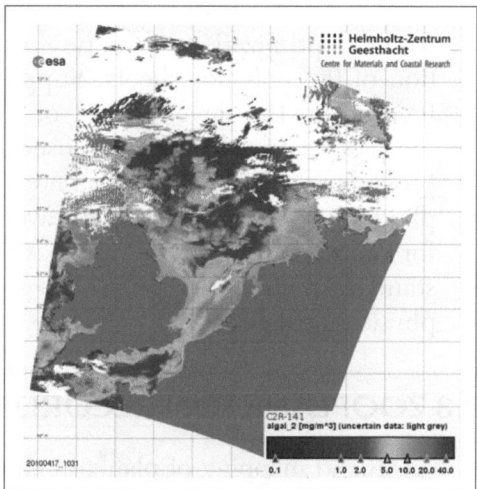

FIGURE 9.21 Chlorophyll concentrations in the sea measured by a satellite.

9.7 UNDERWATER NODES

Within recent years, *underwater nodes* have gained in importance as interfaces for underwater observing networks. They provide the necessary infrastructure (power and data communication) to operate sensors and complex devices at the sea bottom. The objective is to deploy autonomous systems that are deployed in different parts of sea areas and allow a flexible and modular coupling of different near-bottom measurement systems. Thus, a network of long-term operated underwater observatories can be established to investigate processes at the sediment water interface in a sufficient spatial and temporal resolution independent from ship cruises. The main challenge in developing an underwater node is to provide power and broadband data communication to many instruments from different users in a reliable manner: Even if an instrument produces a short circuit by penetrating water the other instruments shall not be influenced. All data will be transported to the user's desk in a transparent way with 100 Mbit/s and each user can control his/her individual instrument by Internet. The first steps were planning and realizing an underwater sampling platform. The newly developed sampling infrastructure was equipped with passive as well as active samplers and the robustness of the installation under field conditions was tested. The implementation of passive samplers started with an optimization of the deployment strategy in terms of size and fixation of the polymer sheets. Supplementary laboratory experiments were focused on the development of cleaning and extraction procedures for the passive sampler material used and on the optimization of the spiking procedure necessary to calculate the sampling rates of the passive sampling devices. Active samplers were deployed and continuously sampled for many months. Laboratory experiments were performed to establish standard procedures for sample preparation and measurements of basic physiological data.

9.8 ZOOPLANKTON RECORDER

Rapid mapping of plankton abundance in combination with taxonomic and size composition will be enabled by the zooplankton recorder [Figure 9.22]. It can image minute objects of sizes below 100 μm with high resolution. Modular components allow the use as devices towed from research vessels, as a component of the ferry box or as a component on unmanned platforms, e.g., piles or underwater nodes. Images of dominant plankton groups can

FIGURE 9.22 Zooplankton recorder.

be classified and objects can be assigned to their respective environmental parameters: depth, temperature, salinity, and oxygen concentration.

9.8.1 Nucleic Acid Biosensor

The surveillance of marine phytoplankton will be greatly facilitated with nucleic acid biosensors. The core of the biosensor is a multi-probe chip that can be used for the simultaneous detection of a variety of algae. A molecular probe as a detection component specifically binds to the target of interest. In turn, an antibody enzyme complex coupled to the signal moiety transforms this detection event into a redox reaction that can be measured as an electrochemical signal, as shown in Figure 9.23. This technique allows rapid detection and counting of microalgae in complex samples. The main steps are automatically carried out in a portable device. Whereas the detection principle has already been verified, the main challenge is to construct a device that reliably filters sea water to concentrate algae cells and "cracks" the cells and transports the resulting fluid to the detector.

9.8.2 Sensors for pH and Alkalinity

Autonomous sensors are required for a comprehensive documentation and characterization of the changes in the marine carbon system. To describe the carbon system, two out of four system parameters must be determined. Alkalinity and pH value are two of these. The pH analysis system adds an acid base indicator dye (m-cresol purple) to a seawater sample. The indicator dye has different extinction coefficients in its acid and in its base state which can be used for spectrophotometric determination of the pH value. If the temperature of the sample is carefully controlled, the precision reaches

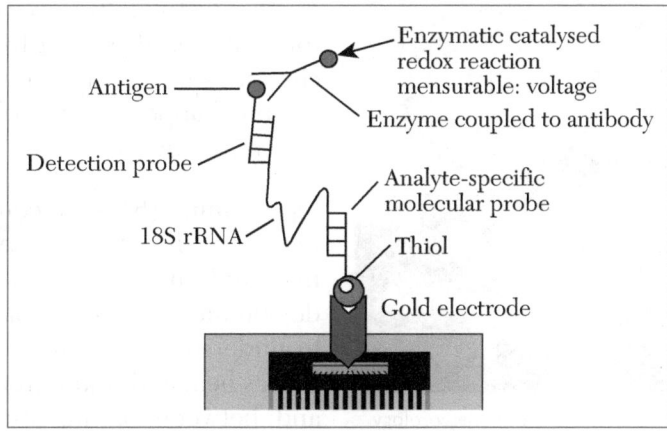

FIGURE 9.23 Principle of a nucleic acid biosensor.

±0.0007 pH units. The accuracy of the determination lies within ±0.01 pH units compared to certified reference material. The advantage of this method is that no drift occurs, hence no calibration of the indicator is needed in the field. Only temperature and salinity must be known accurately. For determination of total alkalinity, a tracer monitored titration with a strong acid (HCl) is performed. Bromocresol green is used as tracer, having similar optical properties such as m-cresol purple. The tracer gives the ability to optically measure the concentration of the indicator and therewith the concentration of the acid. In addition, the pH value can be calculated during the titration procedure which is needed for determination of the alkalinity. Application of these sensors allows monitoring of seawater pH and alkalinity in autonomous underway systems in a high spatial and temporal resolution.

9.9 AIR SEA INTERACTION METEOROLOGY (ASIMET)

The ASIMET (*air sea interaction meteorology*) system is a set of seven very precise sensors that measure how energy and water move between the ocean and atmosphere, shown in Figure 9.24. Those measurements are the raw materials for many calculations about climatic conditions. When climatologists build models of the Earth's climate, they can use ASIMET measurements to check their predictions against reality. The system is used on ships and on buoys anchored in fixed locations in the ocean. In addition to the sensors, the system includes data recorders and a central data processor that sends information back to scientists by satellite. Climatologists are interested in understanding the long-term weather patterns that characterize regions of the Earth or periods of the Earth's history. For example, weather forecasts predict rainstorms, whereas climate studies explain why monsoon seasons happen. Meteorologists explain how hurricanes form.

FIGURE 9.24 Air-Sea Interaction Meteorology.

Operation: ASIMET consists of seven modular sensors of two sets that can be mounted on ocean buoys or ships. Understanding Earth's climate boils down to understanding how the Sun's energy moves heat and water around the planet and between ocean, atmosphere, and land. For oceanographers to balance the checkbook, they need to follow at least three types of exchange, or flux, between air and sea:

1. Heat flux, the movement of energy by two kinds of radiation, shortwave (light), and longwave (heat); plus direct contact between ocean and air (sensible heat); and evaporation or condensation (latent heat)

2. Water flux, the balance between evaporation from the sea surface and precipitation back into the ocean

3. Momentum flux, the transfer of energy from the wind physically pushing against the water

Each must be calculated using precise values for several variables (about a dozen total variables for all three fluxes). The ASIMET system provides those variables at the necessary precision using seven sensors: 1) barometric pressure, 2) relative humidity and air temperature, 3) sea surface temperature and salinity, 4) long-wave radiation, 5) short-wave radiation, 6) wind speed and direction, and 7) precipitation. The seven sensors are individually housed in weatherproof titanium canisters, so they all look fairly similar. The canisters are about 56 cm (22 in) tall and 9 cm (3.5 in) across. Only one sensor (sea surface temperature and salinity) goes in the water; the other six are mounted atop a buoy or high on a ship's bow mast. ASIMET deployments carry two full systems of sensors, data handlers, satellite antennas, and batteries. The duplication ensures there is a backup if one sensor goes wrong and helps with calibrating the data as it comes in. The sensors record data once per minute onto a central data logger. Through the course of their deployment, they also send hourly averages back to researchers via satellites.

Advantages:

1. Reliability: Each deployment uses two sets of sensors that are calibrated both before and after their year at sea. In addition to transmitting home hourly averages, the system collects all the data for the year on flash memory cards. These precautions mean that there are very few gaps in the data record.

2. Mobility: The sensors are modular, making them versatile enough to work on a variety of buoys and ships. If a sensor malfunctions, technicians can replace it without having to dismantle the entire ASIMET system. Individual sensors can be used separately on research projects that don't require the full ASIMET system.

Limitations: Most of the ASIMET system's limitations are imposed by the demanding ocean environment. This is particularly true on buoys, where systems must work perfectly for a year amid crashing waves, crusty salt spray, and seabird poop. ASIMET systems work well in the tropics and subtropics; at higher latitudes the instruments are prone to error from low sun angles, severe weather, and low temperatures. Radiation sensors work best when the Sun is directly overhead. During rough seas, that's only part of the time. Motion sensors on the buoy help track how it rides the swells, but corrections are complicated. Measuring wind direction is also complicated by the way the buoy swings to follow the wind. In light winds, temperature readings can be inflated, and this can bias humidity readings. Fixing these issues, or correcting for them, keeps ASIMET engineers busy between deployments. At present, the ASIMET system does not measure carbon dioxide, a key ingredient in greenhouse warming. Sensors are under development and may be in use on ASIMET systems in the next three to five years.

9.10 BIOMAPER

BIOMAPER is a set of sensors on a long aluminum frame. A research vessel tows the instrument through the water on a specialized tow cable that sends power to the sensors and brings data back to the ship. People use BIOMAPER to learn about phytoplankton and zooplankton over areas that are too large to study with the traditional net and microscope method [Figure 9.25]. Whereas nets can sample areas up to about 5 meters (16 ft) on a side, BIOMAPER can record data from 500 meters (1,640 ft) or more of the water column at a time. The instrument's standard suites of sensors were chosen for studying plankton: a five-frequency sonar system, a video plankton recorder, and an *environmental sensor system* (ESS). The ESS measures water temperature, salinity, oxygen, chlorophyll, and light levels. BIOMAPER is about 3.8 meters (12.5 ft) long, 85 cm (2.8 ft) tall and 55 cm (1.8 ft) wide. It weighs 907 kg (1 ton) in air, or 544 kg (1,200 pounds) in water.

BIOMAPER can be towed at speeds up to 18 km per hour (10 knots). When in "tow-yo" mode as shown in Figure 9.26, the top speed is 11 km per hour (6 knots) and the instrument can climb or fall at up to 10 meters (33 feet) per minute. The sensors produce a combined 1 GB of data per hour. BIOMAPER can operate indefinitely as long as there is power, data storage space, and a team of scientists to watch over the instrument panels. The standard sensor package requires 500 Watts of power, and the tow

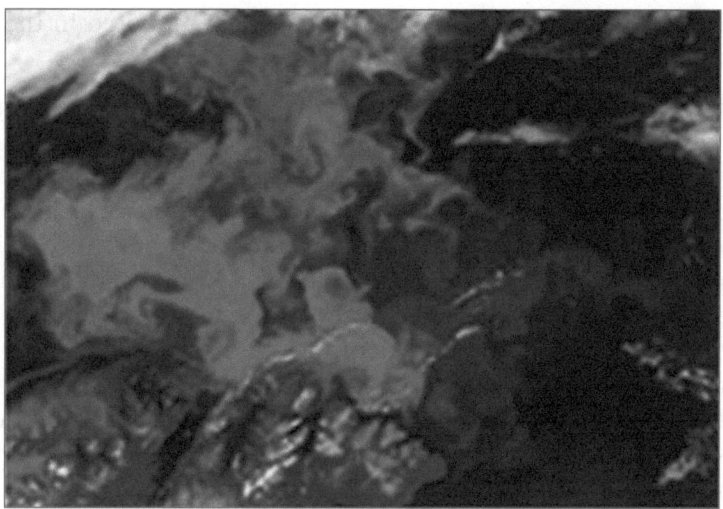

FIGURE 9.25 BIOMAPER for studying huge plankton bloom swirls.

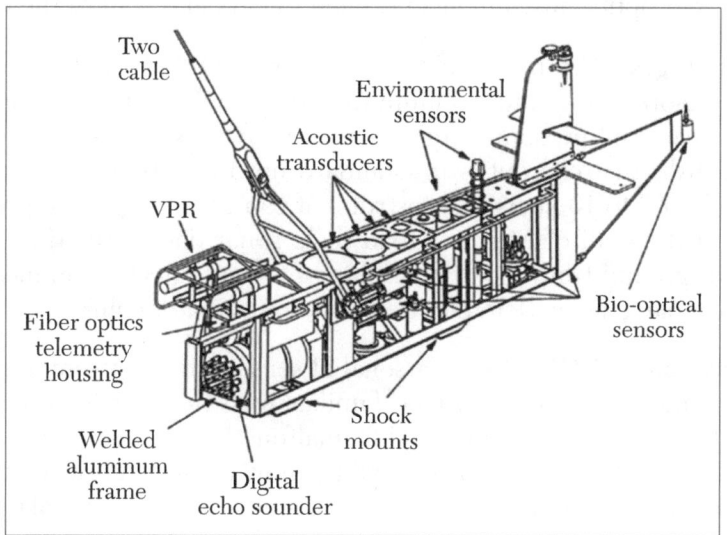

FIGURE 9.26 BIOMAPER sensor placement.

cable can deliver up to 2 kilowatts. BIOMAPER is usually towed to within 2 meters (6.6 ft) of the surface (allowing light sensors to take reference readings) and can then descend to about 500 meters (1,640 ft). At normal settings, the range of the sonar extends about 250 meters (820 ft) above and below BIOMAPER's body. In BIOMAPER, the black apparatus above the nose is a video plankton recorder. Sonar units mounted in the body measure particle size up and down the water column.

BIOMAPER records plankton over large areas by combining the strengths of sonar and video imaging. The sonar data pinpoints small objects in 1-meter increments through several hundred meters of the water column at a time. As a check, the video plankton recorder mounted on the nose images a few centimeters of water at a time. Its detailed images allow scientists to identify the plankton with certainty. Meanwhile, the ESS collects physical data about the ocean water, helping scientists understand relationships between water conditions and ocean life. As BIOMAPER tows through the water, five sonar units transmit sound waves upwards at 43, 120, 200, 420 and 1,000 kHz. Another five units transmit downward at the same frequencies. The different frequencies bounce off objects of different sizes. By timing the echoes the instrument calculates how large and how far away particles are. There is a lower limit to the size of particles that sonar can detect. For studying extremely small phytoplankton, biologists use other optical instruments fastened to bays in the aft of the instrument.

Advantages: The range of the sonar on BIOMAPER lets scientists see much more of the water column than they could with other methods. The sonar can capture some larger creatures (from several-centimeter krill to large fish) that can out-swim standard net tows. By making long tows at speeds of 6 to 10 knots, the instrument can cover regions approaching the size of the Gulf of Maine. Broad scale sonar data verified by small scale video «ground-truthing» and combined with physical ocean measurements help to paint a complete picture of life in the water column.

Limitations: BIOMAPER's arsenal of sensors does present a few practical limitations. Operators must be familiar with the details of all the instruments so they can diagnose and fix malfunctions or poor connections. And before each tow, each sensor must be individually calibrated to make sure it's reporting accurately. The 10 sonar units, the video plankton recorder, the instruments that make up the ESS, and any additional instruments generate lots of data—about 1 GB per hour. Analyzing and comparing readouts among sensors takes a lot of time and computing power. The sonar units have a blind spot that extends 6 meters (21 ft) on either side of the BIOMAPER body (this is one reason for using a tow-yo survey pattern). This means that the video plankton recorder and the sonar are never sampling exactly the same water at the same time.

9.11 GRAVITY CORER

The *gravity corer* allows researchers to sample and study sediment layers at the bottom of lakes or oceans. It got its name because gravity carries it to the bottom of the water body. Recovering sediment cores allows scientists to see the presence or absence of specific fossils in the mud that may indicate climate patterns in the past, such as during the ice ages. Scientists can use this information to improve understanding of the climate system and predict patterns and events in the future. Cores capture a time capsule that, in some cases, can span the past hundreds of thousands and even millions of years. Because sedimentation rates in some areas are quite slow, even a smaller corer a few meters in length may represent thousands of years of particles. These particles are a historical record of condition in the water column and in the atmosphere and can be used to reconstruct past conditions on Earth. To operate the gravity corer, users need a boat with a winch powerful enough to lower and raise the corer [Figure 9.27]—it can be heavy,

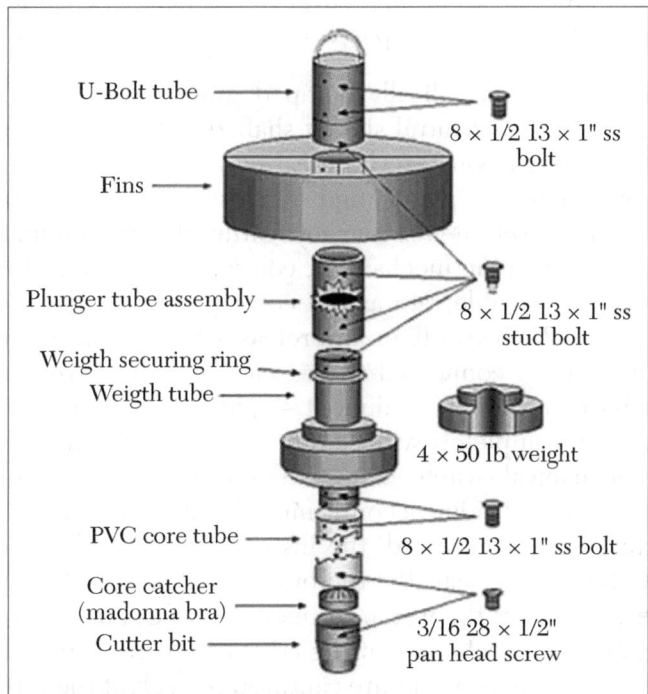

FIGURE 9.27 Schematic drawing of a gravity corer.

with have hundreds to thousands of pounds of lead weights attached. Users also need enough wire rope to reach the bottom of the water body.

Advantages: It is simple, robust, and relatively reliable. There is little cost associated with it, it is extremely easy to use, and requires little maintenance.

Disadvantages: It can be heavy and awkward to deploy and recover. Long cores must be lifted to a horizontal position and maneuvered over the rail to bring aboard the vessel.

9.11.1 Hydraulically Damped Gravity Corer

The *hydraulically damped gravity corer*, like other corers and grab samplers, is used to sample sediments from the ocean floor. The unique aspect of this instrument is that it penetrates the sediment slowly to minimize disturbance of the water/sediment interface. The rate of penetration into the sediment is controlled by a water-filled piston that empties at a selectable rate. This feature allows collection without the bow wave that may occur when samplers approach the bottom swiftly. Cores up to 60 cm long are recovered in muddy sediments. Cores up to about 30 cm are collected in sand.

Operation: The hydraulically damped gravity corer consists of a four-legged frame with a central sliding shaft that holds a heavy weight stand and a clear polycarbonate cylindrical core barrel. The apparatus is constructed of aluminum, plastic, and stainless steel to minimize heavy-metal contamination of sediments. The key feature that differentiates this corer is the hydraulic damping mechanism, consisting of a water-filled piston that empties at a selectable rate as the corer penetrates the mud. When the corer contacts the ocean floor and releases tension on the winch wire, the water-filled piston connected to the sliding shaft controls the speed that the core barrel enters the sediment. Typically, the core barrel takes 10–15 seconds to fill completely with sediment. As the core tube enters the sediment, a mechanical switch changes the sampling rate of a sonar transducer (optical sensor), providing a confirming signal on the ship's depth recorder. When this signal is received, the instrument is winched out of the sediment. A check valve seals the top of the core barrel. When the bottom of the core tube clears the water/sediment interface, a spring-loaded paddle slides against and seals the cutting edge of the core tube. Both sediment and ambient overlying water are captured by seals at the top and bottom of the core tube at pullout.

The core quality and sedimentary features can be observed by visual inspection through the clear core barrel. The core is then removed from the frame, capped and taped at both ends, then stored under refrigeration. To preserve the water/sediment interface, the core should always be secured in a vertical position. Overlying water is left to completely fill the head space above the water sediment for transport and storage. The cylinder can be safely transported by ship or car without resuspension or other disturbance.

Advantages:

- Samples both sediments and overlying water

- Minimal disturbance to water/sediment interface

- Can be used to sample metals and other inorganic material

- Metal barrels can be used for organic geochemistry studies compromised by the use of plastic

Disadvantages:

- Small sampling area and volume

- Not designed to sample living organisms

- More complex to operate

9.12 MARINE MAGNETOMETER

A *magnetometer* is a scientific instrument used to measure magnetic field strength. On land, magnetometers can be used to find iron ore deposits for mining. Under the sea, marine geophysicists, ocean engineers, and nautical archeologists use marine magnetometers to detect variations in the total magnetic field of the underlying seafloor. Usually, the increased magnetization is caused by the presence of ferrous (unoxidized) iron on the seafloor, whether from a shipwrecked boat made of steel or a volcanic rock containing grains of magnetite, a highly magnetic mineral. After corrections are made to measurements of the total magnetic field, scientists can use magnetic data to estimate the age and thickness of volcanic lava flows at mid-ocean ridges and ocean island hot spots; to locate pipelines, undersea cables, and bridge foundations; and to identify important archeological sites.

FIGURE 9.28 Marine magne-tometers.

Operation: Marine magnetometers are gener-ally "fish-type" instruments (so called because they are sleek and only a few meters in length [Figure 9.28]). They are towed at least two-and-a-half ship lengths behind the ship, so that the ship's magnetic field does not interfere with magnetic measure-ments. Marine magnetometers can be scalar, mea-suring the total strength of the magnetic field; or vector, resolving the magnetic field into the vectors of strength, inclination (the angle at which magnetic field lines intersect the surface of the earth, 0° at the equator and 90° at magnetic poles), and declination (the angle the magnetic field makes with geographic north). Marine magnetometers contain a chamber filled with a liquid that is rich in hydrogen atoms, like kerosene or methanol. Electrons dissolved in the liquid are excited by a radio frequency (RF) power source and pass on their energy to the hydrogen atoms' nuclei (protons), altering their spin states. The transfer of energy from electrons to the protons in the hydrogen atoms is called the *Overhauser effect* (after American physicist Albert Overhauser, who discovered it in the early 1950s) and the magnetometers that use the effect are called *Overhauser magnetometers*. Once the protons are spin-ning, the RF power is removed and the protons spiral back to their original alignment with the total geomagnetic field. The frequency of their spiral-ing, or "precession," is measured with a coil and is dependent on a known constant, the *gyromagnetic ratio*, and the total geomagnetic field. Thus, if the frequency is measured, and the constant is known, the total geomag-netic field can be calculated.

Advantages: Overhauser magnetometers are vastly more energy effi-cient than their predecessors, proton precession magnetometers, which relied on excitement of protons by a direct current (DC) source. Over-hauser magnetometers also have faster cycling rates (up to five magnetic measurements taken each second) and higher sensitivities than the older proton precession magnetometers. The power-saving improvement is very important, because Overhauser magnetometers can be mounted on re-motely operated vehicles that have limited battery power, thereby improv-ing the spatial coverage of magnetic mapping surveys. The design of most marine magnetometers makes them light weight, so they are quite simple to load onto the ship and deploy. Surface-towed magnetometers can cover larger areas and are relatively inexpensive. Deep-towed magnetometers can cover detailed areas of the seafloor and have higher sensitivities than surface towed magnetometers.

Disadvantages: Surface-towed magnetometers have lower sensitivities than near-bottom magnetometers. Near-bottom magnetometers are more expensive and cover smaller areas than surface-towed magnetometers.

9.13 OCEAN-BOTTOM SEISMOMETER

Seismometers measure movement in the Earth's crust. About 90% of all natural earthquakes occur underwater, where great pressure and cold make measurements difficult. The *ocean-bottom seismometer* (OBS) was developed for this task. Scientists use seismometer data to calculate the energy released by earthquakes, like the massive one in December 2004 that caused the Indian Ocean tsunami. By using sensitive seismometers to study small earthquakes, researchers are working to predict large earthquakes or volcanic eruptions. Other scientists use seismometers to peer inside the Earth itself. The waves that earthquakes generate get deformed or slowed down as they pass through different materials inside the Earth. Seismometers equipped with precise clocks record the shape and speed of these waves when they arrive. After an earthquake, data from many widespread seismometers help geologists to calculate the structure of Earth's mantle and crust.

Operation: Seismometers work using the principle of inertia. The seismometer body rests securely on the sea floor. Inside, a heavy mass hangs on a spring between two magnets. When the Earth moves, so do the seismometers and its magnets, but the mass briefly stays where it is. As the mass oscillates through the magnetic field it produces an electrical current which the instrument measures. The seismometer itself is a small metal cylinder; the rest of the footlocker-sized OBS consists of equipment to run the seismometer (a data logger and batteries), weight to sink it to the sea floor, a remote controlled acoustic release and flotation to bring the instrument back to the surface.

Two types of OBS: The ground motion caused by earthquakes can be extremely small (less than a millimeter) or large (several meters). Small motions have high frequencies, so monitoring them requires measuring movement many times per second and produces huge amounts of data. Large motions are much rarer, so instruments need to record data less frequently, to save memory space and battery power for longer deployments. Because of this variability, engineers have designed two basic kinds of seismometers: short-period OBS and long-period OBS. Short-period OBSs record high frequency motions (up to hundreds of times per second). They can record small, short period earthquakes and are also useful for studying the outer tens of kilometers of the seafloor. Long-period OBSs record a much broader range of motions, with frequencies of about 10 per second to once or twice a minute. They are used for recording mid-sized earthquakes and seismic activity far from the instrument.

Advantages: Very stable clocks make comparable the readings from many far-flung seismometers. (Without reliable time-stamps, data from different machines would be unusable.) Development of these clocks was a crucial advance for seismologists studying the Earth's interior. After recovering an ocean-bottom seismometer, scientists can offload the instrument's data by plugging in a data cable. This feature saves the task of gingerly disassembling the instrument's protective casing while aboard a rolling ship. The ability to connect a seismometer to a mooring or observatory makes the instrument's data instantly available. This is a huge advantage for geologists scrambling to respond to a major earthquake.

Limitations: Ocean-bottom seismometers are hard to install with pinpoint accuracy (usually they are lowered into place through thousands of meters of water). They can wind up sitting on a cushion of sediment rather than on bedrock. That soft layer can dampen the very tremors the instrument is trying to measure.

Short-period seismometers have short battery lives, so large numbers of them must be set out repeatedly during 30-day cruises. These instruments are designed to be small and light to make deployment and recovery easier.

Seismometers record so much data that storing it requires writing to a disk drive (up to 27 GB), which presents another drain on battery power. The seismometer's data logger and batteries must be protected from the pressures of the deep sea. Engineers house them inside thick glass spheres. To cushion the glass, the spheres then go inside yellow fiberglass "hardhats." A long-term ocean-bottom seismometer designed to work for more than a year is shown in Figure 9.29. The four yellow hardhats protect

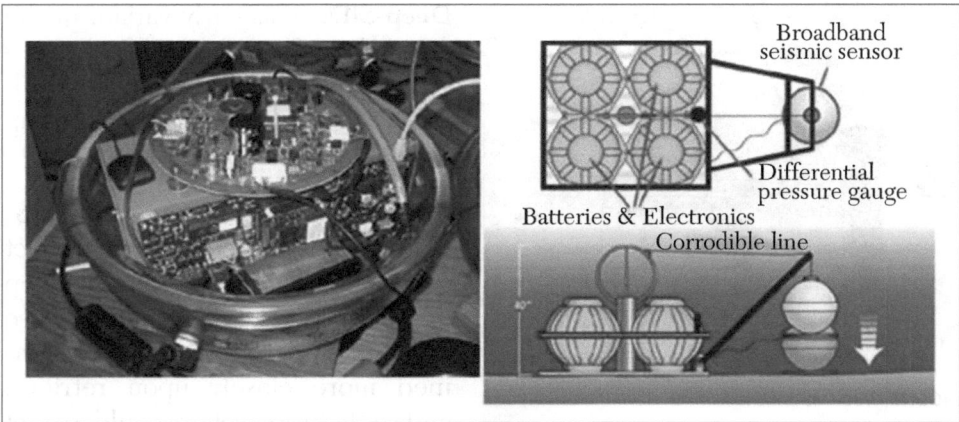

Broadband
seismic sensor

Differential
pressure gauge

Batteries & Electronics

Corrodible line

FIGURE 9.29 Seismometer data logger and long-term OBS.

equipment from the high pressure at depth. A pressure gauge measures earthquake waves in the water. The seismometer is housed in a metal sphere that drops gently into place after the entire instrument has been installed on the sea floor.

9.14 SUBMERSIBLE INCUBATION DEVICE

The *submersible incubation device* (SID) is a robotic mini laboratory that permits researchers at sea to automatically collect and process samples of seawater of precise volume at specified depth to study organisms that live in the water column, on the seafloor, or in hydrothermal vent fluid. Variations of the original SID have been optimized to sample, incubate, and preserve organisms that at precise locations or conditions within the seafloor environment. The original SID was designed to measure a key aspect of ocean ecosystems: how fast single-celled photosynthetic organisms at the heart of the food web convert carbon dioxide into organic carbon. At the time, the standard shipboard method to do that required deploying sample bottles late at night, retrieving them to the sea surface by dawn, incubating the phytoplankton all day under conditions that allowed them to proceed normally with photosynthesis, and then, for hours after dark, preparing the samples for analysis. The objective of SID (and every variant thereafter) was to automate sample collection and processing to give researchers a glimpse into critical microbial functions beneath the surface without exposing collected or processed samples to surface conditions.

FIGURE 9.30 MS-SID.

Deep-SID: This early variant of the original SID was specifically designed to collect and incubate unfiltered seawater samples in depths up to 1,000 meters below the surface.

MS-SID: *Microbial sampler* (MS) SID was designed to not only collect and incubate samples from the deep ocean, but also to filter and preserve organisms so they could be examined more closely upon retrieval and so that researchers could extract RNA from samples. MS-SID can process volumes five times larger than the first SIDs, vastly improving the odds of detecting sparse organisms and their subtle chemical reactions. It can sample a 6-ft thick zone more than two miles below the sea surface, and it can communicate with the surface in real time, permitting scientists to adjust sampling and/or processing protocols based on data from the instrument. MS-SID during recovery in the Mediterranean showing collection chambers filled with yellow preservative is shown in Figure 9.30.

Vent-SID: *Vent-SID* is a variant of the MS-SID that incubates samples of hydrothermal vent fluid at temperatures up to 60°C (140°F). Like MS-SID, it will also filter and preserve organisms in situ for later study and can process large volumes at depth with real-time communication and control of sampling and processing protocols. First deployment of Vent-SID is scheduled for November 2014 at 9N on the East Pacific Rise. Vent-SID undergoing testing at WHOI prior to its first deployment in November 2014 at 9N on the East Pacific Rise is shown in Figure 9.31.

FIGURE 9.31 Vent-SID.

9.15 DEEP OCEAN TSUNAMI DETECTION BUOY

Deep ocean tsunami detection buoys are one of two types of instrument used by the Bureau of Meteorology to confirm the existence of tsunami waves generated by undersea earthquakes. These buoys observe and record changes in sea level out in the deep ocean. This enhances the capability for early detection and real-time reporting of tsunamis before they reach land, which is shown in Figure 9.32.

FIGURE 9.32 The surface component of an operational deep ocean tsunami detection buoy.

Operation: A typical tsunami buoy system comprises two components: the pressure sensor anchored to the sea floor and the surface buoy. The sensor on the sea floor measures the change in height of the water column above by measuring associated changes in the water pressure. This water column height is communicated to the surface buoy by acoustic telemetry and then relayed via satellite to the tsunami warning center. The system has two modes: standard and event. The system generally operates in standard mode, where it routinely collects sea level information and reports via satellite at relatively low frequency transmission intervals (i.e., every 15 minutes). This helps to conserve battery life and hence extend the deployment life. The tsunami buoy is triggered into event mode when the pressure sensor first detects the faster-moving seismic wave moving through the sea floor. It then commences reporting sea level information at one-minute intervals to enable rapid verification of the possible existence of a tsunami. The system returns to standard mode after four hours if no further seismic events are detected.

Determination of location for deployment of tsunami buoys: The best location for deployment of a tsunami buoy is determined by careful consideration of several factors. The tsunami buoy needs to be far enough away from any potential earthquake epicenter to ensure there is no interference between the earthquake signal at the buoy and the sea level signal from the tsunami. On the other hand, the tsunami buoy needs to be close enough to the epicenter to enable timely detection of any tsunami and maximize the lead time of tsunami forecasts for coastal areas. In addition, tsunami buoys must ideally be placed in water deeper than 3,000 meters to ensure the observed signal is not contaminated by other types of waves that have shallower effects (e.g., surface wind waves). International maritime boundaries must also be considered when deploying tsunami buoy systems. Deep-ocean tsunami detection buoy technology was initially developed in the United States by the Pacific Marine Environmental Laboratory (PMEL) of the National Oceanic and Atmospheric Administration (NOAA) as DART (Deep-ocean Assessment and Reporting of Tsunami) buoys [see Figure 3.5]. The latest DART II systems contain two independent and redundant communications systems as back-up. These systems are capable of measuring sea level changes of less than a millimeter in the deep ocean. Two-way communication between the tsunami buoy and the tsunami warning center means that the buoy can be controlled remotely. This two-way communication allows for troubleshooting of the system and also allows people to put the systems into event mode in case of a possible tsunami or for research purposes.

9.16 OCEANOGRAPHC INSTRUMENTS

The following are few more instruments used for measuring the parameters of ocean.

Towed ocean bottom instrument: *Towed ocean bottom instrument* (TOBI) is an instrumented vehicle which is towed close to the bottom of the deep ocean from a ship, and uses sound to form detailed images of the sea floor. TOBI was developed and has been in service since 1990; it is has become one of world's best tools for underwater geological surveys using side scan sonar.

Autonomous underwater vehicles: *Autonomous underwater vehicles* (AUVs) are robot submarines that are used to explore the world's oceans without a pilot or tether. Before launch from the mother ship, the AUV's computers are programmed with instructions of where to go, what to measure, and what depths to go to.

Seismic exploration: While seismic surveys are often used by the oil and gas industries to find reserves, scientists can use the same techniques to look for air bubbles, which can signify the presence of hydrothermal vents. Surveys are a complex operation that requires skilled specialists to operate the equipment.

Corers: Corers are used to collect sediment from the ocean floor and work by pushing or grabbing sediment into containers. There are different types of corer with both tubular and box varieties available. Sediment cores are an expensive and unique resource of immense scientific value, analysis of which can provide clues about climate change, conditions in past oceans, and sedimentary processes.

Dredges: Dredges gather loose rocks sitting on the ocean floor using a technique that has changed little in hundreds of years. They have a chain-link bag with large metal-jawed opening that scoops the contents into the bag. They are lowered to the seabed on a cable and dragged along the bottom for some distance before being brought to the surface. A variety of equipment can be used to get samples of rock that are not loose. If the rock is soft like the mineral rich chimneys of a hydrothermal vent, it may be picked up using grabbers or pincers, as found on the video guided robotic underwater vehicle.

Wax corer or rock chipper: The wax corer is used to collect fresh volcanic glass. This material is the gold standard for volcanic studies as it represents the magmatic liquid rock. The rock chipper will smash into the ground and capture shards of glass in wax containers. When the containers are returned the wax is melted in beakers of hot water and the contents simply fall to the bottom. For deeper rock samples, rock drills can be used to take core samples of bedrock in a similar fashion to sediment cores.

EXERCISES: PART A (ANSWER IN A WORD OR A SENTENCE)

1. _____ instrument uses a strain gauge pressure transducer to sense depth.

2. The abbreviation of SONAR and SOFAR are _____.

3. The echo-sounder is used to measure _____.

4. Give an example of a temperature sensor.

5. _____ is the measure of the quantity of salt in a volume of sea water.

6. _____ instrument is used to measure water clarity.

7. What is a hydrophone?

8. _____ is dropped from airborne platform to record ambient noise and locate ships.

9. Sound speed is a measure of _____.

10. _____ is the instrument used for measuring waves.

11. Tides are measured by _____.

12. An accelerometer is used to measure _____.

13. _____ is the emission of light by living organisms.

14. Seismometer is used to measure _____.

15. _____ are used to collect sediment from the ocean floor.

16. _____ is a scientific instrument used to measure magnetic field strength.

EXERCISES: PART B (ANSWER IN A PAGE)

1. Write about the measurement of sea depth.

2. What are the different methods of measuring temperature?

3. Write about the method of measuring salinity.

4. What are the methods of measuring water clarity?

5. Write about sound and sound speed.

6. What are the different ways of measuring current?

7. Write about seabed sampling.

8. Write a note on research vessels.

9. Write about moorings.

10. Write a note on ferry boxes.

11. Write about the radar Doppler current profiler.

12. What is a nucleic acid biosensor?

13. Write about ASIMET.

14. Write about tsunami detection buoys.

15. How do sensors measure pH and alkalinity?

EXERCISES: PART C (ANSWER WITHIN 2 OR 3 PAGES)

1. Explain the techniques to measure current in detail.

2. Write about oceanographic instrumentation.

3. Write about the ARGO float in detail.

4. Explain in detail about the measurement of hydrographic properties.

5. In detail explain the measurement of dynamic properties.

6. Write the techniques of wave measurement.

7. Write about BIOMAPER in detail.

8. Write about corers.

WEB LINKS

http://www.aoml.noaa.gov/

http://uskess.whoi.edu/

http://www.bodc.ac.uk/

http://www.whoi.edu/instruments/

CHAPTER 10

OCEAN OPTICS AND OCEAN FACTS

This chapter deals with ocean optics and some interesting facts about the ocean.

10.1 INTRODUCTION TO OPTICAL PROPERTIES OF THE OCEAN

The optical properties of sea water are grouped into inherent and apparent properties. *Inherent optical properties* (IOPs) are those properties that depend only upon the medium and therefore are independent of the ambient light. The two fundamental IOPs are the absorption coefficient and the volume scattering function. *Apparent optical properties* (AOPs) are those properties that depend both on the medium (the IOPs) and on the directional structure of the ambient light and display enough regular features and stability to be useful descriptors of a water body. Commonly used AOPs are the *irradiance reflectance*, the *remote sensing reflectance*, and various *diffuse attenuation functions*.

Case 1 waters are those in which the contribution by phytoplankton to the total absorption and scattering is high compared to that by other substances. Absorption by chlorophyll and related pigments therefore plays the dominant role in determining the total absorption in such waters and dissolved organic matter derived from the phytoplankton also contribute to absorption and scattering in case 1 waters. Case 1 water can range from very clear (oligotrophic) to very productive (eutrophic) water, depending on the phytoplankton concentration.

Case 2 waters are everything else, namely, waters where inorganic particles or dissolved organic matter from land drainage contribute significantly to the IOPs, so that absorption by pigments is relatively less important in determining the total absorption. Roughly 98% of the world's open ocean and coastal waters fall into the case 1 category, but near-shore and estuarine case 2 waters are disproportionately important to human interests such as recreation, fisheries, and military operations.

The constituents are identified operationally based upon how we measure their optical properties and often are grouped by like optical properties. For example, the distinction between particulate and dissolved is operationally defined by the filter type/pore size. It is essential to remember 1) that the strict chemical definition is quite different, 2) filter pore size varies, and 3) keep track of pore sizes to ensure closure (i.e., don't define dissolved organic matter by the filtrate of a 0.2 micron pore-sized filter and then measure particulates on a 0.7 micron pore size). Similarly, often all the non-phytoplankton particles are lumped into a single compartment as their optical properties are quite similar. Sometimes all the particulate material is lumped together into *suspended particulate material* (SPM) or part of it into the *particulate organic material* (POM). This is often done when studying a specific biogeochemical property using optics. The differentiation of dissolved and particulate materials (using a filter) does not imply that the dissolved material is organic, although this is most often the assumption. For example, inorganic dissolved substances such as iron oxides (rust) could contribute in certain cases.

10.1.1 Optical Constituents of Seawater

Oceanic waters are a witch's brew of dissolved and particulate matter whose concentrations and optical properties vary by many orders of magnitude, so that ocean waters vary in color from the deep blue of the open ocean, where sunlight can penetrate to depths of several hundred meters, to yellowish brown in a turbid estuary, where sunlight may penetrate less than a meter. The most important optical constituents of sea water can be briefly described as follows.

1. Sea water (water + inorganic dissolved materials)

2. Phytoplankton

3. Colored (or chromophoric) dissolved organic material (CDOM)

4. Bubbles

5. Non-phytoplankton organic particles (sometimes referred to as detritus or tripton)

6. Inorganic particles

Sea water: Water itself is highly absorbing at wavelengths below 250 nm and above 700 nm, which limits the wavelength range of interest in optical oceanography to the near ultraviolet to the near infrared. Water contributes to both absorption and scattering by seawater. In clear ocean waters, water effect on ocean color in the visible cannot be neglected and hence must be taken into account. Temperature and salinity affect both absorption and scattering by water and hence must be taken into account when the optical properties of water are computed.

> Absorption by water: Absorption as a rich structure due to the excitation of the different vibration modes of the water molecule and water absorption is affected by temperature and salinity.

> Elastic scattering by water: Elastic scattering by sea water depends on salinity (~30% increase for range of salinities observed in the oceans), much less so of temperature (~4% between 0 and 26°C) and pressure (~1.3% for an increase in P of 100bar).

> Raman scattering by water: In Raman scattering a fraction of the incident light of wave number υ_o is absorbed and re-emitted at wave number $\upsilon_s = \upsilon_o - \upsilon_r$ where υ_r is the Raman shift of a vibration mode of the water molecule. For water $\upsilon_r = 3400$ cm^{-1}. Raman scattering is used to calibrate the intensity of the source of a LIDAR system as the signal leaving the ocean is proportional to the intensity of the source.

Phytoplankton: Phytoplanktons have a major effect on the ocean color and are one of the primary reasons for studying it. Additionally, these microscopic, single-cell, free-floating organisms possess chlorophyll, pigment that allows them to harvest the sunlight and through the process of photosynthesis produce energy-rich organic material, while releasing oxygen. That makes them the most important primary producers in the ocean, base of the oceanic food web, and an important component of the global carbon cycle. For all these reasons, it is of a great importance to understand phytoplankton abundance and dynamics.

Absorption by phytoplankton: Phytoplankton absorbs sun light and uses this energy to produce energy rich organic material (photosynthesis). Chlorophylls, present in all phytoplankton cells, will cause two dominant peaks in absorption spectra, primary at blue (440 nm) and a secondary peak in red part of the spectra (675 nm). Presence of other pigments (depending on species) will cause the broadening of blue peak and appearance of additional absorption maxima.

Scattering by phytoplankton: Scattering properties of phytoplankton are important since they are directly related to remote sensing reflectance calculations (via backscattering to absorption ratio). Scattering and backscattering coefficients of phytoplankton as well as the volume scattering function are derived from either theoretical models or direct measurements of the above-mentioned properties. They highly depended on size, shape, and refractive index of all components of the phytoplankton cell. Values of phytoplankton scattering coefficients, when compared to the rest of the oceanic particles, are relatively low, based on their high water content and strong absorptive properties. Exception to the rule are coccolithophores—phytoplankton that produces small calcium carbonate scales, that makes them very effective scatterers and allow us to see coccolithophorid blooms from space.

Fluorescence by phytoplankton: A portion of the light absorbed by phytoplankton cell can be emitted at another, longer wavelength; a process referred to as fluorescence. Several phytoplankton pigments (chlorophylls, pheopigments, and phycobilins) have fluorescence, with chlorophyll a fluorescence being the most significant one. Although fluorescence is only a form of energy dissipation of the absorbed light, secondary to photosynthesis, it is still significant enough to be observed from space. Fluorescence from phytoplankton chlorophyll is often expressed this simplified formula (Falkowski and Kiefer (1985):

$$F = PAR[chla]a^*_{pyto} \Phi_f$$

Where PAR is the intensity of light impinging on the cell, [chla] is chlorophyll concentration, a^*_{pyto} is chlorophyll specific phytoplankton absorption coefficient, and Φ_f is the quantum yield of fluorescence the emission efficiency of the cell. Phytoplankton fluorescence and intensity depends (via quantum yield of fluorescence and chlorophyll specific phytoplankton absorption coefficient) on several factors: taxonomic position of algae, pigment content and ratios, photo adaptation, physiological state of phytoplankton, nutrient conditions, and stage of growth.

Dissolved organic compounds: These compounds are produced during the decay of plant matter. In sufficient concentrations, these compounds can color the water yellowish brown; they are therefore generally called *yellow matter* or *colored dissolved organic matter* (CDOM). CDOM absorbs very little in the red, but absorption increases rapidly with decreasing wavelength, and CDOM can be the dominant absorber at the blue end of the spectrum, especially in coastal waters influenced by river runoff.

Colored or chromophoric dissolved organic matter (CDOM): CDOM is an important optical constituent in water often dominating absorption in the blue. It is based on the absorption or fluorescence by material passing through a given filter (most often with pore size of $0.2\mu m$). As such, it is an absorption (or fluorescence) weighted sum of the different dissolved materials in the water. Note that most of the material comprising DOM does not absorb or fluorescence and that there exist inorganic dissolved materials that also absorb (e.g., iron oxides, nitrate) although it is believed that fluorescence is due solely to organic materials. From this discussion, it follows that CDOM is thus not necessarily a good proxy for DOM, particularly in the open ocean.

CDOM absorption: CDOM spectrum is the visible most often described by an exponentially decreasing function:

$$a_g(\lambda) = a_g(\lambda_o)\exp^{-s(\lambda - \lambda^o)}[m^{-1}]$$

Where, s is referred to as the spectral slope and λ_o a reference wavelength. Single bonds, which are most abundant, will absorb short wavelength radiation while resonance of multiple bonds, less abundant, absorb longer wavelength radiation. Since numerically there many more short bonds, the spectrum is higher at short wavelengths. This explanation is consistent with the observation that small values of the spectral slope of CDOM, s are associated with higher molecular weight materials. For visible wavelength the most common values of s appear to be near 0.014nm^{-1}, varying in the visible from 0.007 to 0.026 nm^{-1}.

Elastic scattering by CDOM: CDOM contribution to scattering by seawater is somewhat controversial. By definition, colloids are part of DOM and, if abundant enough, could contribute significantly to scattering (particularly to backscattering) by sea water. However, there is no observational evidence that CDOM contribute significantly to scattering. Thus, currently, CDOM contribution to scattering is most often neglected.

Inelastic scattering by CDOM: One of the primary methods to quantify CDOM is through fluorescence. Since not all dissolve material that absorbs fluorescences, this material is often denoted as FDOM. In general, absorption and fluorescence can vary, but their ratio can vary by orders of magnitude between different locations. The fluorescence of CDOM in the field is often limited to a single excitation/emission band pair. With lab instrumentation, 2-dimensional *excitation emission spectra* (EEMS) are measured and used to characterize the FDOM based on the size and presence of known excitation emission peaks.

Bubbles: Bubbles in the upper ocean are primarily generated by breaking waves. When wind speed exceeds 7 ms^{-1}, field observations have shown that a stratus layer of bubbles forms under the sea surface and persist through continuous supply of bubbles by frequent wave breaking and the subsequent advection by turbulence. As wind subsides, bubbles that have been injected will evolve under the effects of buoyancy and gas diffusion and merge into the background population on time scales of minutes to hours. When wind speeds are lower than 3 ms^{-1}, few waves break. Once formed, bubbles are coated with surfactant material almost instantaneously and the accumulation of organic films onto their surfaces provides a stabilizing mechanism against surface tension pressure and gas diffusion.

Organic particles: Biogenic particles occur in many forms.

Bacteria: Living bacteria in the size range 0.2–1.0 µm can be significant scatterers and absorbers of light, especially at blue wavelengths and in clean oceanic waters, where the larger phytoplankton is relatively scarce.

Phytoplankton: These ubiquitous microscopic plants occur with incredible diversity of species, size (from less than 1 μm to more than 200 μm), shape, and concentration. Phytoplankton are responsible for determining the optical properties of most oceanic waters. Their chlorophyll and related pigments strongly absorb light in the blue and red and thus, when concentrations are high, determine the spectral absorption of sea water. Phytoplankton are generally much larger than the wavelength of visible light and can scatter light strongly.

Detritus: Nonliving organic particles of various sizes are produced, for example, when phytoplankton die and their cells break apart, and when zooplankton graze on phytoplankton and leave cell fragments and fecal pellets. Detritus can be rapidly photo-oxidized and lose the characteristic absorption spectrum of living phytoplankton, leaving significant absorption only at blue wavelengths. However, detritus can contribute significantly to scattering, especially in the open ocean.

Inorganic particles: Particles created by weathering of terrestrial rocks can enter the water as windblown dust settles on the sea surface, as rivers carry eroded soil to the sea, or as currents resuspend bottom sediments. Such particles range in size from less than 0.1 μm to tens of micrometers and can dominate water optical properties when present in sufficient concentrations. Particulate matter is usually the major determinant of the absorption and scattering properties of sea water and is responsible for most of the temporal and spatial variability in these optical properties. A central goal of research in optical oceanography is to understand how the absorption and scattering properties of these various constituents relate to the particle type (e.g., microbial species or mineral composition), present conditions (e.g., the physiological state of a living microbe, which in turn depends on nutrient supply and ambient lighting), and history (e.g., photo oxidation of pigments in dead cells). Bio-geo-optical models have been developed that attempt (with varying degrees of success) to predict the IOPs in terms of the chlorophyll concentration or other simplified measures of the composition of a water body.

10.1.2 Inherent Optical Properties (IOP) Variability

Inherent optical properties (IOPs) depend only on the properties of the medium and its constituents, which include spectral absorption and scattering coefficients—$a(\lambda)$ and $b(\lambda)$, respectively, where λ represents wavelength. The fundamental IOPs are the absorption coefficient and the volume scattering function, as various scattering coefficients (e.g., total, backward) can be determined by integration of the volume scattering function over the appropriate angles. An important characteristic of IOPs is that they are additive. This means that, for a seawater sample containing a mixture of constituents, the absorption and scattering coefficients of the various constituents are independent and the total coefficient can be determined by summation. This fact arises from the definition of IOPs with respect to collimated light. The current methods to measure IOPs can only approximate ideal light field and collection geometry, so corrections are sometimes required to obtain adequate estimates of true IOPs. To explain natural variability in total IOPs and to derive estimates of ecologically and bio–geo-chemically relevant constituents from measured IOPs, it is useful to identify categories of constituents, each of which makes a distinct contribution to the total IOPs. Typically, categories are selected on a combination of operational and functional criteria. For total $a(\lambda)$, for example, contributions from water $a_w(\lambda)$, chromophoric dissolved organic matter (CDOM) $a_{CDOM}(\lambda)$, phytoplankton $a_{ph}(\lambda)$ and non-algal particles (NAP) $a_{NAP}(\lambda)$:

$$a(\lambda) = a_w(\lambda) + a_{CDOM}(\lambda) + a_{ph}(\lambda)\, a_{NAP}(\lambda) \ \text{-----}\ 1$$

Similar summations can be applied to other IOPs such as $b(\lambda)$, backscattering $b_b(\lambda)$ and the beam attenuation coefficient $c(\lambda)$, which is defined as the sum

$$c(\lambda) = a(\lambda) + b(\lambda)$$

It should be emphasized that (eqn.1) represents an example set of constituent categories, and the concept can be generalized to as many levels as practical or important for specific problems; The following sections focus on water constituents and their effects on IOPs.

Absorption CDOM: For optical oceanographers and many marine chemists, CDOM is operationally defined by its passage through a small pore size filter (usually 0.2 µm) and its ability to absorb visible and ultraviolet radiation. CDOM is a poorly characterized portion of the total dissolved organic matter (DOM) pool and there are no routine analytical techniques for chemically quantifying total CDOM; for this reason, CDOM is frequently quantified in terms of its measurable optical properties (i.e. absorption or fluorescence). Especially in coastal waters with substantial riverine inputs, absorption by CDOM can be very high compared with other constituents, and so it influences the quantity and color of light penetrating into and reflecting from the upper ocean. Even in the open ocean, however, absorption by CDOM cannot be neglected. CDOM absorption is characterized by smoothly varying spectral dependence with amplitude tending to increase exponentially towards blue and ultraviolet wavelengths. For this reason, its contribution to absorption is usually represented by an exponential function:

$$a_{CDOM}(\lambda) = a_{CDOM}(\lambda 0) \, e^{-S(\lambda - \lambda 0)}$$

Where $\lambda 0$ represents a reference wavelength and S is the slope of the exponential increase with decreasing wavelength. Both the amplitude and the spectral slope of a_{CDOM} depend on the composition of the dissolved organic matter pool and this in turn depends on a variety of source and sinks processes. Terrestrial input, primarily from riverine sources, is significant; in addition, other processes including photo oxidation (enhanced by CDOM's strong ultraviolet absorption) and local microbial activity can lead to production or loss of different forms of CDOM.

Phytoplankton: As photosynthetic organisms, phytoplankton contain high concentrations of pigments that harvest energy from sunlight. These pigments consist of different chlorophylls and carotenoids present in varying amounts and each type of pigment has spectral properties that lend phytoplankton their characteristically featured absorption spectra. Due to the ubiquitous presence of chlorophylls, absorption peaks at blue (~440 nm) and red wavelengths (~675 nm) are always present, with the blue peak broadened and enhanced by accessory pigments. Pigment compositions, and thus general light absorption characteristics, are partly constrained by phylogeny, but there are also important variations associated with environmental factors that affect growth. Pigment composition and physiological status also affect the amount of light energy absorbed by phytoplankton that is reemitted as fluorescence at red wavelengths. The first order source of variation in light absorption by phytoplankton is total biomass and this, of course, depends on complex ecological and environmental factors that regulate phytoplankton growth and loss rates. Secondary effects with important consequences for how phytoplankton absorb light include variations in intracellular pigment concentration and composition and variations in cell size and shape; size and shape directly impact pigment package effects.

Non-algal particles (NAP): Marine particles besides phytoplankton are also known to absorb light. In natural samples, it is difficult to separate the broad category of NAP into its different contributors, which can include heterotrophic organisms such as bacteria and micrograzers, other organic particles of a detrital nature such as fecal pellets and cell debris, and various mineral particles of both biogenic (e.g., calcite liths and shells) and terrestrial (e.g., clays and sand) origin. In comparison with phytoplankton, much less is known about the optical properties of these particles, but some generalizations have emerged. In coastal and open ocean waters, total NAP absorption tends to exhibit an absorption spectrum that monotonically increases with decreasing wavelength, similar in form to that observed for CDOM. Consequently, a_{CDOM} can be replaced by a_{NAP} in eqn.1 to provide an adequate description of NAP absorption. Because of the similar spectral character of CDOM and NAP, for some applications these pools have been combined into a single class, referred to as colored detrital matter (CDM; which is technically a misnomer as it also includes living and inorganic matter), whose absorption follows eqn.1 with a composite S parameter that will vary with the relative contributions of CDOM and NAP.

Scattering and backscattering: All particulate material whose index of refraction differs from the surrounding medium will scatter light. The amount of scattering is influenced by particle size and shape and by any absorption that occurs within the particle. In contrast to absorption, scattering is not completely characterized simply by specifying its wavelength dependence; it also has angular dependence. Total scattering is summed over all possible scattering angles, but it is also possible to define scattering coefficients over some angular subsets. Backscattering, which is simply scattering integrated over the backward hemisphere with respect to the direction of light incidence, is a quantity that has received a lot of attention due to its importance for ocean color interpretation. As for total scattering, backscattering depends on particle concentration, size, shape, and complex refractive index. Theoretical considerations show that backscattering is generally enhanced relative to forward scattering as particle size decreases, so different particles may dominate total scattering and backscattering in natural waters. For a given wavelength, scattering by polydisperse particles tends to increase with average particle size and with the average real part of the refractive index, so highly refractive mineral particles scatter more light than a population of bacteria of similar concentration and size distribution, for example. Furthermore, the wavelength dependence of the scattering cross section tends to be steeper for smaller particles. Because of its extreme small size (usually defined <0.2 µm) and relatively dilute nature (compared with water molecules, for example), light scattering by CDOM can be neglected.

In general, ocean color remote sensing is a passive remote technique. The sensor, mounted on a satellite, an aircraft or other remote platform, detects the radiometric flux at several selected wavelengths in the visible and near-infrared domains.

The signal received by the sensor is determined by different processes in the water, as well as in the atmosphere [Figure 10.1].

1. Scattering of sunlight by the atmosphere

2. Reflection of direct sunlight at the sea surface

3. Reflection of sunlight at sea surface

4. Light reflected within the water body

Only the portion of the signal originating from the water body contains information on the water constituents; the remaining portion of the signal, which takes up more than 80% of the total signal, has to be assessed precisely to extract the contribution from the water body. There are two strategies to derive oceanic constituents from the signal of ocean color sensor

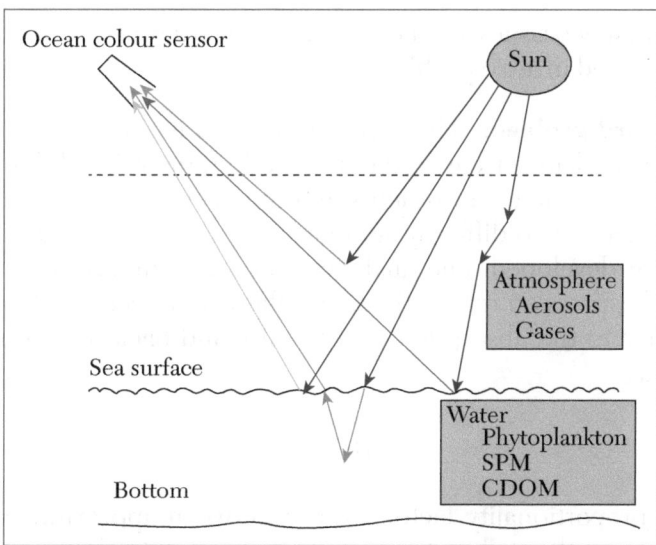

FIGURE 10.1 Sketch of different origins of light received by space-borne sensor.

at *top of atmosphere* (TOA), a one-step method and a two-step method. For the traditionally used two-step method, the water leaving radiance (or reflectance) is firstly derived from the signal at TOA (this procedure is called *atmospheric correction*), and then oceanic constituents are retrieved from water leaving radiance (or reflectance). For the one-step method, oceanic constituents are directly derived from the signal at TOA. The one-step method assumes that radiative transfer in the ocean and atmosphere is coupled. The oceanic constituents and aerosol properties are simultaneously derived from satellite measurements at TOA by using the entire spectrum available to ocean color instruments.

10.2 RETRIEVAL OF OCEANIC CONSTITUENTS FROM OCEAN COLOR MEASUREMENTS

There are three major issues in the retrieval of oceanic constituents from ocean color:

How to quantify the relationship between optically significant oceanic constituents and inherent optical properties (IOPs)?

How do IOPs determine ocean color?

How to obtain oceanic constituents from ocean color measurements?

The first two issues are the so-called forward problem, and the last issue is the so-called inverse problem.

The forward problem: The forward problem is solved by radiative transfer theory. Radiative transfer theory describes the relationship between the IOPs of the oceanic constituents and the ocean color. Based on radiative transfer theory, two different approaches relating the ocean color to IOPs have been developed: one analytical and one numerical. The most-used analytical expression relates the hemispherical reflectance R just below the sea surface to the absorption coefficient a and back scattering coefficient b_b and was

$$R = f \frac{b_b}{a + b_b}$$

The proportionality factor f varies between approximately 0.3 to 0.5, depending on the ambient light field and the optical properties of water.

Another analytical expression relating the remote sensing reflectance to the IOPs of oceanic constituents was derived as

$$R_{rs} = \frac{ft^2}{Qn^2} \frac{b_b}{a + b_b}$$

where t is the transmittance of the air-sea interface, Q is the upwelling irradiance-to-radiance ratio, which is a function of the solar zenith angle and optical properties of water, and n is the real part of the refractive index of seawater. The numerical approach is based on simulations of radiative transfer. It allows including all factors determining the ocean color, i.e., IOPs, rough sea surface, observation geometry, inelastic scattering processes, etc., and has a potential for the development of more advanced retrieval methods. Another advantage is to avoid errors due to eventually poor approximation of the factor Q and the parameter f.

The inverse problem: The determination of the oceanic constituents from ocean color is a parameter estimation problem, where a set of parameters $C = \{c_i, i = 1, ..., I\}$ are estimated from a set of measurements $R = \{r_j, j = 1, ..., J\}$. The functional relationship between measurements and parameters can be expressed as:

$$R = g(C) \ (10.3)$$

Inverting Equation 10.3, one obtains the set of parameters C from the set of measurements R:

$$C = g^{-1}(R) \ (10.4)$$

In this, C represents three different oceanic constituents: pigment, suspended particulate matter and colored dissolved organic matter, while R is either the remote sensing reflectance, defined as the ratio of water leaving radiance to downwelling irradiance or the hemispherical reflectance, defined as the ratio of upwelling to downwelling irradiance, at sea level in J spectral channels.

If g would be a linear function, one could derive the inverse function g^{-1}, and such obtain the oceanic constituents from the measured spectral reflectance. However, the functional relationship between the oceanic constituents and the resulting reflectance is complex and nonlinear. It is therefore mostly impossible to achieve an analytic inversion of g.

The traditional way to overcome this problem is to make assumptions on the functional form of g^{-1} and then to solve Equation 10.4 by regression techniques or other statistical methods. In recent years, *artificial neural networks* (ANN) have been increasingly applied to remote sensing data from ocean observing instruments, among those scatterometers and ocean color sensors. ANN techniques are well suited for solving nonlinear problems. No assumptions on the functions g or g^{-1} defined in Equations 10.3 or 10.4 are required. The training of the ANN requires considerable computational effort, its application is very fast. Therefore, ANN techniques are a promising method to derive oceanic constituents from ocean color data.

10.2.1 Retrieval of Oceanic Constituents from Ocean Color at TOA

Traditionally, the retrieval of oceanic constituents is performed by a two-step process: atmospheric correction followed by a bio-optical algorithm to obtain the desired parameters. The atmospheric correction algorithms which are commonly used are based on the *black pixel assumption*. These algorithms were primarily designed for clear, deep ocean areas. The information about atmospheric aerosols is derived from channels in the red and near infrared (above 670 nm), where the water leaving radiance is close to zero. The derived aerosol information is extrapolated towards the visible channels and the atmospheric contribution is calculated and removed for full spectrum. For the turbid coastal environment, the ocean can no longer assumed to be black in the red and near-infrared because of strong back scattering by suspended materials. Under these conditions, the black pixel assumption is no longer valid for deriving information on atmospheric aerosols. Thus, the algorithms developed for applications to clear ocean waters cannot be easily modified to retrieve water leaving radiance from remote sensing data acquired over the coastal environments.

10.2.2 The Atmospheric Correction Problem

An instrument views the ocean from a satellite or an aircraft measures upwelling radiances. These include contributions by the atmosphere, the water surface, and the water column. The atmospheric contribution L_a comes from solar radiance that is scattered one or more times by atmospheric gases and aerosols into the direction of the sensor. The surface reflected radiance (sun and sky glint) L_r is downwelling solar radiance that is reflected toward the sensor by the water surface. The water leaving radiance L_w comes from light that penetrates the ocean, is changed by the

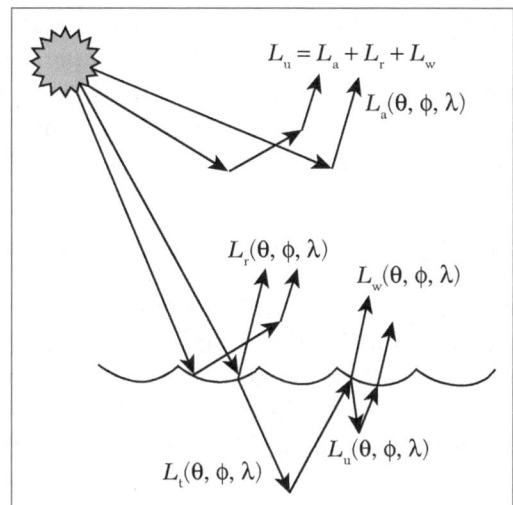

$$L_u = L_a + L_r + L_w$$

$$L_a(\theta, \phi, \lambda)$$

$$L_r(\theta, \phi, \lambda)$$

$$L_w(\theta, \phi, \lambda)$$

$$L_u(\theta, \phi, \lambda)$$

$$L_t(\theta, \phi, \lambda)$$

FIGURE 10.2 Contributions to the total upwelling radiance above the sea surface, L_u. Yellow arrows are the sun's unscattered beam; orange arrows are atmospheric path radiance L_a; red is surface-reflected radiance L_r; green is water-leaving radiance L_w. Thick arrows represent single-scattering contributions; thin arrows illustrate multiple scattering contributions.

absorbing and scattering components in the water, and is then scattered into an upward direction that eventually leaves the sea surface in the sensor direction. Figure 10.2 shows this conceptually.

Radiance reflected by the sea surface contains information about the wave state of the surface, which may be of interest in it or which, for example, may be useful for detection of surface oil slicks. Radiance scattered by the atmosphere along the path between the sea surface and the sensor contains information about atmospheric aerosols and other atmospheric parameters. However, only the water leaving radiance carries information about the water column and bottom conditions. A sensor looking downward measures the total radiance $L_u = L_a + L_r + L_w$ and cannot separate the various contributions to the total. Atmospheric correction refers to the process of removing the contributions by surface glint and atmospheric scattering from the measured total to obtain the water leaving radiance. The atmospheric correction problem becomes even more intimidating when effects of sun and sensor viewing directions, atmospheric conditions, and surface wave state are considered.

Most remote sensing retrieval algorithms are developed and use the remote sensing reflectance $R_{rs} = L_w / E_d$ or an equivalent nondimensional reflectance $\rho = \pi L_w / E_d$. (If L_w were directionally isotropic, $R_{rs} = 1 / \pi$, so is ρ the ratio of the actual R_{rs} to an idealized isotropic remote sensing reflectance. The π carries units of steradian. The use of an apparent optical property like R_{rs} or ρ minimizes the effects of external environmental conditions like sun angle on the magnitude of the spectra.

10.3 OCEAN OPTICS DIP PROBE SPECTROMETERS

A solution that is colored obtains this color by absorbing light of a complementary color in the visible spectrum. Absorbance refers to the amount of light that is absorbed by the sample. Spectrometers measure the percentage of light transmitted at a particular wavelength. The absorbance is then calculated from the percent transmittance by the relationship:

$$\text{Abs} = (2\text{-log } \%\text{transmittance})$$

A spectrophotometer has a source of light that is directed over a specific path. In the Ocean Optics system this path is the entire blue fiber optic cable. The light is produced with a tungsten halogen light source, the blue box in the picture on the right. This light goes into the long blue fiber optic cable to the dip probe seen in Figure 10.3. Fiber-optic cables are widely used in the telecommunications industry because they can transmit light over long distances without intensity losses or signal dispersion. At the open part of the dip probe tip, the light interacts with the sample. The path length of this part of the dip probe is 1 cm in length. Depending on the nature of the molecules in the sample, some of the light will be absorbed. The remaining light is then reflected through the second fiber-optic cable that leads to a detector inside the computer, seen at the right. The detector measures the number of photons of light that reach it and converts this measurement to electrical current. To perform a measurement using a spectrophotometer, the instrument must first be calibrated at 0% and 100% transmittance.

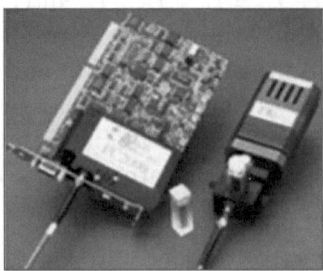

FIGURE 10.3 Fiber optic cable and light source.

At 0% transmittance, all of the light produced by the light source is absorbed and no photons reach the detector. For the ocean optics system, the calibration for 0% transmittance is performed by blocking the light by placing a folded sheet of paper in the slit of the light source. Next the instrument is calibrated for 100% transmittance using a solution called the blank, which contains all sources of absorbance except the analyte of interest. Sometimes the blank solution is a complicated mixture (i.e., urine or motor oil). For other analyses a pure solvent, such as distilled water, is used as the blank. A new blank is measured for every set of analyses. When the dip probe is placed in the blank solution, the percent transmittance is defined to be 100%. Thus when the dip probe is placed in the sample being measured, the analyte of interest is the only thing that contributes to the absorbance.

The lamp of the light source will need to be turned on approximately 15 minutes prior to analysis so that it reaches a constant temperature and the intensity of the light produced is consistent. The software used with the Ocean Optics equipment is called OOIBase32. To verify that the detector is receiving sufficient signal, the intensity of the highest peak should fall between 3,000 and 4,000 counts when the probe is in a blank solution. To check this, make sure scope mode is turned on and the probe is in your blank solution. Your blank solution should contain all sources of absorbance except the analyte of interest. A spectrum should be acquired that looks like the spectrum in Figure 10.4. If the peak intensity is not between 3,000 and 4,000 counts and yet a spectrum appears, adjust the integration time, shorter to reduce the counts or longer to increase the counts so that the peak intensity lies within this range.

Making a measurement: An absorbance measurement of the sample of interest can be made after the dark spectrum and a reference blank spectrum have been stored and the instrument is calibrated. To make the absorbance measurement for a sample, place the dip probe tip in the sample of interest.

1. Make sure no bubbles are present in the tip area

2. Under Spectrum, select Absorbance mode.

3. Under File - Save - Processed. Name the spectrum.

4. Open the saved spectrum by selecting File - Open - Processed

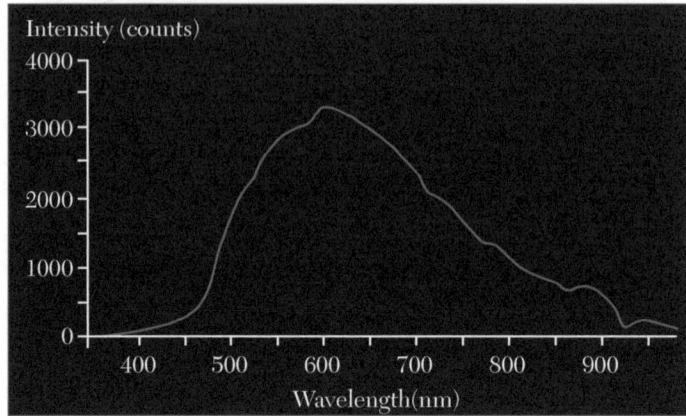

FIGURE 10.4 Spectrum.

When done with the spectrophotometer, turn off the light source and close the software.

10.3.1 Spectrometers

Ocean Optics revolutionized the spectrometer field when it introduced the *diffraction grating based spectrometer* using a *charged coupled device* (CCD) for light collection. Combined with fiber optic technology, the Ocean Optics equipment was, and continues to be, a powerful and relatively inexpensive research tool. The spectrometers are very compact (about the size of a deck of playing cards) and light weight. These spectrometers utilize USB technology, making them ideal for field use (in fact, many researchers have their spectrometers in underwater housings for work with corals in their natural habitats).

As mentioned, these spectrometers utilize diffraction gratings to split incoming light into its spectral components. This diffracted light falls upon the CCD array and specialized software analyzes and reports spectral characteristics. No single diffraction grating is efficient for broadband analyses (200–1,100 nm). Ocean Optics offers a choice of 14 gratings for the USB4000. Analyses of UV (down to ~220 nm) and visible radiation will require a specific grating, while work with visible wavelengths and near infrared will call for a different grating. At a minimum, one should decide what part of the spectrum is of interest for analysis. Even with the proper grating, the reported spectral quality is not 100% correct. A grating is efficient over a given spectral range (Ocean Optics uses 30% efficiency as the cutoff). This means these spectrometers generally underreport some wavelengths. It is most efficient at 500 nm and underreports other wavelengths, particularly blue and red wavelengths.

Another important consideration is of expected light intensity. Six variously sized apertures (called *slits*) are available for the USB4000 (5, 10, 25, 50, 100, 200 μ and "none"—the fiber optic size acts as the light regulating mechanism). Low light applications (such as fluorescence measurements) require larger slits (say, 200 μ), while higher light intensity requires a smaller slit. Optical resolution is a function of slit width and holographic grating. An optical resolution of just a fraction of a nanometer is possible. However, the optical resolution will be generally between 2–10 nm. These spectrometers are compatible with USB 1.1 and USB 2.0; use of a serial port is possible with the available 5v power supply.

Fiber optics: It is not absolutely necessary to use fiber optic patch cords if the goal is simply to measure lamp spectra—one can merely remove the protective cover from the input aperture and point the spectrometer at the light source. However, fiber optic cords do offer advantages in that they can be tightly attached to the aperture, thus protecting the internal works of the spectrometer (the thought of a drop of saltwater or debris entering the spec housing is frightening). The fiber optic cords also offer other advantages. While it is apparent that cords are a must for connecting optional accessories, it should be noted that the size (diameter) of the cord is also a critical consideration. Very simply, the larger the diameter of the fiber optic cord, the more light is transmitted. This is useful to know if high light intensity saturates the CCD array and causes the reported intensity to be above the maximum allowed. Use of a smaller diameter cord could attenuate (weaken) the signal, thus allowing measurement. Fibers are available in the following diameters (in μ): 8 (VIS/NIR only, range of 450–1,000 nm), 50, 100, 200, 300, (for use with UV <250 nm) 400, 600, and 1,000. Fibers are usually 2 meters in length, and custom lengths are available.

Software and applications:

SpectraSuite: SpectraSuite is Java-based software that operates with Windows 98/Me/2000/XP, Mac OSX, and Linux w/USB port. Light measurements are usually reported as *counts* (a generic term) but it is possible to measure absolute irradiance. Graphical data can be copied, and numerical data exported to spreadsheets such as Excel. Integration time is programmable, as are functions such as averaging and boxcar smoothing. Two spectral charts can be open at one time thus allowing simultaneous use of two spectrometers (especially useful when observed fluorescence excitation and emission wavelengths).

OOISensors: Software specifically for use with fluorescent pH and dissolved oxygen probes.

OOIIrrad: Software for measure of relative and absolute irradiance.

10.3.2 Ocean Optics Visible Spectrophotometer

A spectrophotometer is an instrument capable of measuring the absorbance or transmittance of a sample at a selected wavelength or range of wavelengths. A tungsten light bulb or other light source in the instrument produces *white light*. The light is focused through the sample and a diffraction grating then disperses the wavelengths from the lamp's continuous spectrum. The light that has passed through the sample (i.e., the transmitted light) and been dispersed strikes an array of detectors: one for each wavelength. These detectors record the amount of transmitted light at each wavelength. The signal given by each detector is used to calculate the absorbance at each wavelength; the computer displays the signal as a plot of absorbance versus wavelength. This graph is called the spectrum of the sample. Simple color meters use red, green, and blue filters in combination with diodes or sensor pixels for measurement. More advanced systems use tri-stimulus filters. These work well for incandescent light sources, but are less accurate for LEDs. Handheld color meters may measure up to 20 wavelength bands, but this is not enough for research or high accuracy measurements. To detect small color changes, very high color resolution is necessary. By capturing the complete spectrum, the color measurement made by a spectrometer allows careful and detailed analysis of data. Color meters and analyzers based on filters or detection over specific bands simply leave a lot of information on the table to better inform the color measurement.

Some color analyzers are also strongly dependent on lighting conditions, since objects tend to appear different colors under different illumination. With the right lighting, two objects can appear to be identical in color even if the reflected spectral power distributions differ, an effect called *metamerism*. If the lighting changes, however, the colors can look significantly different. This makes controlled lighting conditions essential to consistent results. When color measurements are made with a spectrometer, a full reflected or emissive spectrum is the starting point for all calculations. That allows the data to be analyzed in different ways, and even recalculated at a later date to change the observer, the illuminant, or the color space. It offers maximum flexibility with the same high accuracy as if the calculation had been performed that way initially.

10.4 INTERESTING FACTS ABOUT THE OCEAN

Area: about 140 million square miles (362 million sq km), or nearly 71% of the Earth's surface.

Average depth: 12,200 ft (3,720 meters).

Deepest point: 36,198 ft (11,033 m) in the Mariana Trench in the western Pacific.

- If you could evaporate all the water out of all the oceans and spread the resulting salt over the Earth, you would have a 500-ft layer covering everything.

- Do you know that life in the ocean varies as we go deep? Plants grow to a depth of about 107 meters. Fish color changes; fish living near surface are often blue, green, or violet. In the twilight zone, which is 180 meters down, fish are silver or light colored. Many fish living 3,000 meters down in the dark ocean waters have their own lights.

- The Red Sea in the Indian Ocean has the saltiest water; it is also known as Dead Sea as its water is so salty that nothing can remain alive in it.

- The Dead Sea is so salty because it is surrounded by a hot desert whose intense heat causes sea water to evaporate faster, and thus as much salt remains in the sea as water goes into air. The Dead Sea exerts a lot of upwards force due to its large quantity of salt, so people remain afloat or can swim with no effort.

- The Pacific is the largest and deepest ocean in the world.

- The Mercury, Gemini, and Apollo spacecraft landed in the Atlantic and Pacific oceans when they returned to earth.

- The Indian Ocean is the warmest ocean. The temperature of surface water sometimes touches 36.6 degrees (C).

- The scientists who specialize in study of oceans are called oceanographers.

- "Pacific" means "peaceful." When Europeans first found it, they found it very calm and peaceful, so they named it "Pacific."

- An estuary is a place where a river flows into the sea.

- Mangroves are trees and shrubs that grow on sea shores and estuaries. Their roots can breathe in oxygen.

- Splash zones are parts of the beaches which are covered by water as the tide come.

- Tsunami is a Japanese word meaning "high sea-waves."

- Buoys are colored metal floats which are anchored to the sea bed, used to warn sailors about dangers of rocks, sand banks, wreckage, etc.

- A knot is a measure of speed at sea. One knot equals 1.85 kilometers per hour.

- Scuba diving means diving under water with the help of scuba equipment for under-water breathing.

- Salt is produced by evaporating seawater. This is done by flooding salt pans or salt farms with seawater and allowing evaporation by the Sun to occur. Salt is left behind by evaporating seawater.

- The world's oceans contain enough water to fill a cube with edges over 1,000 kilometers (621 miles) in length.

- Ocean tides are caused by the Earth rotating while the Moon and Sun's gravitational pull acts on ocean water.

- While there are hundreds of thousands of known marine life forms, there are many that are yet to be discovered. Some scientists suggest that there could be millions of marine life forms out there.

- Oceans are frequently used as a means of transport, with various companies shipping their products across oceans from one port to another.

- The largest ocean on Earth is the Pacific Ocean; it covers around 30% of the Earth's surface.

- The Pacific Ocean contains around 25,000 different islands, many more than are found in Earth's other oceans.

- The Pacific Ocean is surrounded by the Pacific Ring of Fire, a large number of active volcanoes.

- The second-largest ocean on Earth is the Atlantic Ocean; it covers over 21% of the Earth's surface.

- The Atlantic Ocean's name refers to Atlas of Greek mythology.

- The Bermuda Triangle is in the Atlantic Ocean.

- The third-largest ocean on Earth is the Indian Ocean; it covers around 14% of the Earth's surface.

- During winter the Arctic Ocean is almost completely covered in sea ice.

- While some disagree on whether it is an ocean or just part of larger oceans, the Southern Ocean includes the area of water that encircles Antarctica.

- World Oceans Day is June 8.

- More than 97% of all our planet's water is contained in the ocean.

- The top 10 ft of the ocean hold as much heat as our entire atmosphere.

- The average depth of the ocean is more than 2.5 miles.

- The oceans provide 99% of the Earth's living space—the largest space in our universe known to be inhabited by living organisms.

- More than 90% of this habitat exists in the deep sea known as the abyss.

- Less than 10% of this living space has been explored by humans.

- Mount Everest (the highest point on the Earth's surface at 5.49 miles) is more than 1 mile shorter than the Challenger Deep (the deepest point in the ocean at 6.86 miles).

- The longest continuous mountain chain known to exist in the universe resides in the Atlantic Ocean at more than 40,000 miles long.

- The Monterey Bay Submarine Canyon is deeper and larger in volume than the Grand Canyon.

- The average temperature of the oceans is 2°C, about 39°F.

- Water pressure at the deepest point in the ocean is more than 8 tons per square inch, the equivalent of one person trying to hold 50 jumbo jets.

- The Gulf Stream off the Atlantic seaboard of the United States flows at a rate nearly 300 times faster than the typical flow of the Amazon River, the world's largest river.

- The world's oceans contain nearly 20 million tons of gold.

- The color blue is least absorbed by seawater; the same shade of blue is most absorbed by microscopic plants, called phytoplankton, drifting in seawater.

- A new form of life, based on chemical energy rather than light energy, resides in deep-sea hydrothermal vents along mid-ocean ridges.

- A swallow of seawater may contain millions of bacterial cells, hundreds of thousands of phytoplankton, and tens of thousands of zooplankton.

- The blue whale, the largest animal on our planet ever (exceeding the size of the greatest dinosaurs) still lives in the ocean.

- The gray whale migrates more than 10,000 miles each year, the longest migration of any mammal.

- The Great Barrier Reef, measuring 1,243 miles, is the largest living structure on Earth. It can be seen from the Moon.

- More than 90% of the trade between countries is carried by ships and about half the communications between nations use underwater cables.

- More oil reaches the oceans each year from leaking automobiles and other non-point sources.

- Fish supply the greatest percentage of the world's protein consumed by humans.

- Most of the world's major fisheries are being fished at levels above their maximum sustainable yield; some regions are severely overfished.

- The Grand Banks, the pride of New England fishing for centuries, are closed due to overfishing.

- Eighty percent of all pollution in seas and oceans comes from land-based activities.

- Three-quarters of the world's mega-cities are by the sea.

- By 2010, 80% of people will live within 60 miles of the coast.

- Plastic waste kills up to 1 million sea birds, 100,000 sea mammals, and countless fish each year. Plastic remains in our ecosystem for years harming thousands of sea creatures every day.

- Over the past decade, an average of 600,000 barrels of oil a year has been accidentally spilled from ships.

- Tropical coral reefs border the shores of 109 countries, most of which are among the world's least developed. Significant reef degradation has occurred in 93 countries.

- Although coral reefs comprise less than 0.5% of the ocean floor, it is estimated that more than 90% of marine species are directly or indirectly dependent on them.

- There are about 4,000 coral reef fish species worldwide, accounting for approximately a quarter of all marine fish species.

- Nearly 60% of the world's remaining reefs are at significant risk of being lost in the next three decades.

- The major causes of coral reef decline are coastal development, sedimentation, destructive fishing practices, pollution, tourism, and global warming.

- Less than 0.5% of marine habitats are protected, compared with 11.5% of global land area.

- The High Seas—areas of the ocean beyond national jurisdiction—cover almost 50% of the Earth's surface. They are the least protected part of the world.

- Although there are some treaties that protect ocean-going species such as whales, as well as some fisheries agreements, there are no protected areas in the High Seas.

- Studies show that protecting critical marine habitats—such as warm-and cold-water coral reefs, seagrass beds, and mangroves—can dramatically increase fish size and quantity.

- More than 3.5 billion people depend on the ocean for their primary source of food. In 20 years, this number could double to 7 billion.

- Populations of commercially attractive large fish, such as tuna, cod, swordfish, and marlin have declined by as much as 90% in the past century.

- Each year, illegal longline fishing, which involves lines up to 80 miles long, with thousands of baited hooks, kills over 300,000 seabirds, including 100,000 albatrosses.

- As many as 100 million sharks are killed each year for their meat and fins, which are used for shark fin soup. Hunters typically catch the sharks, de-fin them while alive, and throw them back into the ocean, where they either drown or bleed to death.

- Global by-catch—unintended destruction caused by nonselective fishing gear, such as trawl nets, longlines, and gillnets—amounts to 20 million tons a year.

- The annual global by-catch mortality of small whales, dolphins, and porpoises alone is estimated to be more than 300,000 individuals.

- Ninety-four percent of life on Earth is aquatic. That makes us land-dwellers a very small minority.

- About 70% of the planet is ocean, with an average depth of more than 12,400 feet. Given that photons (light) can't penetrate more than 330 feet below the water's surface, most of our planet is in a perpetual state of darkness.

- Because the architecture and chemistry of coral is so like human bone, coral has been used to replace bone grafts in helping human bone to heal quickly and cleanly.

- The deep sea is the largest museum on Earth: There are more artifacts and remnants of history in the ocean than in all the world's museums, combined.

- We have only explored less than 5% of the Earth's oceans. In fact, we have better maps of Mars than we do of the ocean floor.

- The longest mountain range in the world is under water. Called the Mid-Oceanic Ridge, this chain of mountains runs through the middle of the Atlantic Ocean and into the Indian and Pacific oceans. It is more than 35,000 miles long, has peaks higher than those in the Alps and comprises 23% of the Earth's total surface.

- We didn't send divers down to explore the Mid-Ocean Ridge until 1973—four years after Neil Armstrong and Buzz Aldrin walked on the moon—when a French-American crew of seven entered the 9,000-foot-deep Great Rift in the French submersible *Archimede*.

- The ocean boasts an array of unusual geographic features, such as pillars that reach several stories high and chimneys that send up sulphuric acid. In the ocean-floor neighborhood of the Gulf of Mexico, brine pools mark the floor, along with underwater volcanoes that spew mud and methane, rather than lava.

- These wonderful formations aren't barren. Underwater hot springs that shoot water that's 650°F—hot enough to melt lead—boast a profusion of life, from 10-ft tall tubeworms to giant clams that function without digestive systems.

- The part of the ocean farthest from land lies in the South Pacific and is known as Point Nemo or "The Pole of Inaccessibility."

The Bermuda Triangle

- Located in the Atlantic Ocean, the Bermuda Triangle falls between Bermuda, Puerto Rico, and Florida as shown in above figure.

- The Bermuda Triangle has long been believed to be the site of a number of mysterious plane and boat incidents have occurred.

- While it has become part of popular culture to link the Bermuda Triangle to paranormal activity, most investigations indicate bad weather and human error are the more likely culprits.

- Research has suggested that many original reports of strange incidents in the Bermuda Triangle were exaggerated and that the actual number of incidents in the area is similar to that in other parts of the ocean.

- While its reputation may scare some people, the Bermuda Triangle is part of a regularly sailed shipping lane, with cruise ships and other boats also frequently sailing through the area.

- Aircraft are also common in the Bermuda Triangle, with both private and commercial planes commonly flying through the airspace.

- Stories of unexplained disappearances in the Bermuda Triangle started to reach public awareness around 1950 and have been consistently reported since then.

- Unverified supernatural explanations for Bermuda Triangle incidents have included references to UFO's and even the mythical lost continent of Atlantis.

- Other explanations have included magnetic anomalies, pirates, deliberate sinkings, hurricanes, gas deposits, rough weather, huge waves, and human error.

- Some famous reported incidents involving the Bermuda Triangle include the *USS Cyclops* and its crew of 309 that went missing after leaving Barbados in 1918; the TBM Avenger bombers that went missing in 1945 during a training flight over the Atlantic; a Douglas DC-3 aircraft containing 32 people that went missing in 1958, with no trace of the aircraft ever found; and a yacht found in 1955 that had survived three hurricanes but was missing all its crew.

Coastlines: The total length of the world's coastlines is about 315,000 miles, enough to circle the equator 12 times. As coastal zones become more and more crowded, the quality of coastal water will suffer, wildlife will be displaced, and the shorelines will erode. Sixty percent of the Pacific and 35% of the Atlantic Coast shoreline are eroding at a rate of a meter every year. More than half the world's population live within 100 km or 60 miles of the coast. This is more than 2.7 billion people. Rapid urbanization will lead to more coastal mega-cities containing 10 million or more people. By the end of the millennium, 13 out of 15 of the world's largest cities will be located on or near the coast. Growing population in coastal areas leads to more marine pollution and distribution of coastal habitats. Some 6.5 million tons (6,500,000,000 kilo) of litter finds its way into the sea each year.

Fisheries: The sea provides the biggest source of wild or domestic protein in the world. Each year some 70 to 75 million tons of fish are caught in the ocean. Of this amount, around 29 million tons is for human consumption. Global fish production exceeds that of cattle, sheep, poultry, or eggs. Fish can be produced in two ways: by capture and by aquaculture. Total production has grown 34% over the last decade. The largest numbers of fish are in the Southern Hemisphere due to the fact that these waters are not largely exploited by man. Fifteen out of seventeen of the world's largest fisheries are so heavily exploited that reproduction can't keep up, with the result that many fish populations are decreasing rapidly. Species of fish endangered by overfishing are: tuna, salmon, haddock, halibut, and cod. In the 19th century, codfish weighing up to 200 pounds used to be caught. Nowadays, a 40-pound cod is considered a giant. Reason: overfishing.

Rising sea level: The sea level has risen an average of 4–10 inches (10–25 cm) over the past 100 years and scientists expect this rate to increase. Sea levels will continue rising even if the climate has stabilized, because the ocean reacts slowly to changes. Ten thousand years ago the ocean level was about 330 ft (110 meters) lower than it is now. If the entire world's ice melted, the oceans would rise 200 ft (66 meters).

Volcanic activity: Ninety percent of all volcanic activity on Earth occurs in the ocean. The largest known concentration of active volcanoes (approximately 1,133) on the sea floor is in the South Pacific.

Density: The density of ocean water varies. It becomes denser as it becomes colder, right down to its freezing point of -1.9°C. (This is unlike fresh water, which is most dense at 4°C, well above its freezing point.)

Water temperature: Under the enormous pressures of the deep ocean, seawater can reach very high temperatures without boiling. A water temperature of 400°C has been measured at one hydrothermal vent. The average temperature of all ocean water is about 3.5°C. Almost all deep ocean temperatures are only a little warmer than freezing (39°F).

Ice: Antarctica has as much ice as the Atlantic Ocean has water. Ten percent of the earth's surface is covered with ice. The Arctic Ocean is the smallest ocean, holding only 1% of the Earth's seawater. This is still more than 25 times as much water as all rivers and freshwater lakes. The average thickness of the Arctic ice sheet is about 9–10 ft, although there are some areas as thick as 65 ft. In the unlikely event that all the polar ice was to melt, the sea level all over the world would rise 500–600 ft. As a result, 85–90% of the Earth's surface would be covered with water as compared to the current 71%. The United States would be split by the Mississippi Sea, which would connect the Great Lakes with the Gulf of Mexico. The Arctic produces 10,000 to 50,000 icebergs annually. The amount produced in the Antarctic regions is inestimable. Icebergs normally have a four-year lifespan; they begin entering shipping lanes after about three years.

Carbon dioxide absorption: Oceans absorb between 30% and 50% of the carbon dioxide produced by burning fossil fuel. Carbon dioxide is transported downwards by plankton. Any change in the temperature of the ocean water, influences the ability of plankton to take up carbon dioxide. This has consequences for the ecosystem, because plankton forms the base of the food web.

Reefs: Over 60% of the world's coral reefs are threatened by pollution, sedimentation, and bleaching due to rising water temperatures caused by global warming. Global Coral Monitoring Network (GCRMN) states that currently 27% of all coral reef worldwide has disappeared and by 2050 only 30% will be left.

Rubbish/contamination: In one year, three times as much rubbish is dumped into the world's oceans as the weight of fish caught. A single quart of motor oil can contaminate up to 2 million gallons of drinking water.

Oil: Oil is one of the ocean's greatest resources. It gives us heat for our homes, endless consumer products, and the ability to run the engines of cars, planes, and boats for auto transport all over the world. Nearly one-third of the world's oil comes from offshore fields in our oceans which, as we've seen, can have devastating effects on our ocean's ecosystems. The transport of ocean oil from the Arabian Gulf, the North Sea, and the Gulf of Mexico reaches all corners of the globe daily. Oil was also born from the sea. Millions of years ago, countless marine microscopic plants (phytoplankton) and animals (zooplankton) lived in the ancient seas as they do today. As they died, the skeletal remains of these tiny organisms settled to the sea floor, mixed with mud and silt, and over millions of years, formed organic-rich sedimentary layers. Other sediments continued to be deposited and further buried the organic-rich sediment layer to depths of thousands of feet, compressing the layers into a rock that would become the source for oil. Over the years, as the depth of the burial increased, pressure increased, along with the temperature. Under such conditions, and over long periods of time, the original skeletal remains of phytoplankton and zooplankton changed, breaking down into simpler substances called hydrocarbons—compounds of hydrogen and carbon. This process continues, although it will be millions of years before the next batch of oil is done cooking. Refined oil is also responsible for polluting the ocean.

Salinity: Some scientists estimate that the oceans contain as much as 50 quadrillion tons (50 million billion tons = 50,000,000,000,000,000) of dissolved solids. If the salt in the ocean could be removed and spread evenly over the Earth's land surface, it would form a layer more than 500 ft (166 meters) thick, about the height of a 40-story office building. The ocean's principal dissolved solids are sodium salts (sodium chloride or common salt), calcium salts (calcium carbonate or lime, and calcium sulfate), potassium salts (potassium sulfate), and magnesium salts (magnesium chloride, magnesium sulfate, and magnesium bromide). Atlantic sea water is heavier than Pacific sea water due to its higher salt content. The freezing point of sea water depends on its salt content. Typical ocean water has about 35 grams of salt per liter and freezes at −19°C.

Seawater's inorganic salt components:

Chloride	Cl-	55.04%
Sodium	NA+	30.61%
Sulfate	SO4--	7.68%
Magnesium	Mg++	3.69%
Calcium	Ca++	1.16%
Potassium	K+	1.16%
Carbonic Acid	HCO3-	0.41%
Bromine	Br-	0.19%
Boric Acid	H3Bo3	0.07%
Strontium	Sr++	0.04%
Total		99.

Desalination: Arabian Gulf reverse osmosis plants treat 500,000,000 gallons of sea water to obtain 100,000,000 gallons of fresh water. Daily over 500,000,000 gallons of seawater must be heated to extremely high temperatures. Mixed with toxic chemicals, the seawater is injected under high pressure through a series of membrane filters. Only 100,000,000 gallons of fresh water is generated. The 5:1 ratio of this highly inefficient process means 400,000,000 gallons of untreated water are returned to the sea each day. The higher temperature of the discharged water causes environmental problems. Worse, the super-heated brine discharge has significantly higher levels of total dissolved solids and toxic chemicals are mixed in with it. This pollution is usually discharged back into the sea.

The 10 largest territorial powers (in million sq km):

Country	Land Area	Sea Claims	Total Area
1. Australia	7,700,000	28,500,000	36,200,000
2. Russia	17,100,000	21,500,000	38,600,000
3. USA	9,400,000	20,000,000	29,400,000
4. Canada	9,900,000	12,400,000	22,300,000
5. China	9,600,000	11,400,000	21,000,000
6. Brazil	8,500,000	11,000,000	19,500,000

(Continued)

Country	Land Area	Sea Claims	Total Area
7. France	500,000	6,000,000	6,500,000
8. Indonesia	1,900,000	6,000,000	7,900,000
9. India	3,200,000	5,700,000	8,900,000
10. New Zealand	300,000	5,500,000	5,800,000

An estimated 50–80% of all life on earth is found under the ocean surface and the oceans contain 99% of the living space on the planet. Eighty-five percent of the area and 90% of the volume constitute the dark, cold environment we call the deep sea. The average depth of the ocean is 3,795 meters. The average height of the land is 840 meters. The oceans contain 97% of the Earth's water. Less than 1% is fresh water, and 2–3% is contained in glaciers and ice caps (and is decreasing).

The speed of sound in water is 1,435m/sec—nearly five times faster than the speed of sound in air. The highest tides in the world are at the Bay of Fundy, which separates New Brunswick from Nova Scotia. At some times of the year the difference between high and low tide is 16.3 meters, taller than a three-story building. Earth's longest mountain range is the Mid-Ocean Ridge more than 50,000km in length, which winds around the globe from the Arctic Ocean to the Atlantic, skirting Africa, Asia, and Australia, and crossing the Pacific to the west coast of North America. It is four times longer than the Andes, Rockies, and Himalayas combined. The pressure at the deepest point in the ocean is more than 11,318 tons/sq meter.

The largest recorded tsunami measured 60 meters above sea level, caused by an 8.9 magnitude earthquake in the Gulf of Alaska in 1899 traveling at hundreds of km/hr.

The average depth of the Atlantic Ocean, with its adjacent seas, is 3,332 meters; without them it is 3,926 meters. The greatest depth, 8,381 meters, is in the Puerto Rico Trench. The Pacific Ocean, the world's largest water body, occupies a third of the Earth's surface. The Pacific contains about 25,000 islands (more than the total number in the rest of the world's ocean combined), almost all of which are found south of the equator. The Pacific covers an area of 179.7 million sq km. The Kuroshio Current, off the shores of Japan, is the largest current. It can travel between 40–121 km/day at 1.6–4.8 km/hr, and extends some 1,006 meters deep. The Gulf Stream is close to this current's speed. The Gulf Stream is a well-known current of warm water in the Atlantic Ocean. At a speed of 97km/day, the Gulf Stream moves a hundred times as much water as all the rivers on earth and flows at a rate 300 times faster than Amazon, which is the world's largest river.

Earth's oceans are unique in the universe—as far as we know: Earth is the only known planet or moon to have large bodies of liquid water on its surface. Our planet lies in the "Goldilocks" zone—not too hot, not too cold, and with enough atmospheric pressure to prevent liquid surface water from evaporating into space. Although we don't yet know of any other planets or moons with liquid water oceans, it's likely that they do exist and we just haven't found them. In our own solar system, there is growing evidence that the planet Mars may have liquid water not on the surface but underground. There is also strong evidence that liquid oceans may be hidden beneath the thick icy surfaces of three of Jupiter's moons (Europa, Callisto, and Ganymede) and two of Saturn's moons (Titan and Enceladus).

For every species of marine life we know of, at least another three are yet to be discovered: Our oceans teem with life ranging from the blue whale—the biggest animal on Earth—to tiny microbes. But nobody knows exactly how many different species live in this environment. There is no data for around 20% of the ocean's volume. The Census of Marine Life, a 10-year international project to identify life in our oceans, found nearly 250,000 species. But scientists believe a least a million species of marine life could be out there, and that's not counting the tens or even hundreds of millions of kinds of microbes that make up the majority of marine life. What we do know is that ocean life survives in the most extreme environments. Scientists have found life that can survive in temperatures that melt lead, where seawater freezes into ice, or there's no light or oxygen. In fact, the dark ocean zone between 1,000 and 5,000 meters known as the abyssal zone has a far greater range of marine life than we once thought.

Water takes around 1,000 years to travel all the way around the whole globe: The oceans not only have waves, tides, and surface currents—they also have a constantly moving system of deep-ocean circulation driven by temperature and salinity. Known as the global ocean conveyor belt or thermohaline current (thermo = temperature, haline = salinity), this deep ocean current gets one of its starts in the polar region near Norway. As sea ice forms, the water left behind becomes saltier and denser and begins to sink, making room for warmer and less dense incoming surface water, which in turn eventually becomes cold and salty enough to sink. The cold dense water flows along the ocean bottom all the way from the Northern Hemisphere to the Southern Ocean, where it merges with more cold dense water from Antarctica and is swept into the Indian and Pacific Oceans as shown in below figure. Eventually it mixes with warmer water and rises to the surface before finding its way back to the Atlantic. It can take 1,000 years to complete this cycle.

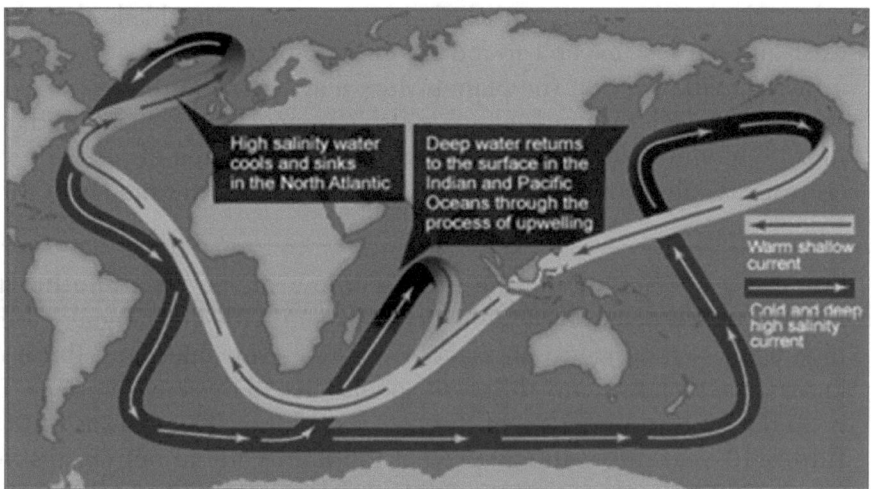

Half of all the oxygen we breathe is produced in the ocean: Some of this oxygen is produced by sea weeds and sea grasses, but the vast majority of the oxygen is produced by phytoplankton, microscopic single-celled organisms that have the ability to photosynthesize. These tiny creatures live in the surface layer of the ocean (and in lakes and rivers) and form the very base of the aquatic food chain. During photosynthesis, phytoplankton removes carbon dioxide from sea water and release oxygen. The carbon becomes part of their bodies.

Oceans hold around 50 times more carbon than the atmosphere

Cold water can dissolve much more CO_2 than warm water, so the cold polar regions are net absorbers of CO_2. But as the cold water finds its way to warmer tropical areas, the oceans release CO_2 back into the atmosphere. The equatorial Pacific is thought to be the biggest single natural source of

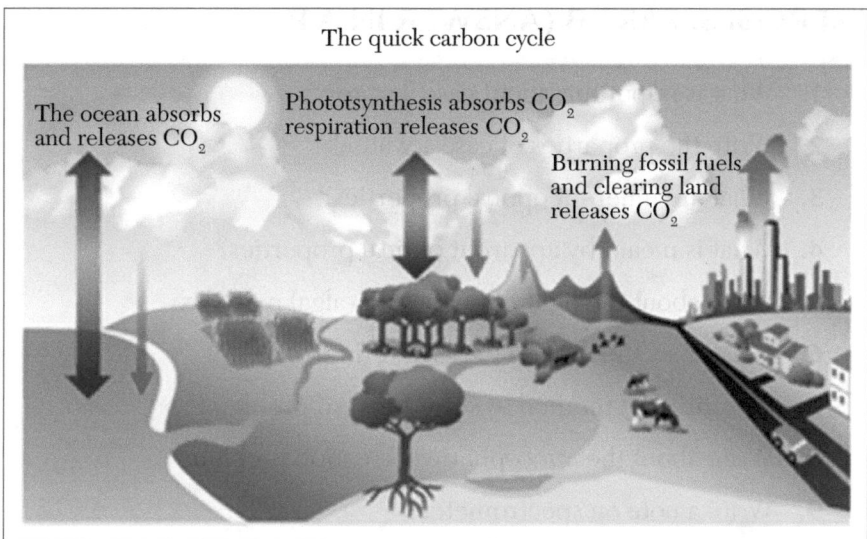

CO_2 in the atmosphere. Most of this carbon is exchanged with the atmosphere on a timescale of several hundred years.

Prior to the industrial revolution, the uptake and release of CO_2 on land and ocean was in a dynamic equilibrium. Since then, the oceans are thought to have absorbed about half of the carbon dioxide released from the burning of fossil fuels, with the rest remaining in the atmosphere.

EXERCISES: PART A (ANSWER IN A WORD OR A SENTENCE)

1. What are the two different types of optical properties of sea water?

2. What are the wavelengths that light absorbs highly?

3. What is Raman scattering by water?

4. _____ is an instrument capable of measuring the absorbance or transmittance of a sample at a selected wavelength.

5. What is the software used to measure relative and absolute irradiance?

6. _____ software is used with fluorescent pH and dissolved oxygen probes.

EXERCISES: PART B (ANSWER IN A PAGE)

1. What is important about case 2 waters?

2. Write the properties of case 1 water.

3. What are inherent optical properties?

4. What is meant by apparent optical properties?

5. Write about the light effect on non-algal particles.

6. What is scattering and backscattering?

7. Describe the method to solve forward problem.

8. Write about the atmospheric correction problem.

9. Write a note on spectrometers.

10. Write about the ocean optics visible spectrophotometer.

EXERCISES: PART C (ANSWER WITHIN 2 OR 3 PAGES)

1. Write the optical constituents of sea water.

2. Explain the absorption, scattering, and fluorescence by phytoplankton.

3. What are the effects of optic properties on CDOM?

4. In detail, write about the retrieval of oceanic constituents from ocean color measurements at sea level.

5. Describe the operation of ocean optics dip probe spectrometers.

WEB LINKS

http://www.oceanopticsbook.info/

INDIAN SATELLITES FOR OCEAN MONITORING

Satellite	Launch date	Remarks
IRS-1A	17 March 1988	Earth observation satellite. First operational remote sensing satellite.
IRS-1B	29 August 1991	Earth observation satellite. Improved version of IRS-1A.
IRS-P2	15 October 1994	Earth observation satellite. Launched by second developmental flight of PSLV. Mission accomplished after three years of service in 1997.
IRS-1C	29 December 1995	Earth observation satellite. Launched from Baikonur Cosmodrome.
IRS-P3	21 March 1996	Earth observation satellite. Carries remote sensing payload and an X-ray astronomy payload. Launched by third developmental flight of PSLV.
IRS-1D	29 September 1997	Earth observation satellite. Same as IRS-1C.

Continued

Satellite	Launch date	Remarks
Oceansat-1 (IRS-P4)	26 May 1999	Earth observation satellite. Carries an Ocean Color Monitor (OCM) and a Multi-frequency Scanning Microwave Radiometer (MSMR).
RESOURCESAT-1 (IRS-P6)	17 October 2003	Earth observation/remote sensing satellite. Intended to supplement and replace IRS-1C and IRS-1D.
CARTOSAT-1	5 May 2005	Earth observation satellite. Provides stereographic in-orbit images with a 2.5-meter resolution.
CARTOSAT-2	10 January 2007	Advanced remote sensing satellite carrying a panchromatic camera capable of providing scene-specific spot images.
CARTOSAT-2A	28 April 2008	Earth observation/remote sensing satellite. Identical to CARTOSAT-2.
Oceansat-2 (IRS-P4)	23 September 2009	Gathers data for oceanographic, coastal and atmospheric applications. Continues mission of Oceansat-1.
CARTOSAT-2B	12 July 2010	Earth observation/remote sensing satellite. Identical to CARTOSAT-2A.
RESOURCESAT-2	20 April 2011	RESOURCESAT-2, ISRO's eighteenth remote-sensing satellite, followed RESOURCESAT-1. PSLV-C16 placed three spacecraft with a total payload mass of 1404 kg—RESOURCESAT-2 weighing 1206 kg, the Indo-Russian YOUTH-SAT weighing 92 kg and Singapore's X-SAT weighing 106 kg—into an 822 km polar Sun Synchronous Orbit (SSO).
RISAT-1	26 April 2012	RISAT-1, first indigenous all-weather Radar Imaging Satellite (RISAT-1), whose images will facilitate agriculture and disaster management weighs about 1858 kg.
SARAL	25 February 2013	SARAL (Satellite with *ARGOS* and *AL-TIKA*) is a joint Indo-French satellite mission for oceanographic studies.

Continued

Satellite	Launch date	Remarks
IRNSS-1A	1 July 2013	IRNSS-1A is the first satellite in the Indian Regional Navigation Satellite System (IRNSS). It is one of the seven spacecraft constituting the IRNSS space segment.
IRNSS-1B	4 April 2014	IRNSS-1B is the second satellite in the Indian Regional Navigation Satellite System (IRNSS).
IRNSS-1C	10 November 2014	IRNSS-1C is the third satellite in the Indian Regional Navigation Satellite System (IRNSS).
IRNSS-1D	28 March 2015	IRNSS-1D is the fourth satellite in the Indian Regional Navigation Satellite System (IRNSS).

TABLE A1.1 Indian Satellites Monitoring the Ocean

Indian Remote Sensing satellites (IRS) are a series of Earth Observation satellites, built, launched and maintained by Indian Space Research Organisation. The IRS series provides many remote sensing services to India. Indian Remote Sensing (IRS) is a satellite program to support the national economy in the areas of water resources, forestry and ecology, geology, water sheds, marine fisheries, and coastal management. IRS system is the largest constellation of remote sensing satellites for civilian use in operation today in the world, with 12 operational satellites. All these are placed in polar Sun-synchronous orbit and provide data in a variety of spatial, spectral, and temporal resolutions. The Indian Remote Sensing Program completed 25 years of successful operations on March 17, 2013.

Data from Indian Remote Sensing satellites are used for various applications of resources survey and management under the National Natural Resources Management System (NNRMS). Following is the list of those applications:

- Space Based Inputs for Decentralized Planning (SIS-DP)

- National Urban Information System (NUIS)

- ISRO Disaster Management Support Program (ISRO-DMSP)

- Biodiversity Characterizations at landscape level

- Preharvest crop area and production estimation of major crops

- Drought monitoring and assessment based on vegetation conditions

- Flood risk zone mapping and flood damage assessment

- Hydro-geomorphological maps for locating underground water resources for drilling wells

- Irrigation command area status monitoring

- Snow-melt run-off estimates for planning water use in downstream projects

- Land use and land cover mapping

- Urban planning

- Forest survey

- Wetland mapping

- Environmental impact analysis

- Mineral prospecting

- Coastal studies

- Integrated Mission for Sustainable Development (initiated in 1992) for generating locale-specific prescriptions for integrated land and water resources development in 174 districts.

The initial versions are composed of the 1 (A, B, C, D). The later versions are named based on their area of application, including OceanSat, CartoSat, and ResourceSat. Some of the satellites have alternate designations based on the launch number and vehicle (P series for PSLV). Data from IRS are available to its users through NRSC Data Centre and through Bhuvan Geoportal of ISRO. NRSC data center provide data through its purchase process while Bhuvan Geoportal provides data in free and open domain.

IRS Launch Plans

RESOURCESAT-3: Resourcesat-3 will carry more advanced LISS-III-WS (Wide Swath) Sensor having similar swath of RESOURCESAT-2 and revisit capability as advanced wide field sensor (AWiFS), thus overcoming any spatial resolution limitation of AWiFS. Satellite would also carry atmospheric correction sensor (ACS) for quantitative interpretation and geophysical parameter retrieval.

CARTOSAT-3: A continuation of Cartosat series, it will have a resolution 30 cm and 6 km swath suitable for cadastre and infrastructure mapping and analysis. It would also enhance disaster monitoring and damage assessment.

OCEANSAT-3: Oceansat-3 would carry thermal IR sensor, 12 channels ocean color monitor, scatterometer, and passive microwave radiometer. IR sensor and ocean color monitor would be used in the analysis for operational potential fishing zones. Satellite is mainly for ocean biology and sea state applications. The National Remote Sensing Centre (NRSC) at Hyderabad is the nodal agency for reception, archival, processing, and dissemination of remote sensing data in the country. NRSC acquires and processes data from all Indian remote sensing satellites like Cartosat-1, Cartosat-2, Resourcesat-1, IRS-1D, Oceansat-1, and Technology Experiment Satellite as well as foreign satellites like Terra, NOAA, and ERS.

Ocean color monitor (OCM) on Indian Remote Sensing Satellite IRS-P4: IRS-P4 is the first Indian satellite envisaged to meet the data requirements of the oceanographic community. The payload to be flown on-board IRS-P4 are: (a) OCM (Ocean Color Monitor) operating in eight narrow spectral bands in the visible/near-infrared region of the electromagnetic spectrum and with high revisit time (2 days), and (b) MSMR (multi-frequency scanning microwave radiometer) operating in microwave bands 6.6, 10.65, 18, and 21 GHz in dual polarization mode. The multi-frequency scanning microwave radiometer is envisaged to provide information on physical oceanographic parameters such as sea surface temperature, wind speed and atmospheric water vapor. The IRS-P4 spacecraft will be a polar orbiting satellite in sun synchronous orbit with nominal altitude of 720 km, providing revisit time of 2 days for OCM. The main features of the OCM instrument are outlined in Table A1.2.

The OCM is the first instrument to take advantage of pushbroom technology for achieving higher radiometric performance and higher spatial resolution while maintaining a large swath to provide high revisit time for ocean observations. Unlike SeaWiFS or OCTS, the instrument does not have common collecting optics coupled to a scan mechanism for realizing the wide swath. The instrument design is an extension of the imaging

Parameter		Specification
1. IGFOV at nominal altitude (m)		360×250
2. Swath (km)		>1420
3. No. of spectral bands		8
4. Spectral range (nm)		402–885
5. Spectral band	Central wavelength (bandwidth) in nm	Saturation radiance (mw cm^{-2} sr^{-1} μm^{-1})
C1	414 (20)	35.5
C2	442 (20)	28.5
C3	489 (20)	22.8
C4	512 (20)	25.7
C5	557 (20)	22.4
C6	670 (20)	18.1
C7	768 (40)	0.0
C8	867 (40)	17.2
6. Quantization bits		12
7. Camera MTF (at Nyquist frequency)		>0.2
8. Data rate (Mbits s^{-1})		20.8
Along Track Steering		+20, 0, 20

TABLE A1.2 OCM Instrument Features

concept adopted for IRS LISS payloads. The instrument has separate wide angle optics and a linear array CCD detector for each of the eight spectral channels. Individual and separate chains for each of the channels enables optimization of the performance of one channel without interdependence and hence the need to compromise the performance of other channels. The issue of spectral response variation with large incidence angles is overcome by the choice of telecentric design and use of a spectral selection filter close to the linear array detector. The pushbroom approach has enabled the use of a 12 bit digitizer to cover the instruments dynamic range. The instantaneous geometric field of view of the pixel is 360 meters across track and the sampling interval along track is 250 meters. The instrument is mounted on a mechanism to provide tilt in the along track direction to avoid sun glint.

The spectral bands for IRS-P4 OCM have been selected mindful of the optical properties of phytoplankton pigments (principally chlorophyll-a), inorganic suspended matter, and yellow substance, and the requirements of spectral bands for atmospheric correction. The first spectral band centered at 414 nm is selected primarily for discriminating Gelbstoffe or yellow substance from viable phytoplankton pigment. The band at 443 nm is close to the absorption maximum of chlorophyll, which is centered at approximately 435 nm, but it has been selected because its location minimizes interference from a Fraunhoffer absorption line at 435 nm. This band is used along with the 557 nm band for determining color boundaries, low chlorophyll concentrations, and diffuse attenuation coefficient. The third band, at 489 nm, along with a fourth channel at 512 nm would allow the use of multi-band spectral curvature algorithms and other second derivative algorithms to be applied to derive chlorophyll concentrations in coastal or Case-II waters. The 512 nm band along with a 557 nm channel would also be useful in deriving higher chlorophyll concentrations in Case 1 waters. The spectral band at 557 nm is used as a hinge point for determining chlorophyll concentration and water optical properties such as diffuse attenuation coefficient. The band at 670 nm is sensitive to backscattering from suspended matter in coastal waters, and is useful in quantifying suspended matter along with the channel at 557 nm. The spectral bands at 768 nm and 867 nm are used in atmospheric correction procedures.

The OCM instrument mounted on IRS-P4 launched by PSLV and placed in a polar sun synchronous 720 kilometer altitude orbit. Equatorial crossing is at 12 noon ±20 min, descending node. The satellite has provision for data recording onboard and will also transmit real time data to ground stations in X band. The ground station at Hyderabad, India will acquire data over the Indian subcontinent and the adjacent Arabian Sea and the Bay of Bengal. IRS-P4 OCM data may also be acquired by other ground stations with suitable augmentation / modification.

The IRS-P4 OCM data would be extremely useful for estimation of phytoplankton in oceanic/coastal waters, detection and monitoring of phytoplankton blooms, coastal upwelling, suspended sediment dynamics, location of fronts, identification of water mass boundaries, and oil pollution.

With additional input from other sensors as well as conventional data, IRS-P4 OCM data will provide detailed information on the coastal region owing to its increased spatial resolution. The information on pigments, in conjunction with sea surface temperature, will greatly assist in identification of potential fishery zones in coastal and oceanic waters. The potential end users of the OCM data products include fisheries management, marine industries, environmental management, and studies related to the estimation of primary productivity in the oceanic basins. IRS-P4 OCM, along with other ocean color sensors such as IRS-P3 MOS, SeaWiFS, MERIS and MODIS will assist the ocean color community in filling data gaps, and can also be used for the inter-calibration of different ocean color sensors.

Ocean monitor satellite launched by India: India's workhorse Polar Satellite Launch Vehicle lifted seven satellites into orbit, bolstering global ocean research, space surveillance, and taking miniature technology to new heights. The mission's seven payloads were deployed in orbit 490 miles above Earth in less than 22 minutes, wrapping up the PSLV's 23rd mission and its 19th success in a row. Among the rocket's passengers: the first asteroid-hunting satellite, a French-Indian ocean research craft, a small spacecraft built around a smartphone, a Canadian space surveillance satellite, two Austrian mini-telescopes, and a CubeSat built by students in Denmark.

The 900-pound SARAL satellite, equipped with a Ka-band altimeter to measure the height of ocean waves, separated first from the Indian booster and unfurled its solar panels moments later. Jointly developed by France and India, the SARAL mission will bounce radar waves off ocean and ice surfaces to measure topography, pulling back the curtain on ocean circulation and giving scientists insights into its role in global climate.

The radar signal will measure the height of waves with an accuracy of just a few inches, a feat similar to measuring the thickness of paper lying on the ground from the top of a skyscraper, according to scientists. SARAL joins the U.S.-French Jason 2 satellite, which also measures ocean topography from orbit. And SARAL's high-frequency Ka-band radar offers twice the spatial resolution of Jason 2's altimeter, giving researchers better data in coastal zones. SARAL will collect data over ice sheets. Sea-surface terrain can be used to chart currents, water temperatures, tides, and ocean eddies.

Forecasters use ocean topography data in computer models predicting weather and climate on time scales ranging from a few days to more than a year. SARAL also carries a communications package named ARGOS to collect observations from a network of ocean buoys and ground stations providing in situ data on wave height, period, water and air temperature, and other conditions.

Oceansat-2: Oceansat-2 is an Indian satellite designed to provide service continuity for operational users of the Ocean Color Monitor (OCM) instrument on Oceansat-1. It will also enhance the potential of applications in other areas. The main objectives of OceanSat-2 are to study surface winds and ocean surface strata, observation of chlorophyll concentrations, monitoring of phytoplankton blooms, study of atmospheric aerosols, and suspended sediments in the water.

Oceansat-2 is ISRO's second in the series of Indian Remote Sensing satellites dedicated to ocean research, and will provide continuity to the applications of Oceansat-1 (launched in 1999). Oceansat-2 will carry three payloads including an Ocean Color Monitor (OCM-2), similar to the device carried on Oceansat-1. Data from all instruments will be made available to the global scientific. Oceansat-2 was launched from Satish Dhawan Space Centre on 23 September 2009 using PSLV-C14 .The mission objectives of Oceansat-2 are to gather systematic data for oceanographic, coastal, and atmospheric applications.

Oceansat-2 will carry two payloads for ocean related studies, namely, the Ocean Color Monitor (OCM) and Ku-band pencil beam scatterometer. An additional piggy-back payload called ROSA (radio occultation sounder for atmospheric studies) developed by the Italian Space Agency (ASI) is also included. The major applications of data from Oceansat-2 are identification of potential fishing zones, sea state forecasting, coastal zone studies, and inputs for weather forecasting and climatic studies. The scientific payload contains three instruments. Two are Indian and one is from the Italian Space Agency.

- Ocean Color Monitor (OCM) is an 8-band multi-spectral camera operating in the visible–near IR spectral range. This camera provides an instantaneous geometric field of view of 360 meter and a swath of 1420 km. OCM can be tilted up to + 20 degree along track.

- Scanning Scatterometer (SCAT) is an active microwave device designed and developed at ISRO/SAC, Ahmedabad. It will be used to determine ocean surface level wind vectors through estimation of radar backscatter. The scatterometer system has a 1-meter parabolic dish antenna and a dual feed assembly to generate two pencil beams and is scanned at a rate of 20.5 rpm to cover the entire swath. The Ku-band pencil beam scatterometer is active microwave radar operating at 13.515 GHz providing a ground resolution cell of 50 x 50 km. It consists of a parabolic dish antenna of 1-meter diameter which is offset mounted with a cant angle of about 46° with respect to the Earth viewing axis. This antenna is continuously rotated at 20.5 rpm using a scan mechanism with the scan axis along the +ve Yaw axis. By using two offset feeds at the focal plane of the antenna, two beams are generated which will conically scan the ground surface. The back scattered power in each beam from the ocean surface is measured to derive wind vector. It is an improved version of the one on Oceansat-1. The inner beam makes an incidence angle of 48.90° and the outer beam makes an incidence angle of 57.60° on the ground. It covers a continuous swath of 1400 km for inner beam and 1840 km for outer beam respectively. The inner and outer beams are configured in horizontal and vertical polarization respectively for both transmit and receive modes. The aim is to provide global ocean coverage and wind vector retrieval with a revisit time of 2 days.

- Radio occultation sounder for atmospheric studies (ROSA) is a new GPS occultation receiver provided by ASI (Italian Space Agency). The objective is to characterize the lower atmosphere and the ionosphere, opening the possibilities for the development of several scientific activities exploiting these new radio occultation data sets.

An India Space Research Organization PSLV rocket (Polar Satellite Launch Vehicle) launched seven satellites from the Satish Dhawan Space Centre in Sriharikota, India, on Feb. 25, 2013. The rocket carried an ocean-monitoring satellite for India, two tiny space telescopes and an asteroid-hunting spacecraft built by the Canadian Space Agency among its payloads.

The ISRO (Indian Space Research Organization) spacecraft OceanSat-2 is envisaged to provide service continuity for the operational users of OCM (Ocean Color Monitor) data as well as to enhance the application potential in other areas. OCM is flown on IRS-P4/OceanSat-1, launched May 26, 1999. The main objectives of OceanSat-2 are to study surface winds and ocean surface strata, observation of chlorophyll concentrations, monitoring of phytoplankton blooms, study of atmospheric aerosols, and suspended sediments in the water.

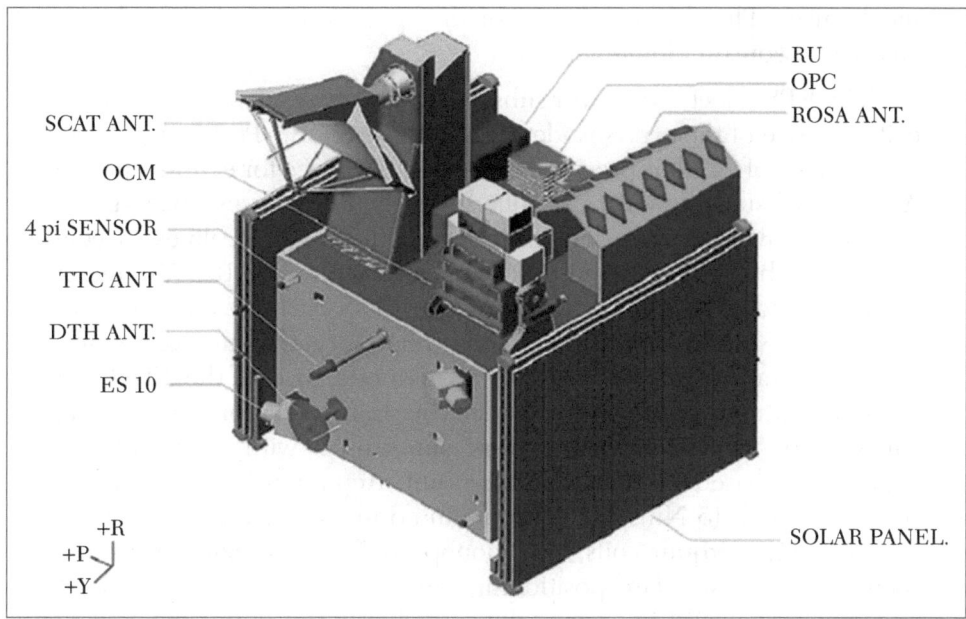

FIGURE A1.1 Oceansat-2.

Coverage of applications:

- Sea-state forecast: waves, circulation, and ocean MLD (mixed layer depth)

- Monsoon and cyclone forecast—medium and extended range

- Observation of Antarctic sea ice

- Fisheries and primary production estimation

- Detection and monitoring of phytoplankton blooms

- Study of sediment dynamics

Spacecraft: OceanSat-2 is a three-axis stabilized spacecraft configured around the proven IRS bus along with improved mission specific subsystem designs. The main structure is made up of a CFRP (carbon fiber reinforced plastic) composite cylinder with a PSLV interface ring. Three deployment mechanisms are included: 1) solar panel auto deployment after separation from the launcher, 2) OCM hold down release tilt mechanism, and 3) OSCAT antenna hold down release mechanism. The thermal design of the spacecraft employs both passive and active control elements.

The EPS (electrical power subsystem) uses two solar arrays with silicon cells, the size of the arrays is identical to those of IRS-P6/P5. A power storage capacity of 24 Ah is provided by 2 NiCd batteries for eclipse operations. All onboard subsystems are supplied with two raw buses of 28–42 V, and DC-DC converters are used to derive required voltage lines. A centralized BMU (bus management unit), designed with a MAR31750 microprocessor, provides the service functions for AOCS (attitude and orbit control subsystem), sensor processing, TT&C (telemetry, tracking, & command), auto-temperature control, and for PSK demodulation of the TT&C uplink carrier. Attitude sensing is provided by Earth horizon sensors, digital sun sensors, tri-axial magnetometers, sun sensors with a FOV of 4π, and a gyroscope-based inertial reference unit. Actuation is provided by four reaction wheels (5 Nms, 0.1 Nm) mounted in a tetrahedral configuration, two magnetic torquer coils, and monopropellant hydrazine thrusters. An 8-channel SPS (standard positioning service) GPS receiver provides both position and velocity, improving the overall orbit determination accuracy.

The payload data handling system is a new design; it is configured to transmit OCM and scatterometer data on a single carrier with QPSK modulation. The OCM data will be transmitted on the I-channel, while the OSCAT/ROSA data will be transmitted on the Q-channel. An indigenous onboard SSR (solid-state recorder) of 64 Gbit capacity is used to record the processed data of OSCAT and ROSA continuously; the OCM data is being recorded per requirement.

RF communications: The payload telemetry data transmission system is configured using SSPAs (solid state power amplifier) with a conventional X-band antenna. In addition, there is a TT&C subsystem in S-band for spacecraft control. The ground segment elements for OceanSat-2 include the SCC (spacecraft control center) at ISTRAC (ISRO Telemetry, Tracking and Command Network), Bangalore; the payload data reception station at NRSA (National Remote Sensing Agency) Shadnagar; data processing, data product generation and dissemination to users at NRSA, Balanagar, Hyderabad; data product software development at SAC (Space Application Center), Ahmedabad; and development of mission software, flight dynamics software and mission management at ISAC (ISRO Satellite Center), Bangalore, India.

Spacecraft bus structure	CFRP: Al honeycomb sandwich cylinder with Al honeycomb panels
Thermal system	Passive/semi-active thermal control with paints, blankets, OSRs and closed-looped auto temperature controllers
Thermal control	- Payloads: 15 ± 2°C for OCM, 5 - 45°C for OSCAT - Battery: 5 ± 5°C - Electronics: 0 to 40°C
Mechanisms	- Solar panel deployment - OCM hold down and release & OCM tilt - OSCAT antenna hold down & release
EPS (Electrical Power Subsystem)	- Solar panels: 3 on either side of S/C, sun tracking, area = 15.12 m² - Power: 1360 W, EOL, oriented normal to sun - Battery: 2 × 24 Ah NiCd batteries - Electronics: Two raw buses (28–42 V)
AOCS: Pointing accuracy Location accuracy Drift rate	- Sensors: Earth sensors, sun sensors, magnetometers, sun sensors, IRU - Actuators: Reaction wheels, magnetic torques, reaction control thrusters - ±0.10° (pitch & roll), ±0.15° (yaw) - 100-150 m (using SPS in autonomous mode) - < 3.0×10^{-4} °/s
Data Handling Subsystem	- Data rate: 42.4515 Mbit/s - RF communications: QPSK modulated transmitter - X band frequency: 8300 MHz
TT&C Subsystem	- Uplink: PCM/PSK/PM modulation, 4 kbit/s, time tag command facility - Downlink: PCM/PSK/PM modulation, 4 kbit/s (real time); 16 kbit/s (playback) - Transponder: Uplink frequency = 2071.875 MHz, downlink = 2250 MHSz
pacecraft mass, design life	970 kg, 5 years

TABLE A1.3 Overview of Spacecraft Parameters.

Orbit: Sun-synchronous near circular orbit, altitude ~720 km, inclination = 98.28°, period = 99.31 min, the LTAN (local time on ascending node) is at 12:00 hours ±10 minutes, revisit cycle of 2 days. The OceanSat-2 tracking system is S-band tone ranging from ISTRAC (ISRO Telemetry Tracking and Command Network) ground stations. The ranging system is CORTEX. Tracking measurements are two-way range, Doppler, and angles (azimuth and elevation).

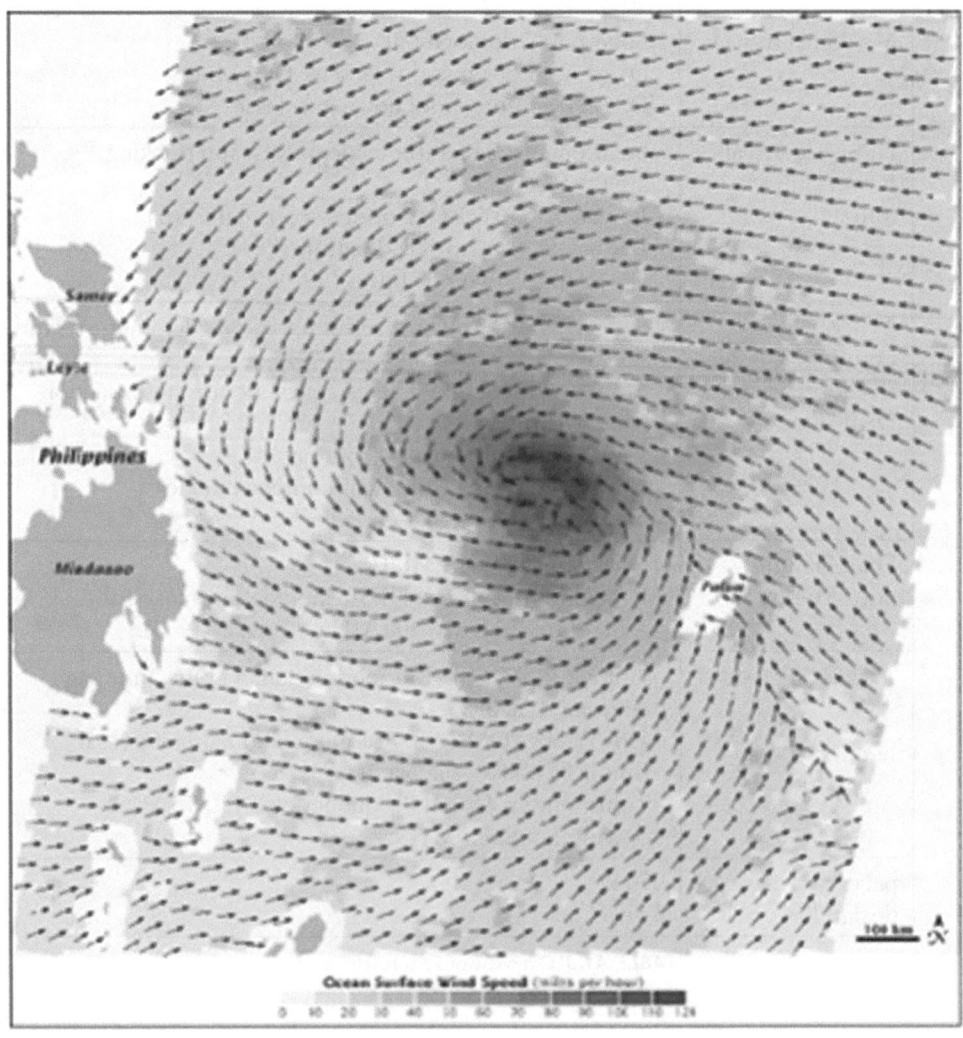

FIGURE A1.2 Assessing Haiyan's Winds with OSCAT on OceanSat-2.

On November 6, 2013, the OSCAT instrument of OceanSat-2 measured Haiyan's surface winds as shown in Figure A1.3. The arrows indicate wind direction while the colors indicate wind speed, with darker shades of purple indicating stronger winds (the strongest ones are shown in red). As is typical of cyclones in the Northern Hemisphere, the area of strongest winds was northeast of the storm center.

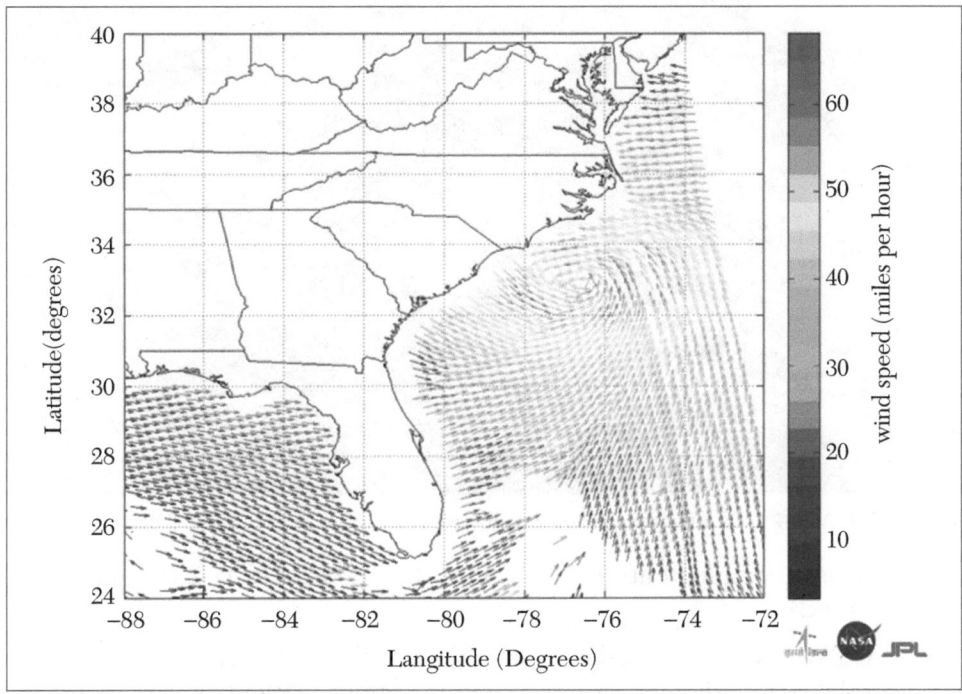

FIGURE A1.3 NASA-ISRO OSCAT image shows Irene's winds before landfall on Aug. 27, 2011.

The satellite image of Hurricane Irene (Figure A1.3), showing the storm's ocean surface wind speed and direction, was acquired at 1:07 a.m. EDT on Aug. 27, 2011, approximately six hours before it hit the North Carolina coast. The data are provided courtesy of the Indian Space Research Organization (ISRO) from the OSCAT instrument on ISRO's OceanSat 2 spacecraft, launched in September 2009. Wind vector data processing was performed at NASA/JPL, Pasadena, CA. The OSCAT winds are obtained at a resolution of 25 km × 25 km and do not resolve the hurricane's maximum wind speeds, which occur at much finer scales. All three payloads on board the OceanSat-2 have been successfully turned on providing good quality data, as shown in Figure A1.4.

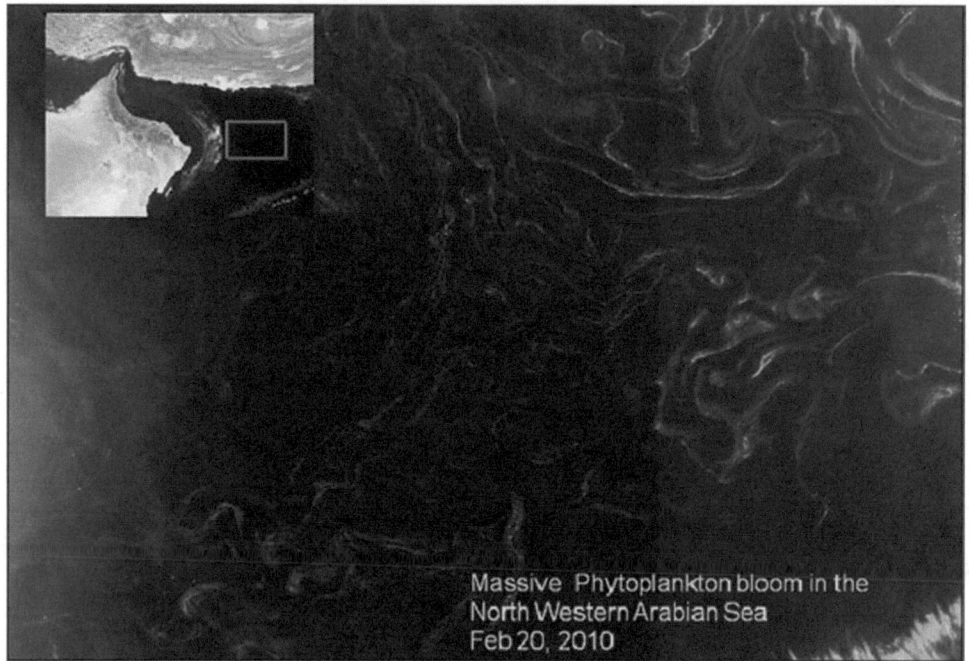

FIGURE 11.4 OCM-2 image of phytoplankton bloom in the Arabian Sea and OCM-2LAC data.

Sensor complement: (OCM-2, OSCAT, ROSA) OCM-2 (Ocean Color Monitor-2):

OCM-2 is an improved version of the one flown on OceanSat-1. OCM-2 is a solid-state radiometer providing observations in eight spectral bands in the VNIR region. The instrument employs pushbroom scanning technology with linear CCD detector arrays (191 6K CCD) of 6,000 elements (3,730 active detectors in the center are used to cover the image field, the rest are used to correct for dark current). A swath width of 1,420 km is provided. An along track instrument tilt capability of ±20° is provided to avoid sun glint. OCM optics is based on one lens per band (wide angle telecentric lens design, refractive system). The ground resolution is 360 meters in the along-track and 236 meters in the cross-track direction. The processing electronics consists of a video processor, timing logic, and interface circuits. An onboard calibration scheme, using light emitting diodes (LEDs) mounted near each CCD, is incorporated to study long-term stability of the radiometric performance.

Two modes in ground resolution are supported:

LAC (Local Area Coverage): 360 m (cross-track) × 236 m (along-track)

GAC (Global Area Coverage): 1 km

Parameter	Value	Parameter	Value
Spectral range (8 bands)	0.4 – 0.9 μm (VNIR)	SNR	> 512 (saturation)
Scan plane tilt	±20°, to avoid sun glitter	Effective focal length	20 mm
Camera MTF	>20% at Nyquist frequency	Absolute radio-metric accuracy	< 10%
IFOV at nadir (spatial resolution)	LAC: 360 m × 236 m GAC: 1 km	No of CCD elements	6000
FOV (swath)	1420 km (±43°)	CCD element size	10 μm × 7 μm
Tilt capability	±20° along track	Integration time	52.4 ms
Data quantization	12 bit	Exposure levels (gain)	16
Data rate (real-time)	20.8 Mbit/s	Onboard calibra-tion	2 LEDs per band
Instrument mass	78 kg	Instrument power	134 W

TABLE 11.4 Specification of the OCM Instrument

The configuration of the OCM payload is identical to the one flown in IRS-P4 (OceanSat-1) except that the spectral band is modified for band 6 and band 7. For band 6, the center wavelength is shifted from 670 nm to 620 nm to improve the reflectance from suspended sediments; for band 7, the center wavelength is shifted from 760 nm to 740 nm to avoid oxygen absorption. However, the bandwidth remains same in both cases.

OSCAT (OceanSat-2 Scanning Scatterometer): OSCAT is an active microwave device designed and developed at ISRO/SAC, Ahmedabad. The objective is to monitor ocean surface wind speed and directions. The instrument is a pencil beam wind scatterometer operating at Ku-band of 13.515 GHz. OSCAT is being utilized for the estimation of the radar backscattered power and subsequent local and global wind vector (velocity magnitude and direction) retrieval over the ocean, from the normalized radar cross-section (σ^o), for cell resolution grids of 25 × 25 km over a swath of 1400 km. The aim is to provide global ocean coverage and wind vector retrieval with a revisit time of 2 days.

The scanning configuration of OSCAT, similar in design to Seawinds of NASA, offers the advantages like simpler onboard payload, better radar backscatter cross section (σ^o) measurement and directional accuracy, continuous and wider swath with no nadir gaps, less complex signal processing and reduced data rates, smaller and lighter onboard instrument and simplified wind retrieval model compared to conventional multiple fan beam scatterometers.

The OSCAT onboard processing requirements are:

- Digital IQ demodulation and decimation
- Doppler shift computation for received return echo
- Doppler frequency compensation
- Reference chirp generation and de-chirping of echo returns
- Multiple 1 K complex FFTs of the de-chirped data
- Binning for estimation of signal+noise energy for every pulse
- Noise filtering and binning for noise-only estimation for every pulse
- Formatting of processed and payload and spacecraft auxiliary data
- Optional formatting and transmission of sensor raw data for selected acquisitions over Indian visibility regions.

The OSCAT payload design includes many new elements; it consists of an antenna, rotary joint, scan mechanism and switch assembly, transmitter, receiver, frequency generator, internal calibration unit, and digital subsystems, that is to say: DCG (digital chirp generator), DACS (data acquisition and compression subsystem), and payload controller. the frequency generator provides coherent reference frequencies for other onboard units and generates LFM (linear frequency modulated) pulses for transmission.

The OSCAT parabolic dish antenna has a diameter of 1 meter, which is offset-mounted with a cant angle of 46° with respect to the yaw axis (earth viewing axis). The antenna is continuously rotated at 20.5 rpm

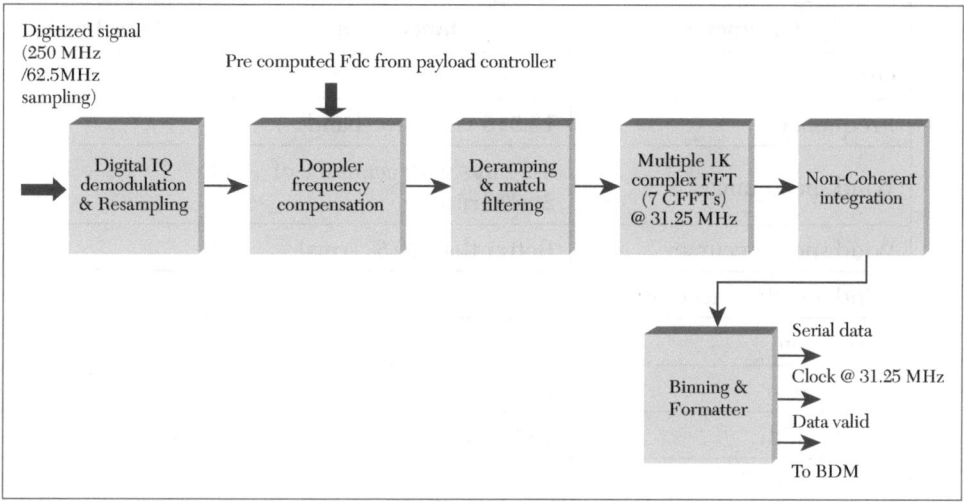

FIGURE A1.5 Onboard signal processor implementation of OSCAT.

using a DC motor with the scan axis along the +velocity yaw axis. By using two offset feeds at the focal plane of the antenna, two beams (inner beam and outer beam) are being generated which scan the ground surface in a conical fashion.

The antenna consisting of two off-axis near prime focus feeds along with a 1-meter paraboloid reflector creates inner and outer beams, which operate in an interleaved manner with an effective PRF (pulse repetition frequency) of 100 Hz each. The antenna, is conically scanned about the positive yaw axis at 20.5 rpm, by an appropriate scan mechanism. The received signal is amplified and down converted to generate the IF signal. This IF received signal from the receiver is fed to the onboard digital system for subsequent digitization, digital I/Q demodulation, and signal processing. The raw and processed data are fed to the spacecraft data handling unit for ground transmission.

The two pencil beams, inner and outer, result in a constant angle of incidence for both beams; this allows σ° measurements at multiple (4 or 2) azimuth angles for the same point on the ocean surface. Each point in the inner swath is viewed twice at different azimuth angles by both beams. The region between the inner and outer swath is subjected to two measurements by only the outer beam, and the wind vector there can only be determined with a directional ambiguity of 180°.

Parameter	Inner beam	Outer beam
Orbiting altitude	720 km	
Frequency	13.515 GHz (Ku-band)	
Wind speed range	4-24 m/s with accuracy of 2 m/s (rms)	
Wind speed accuracy	Better than 20 % (rms)	
Wind direction accuracy	20° (rms)	
Resolution	50 km × 50 km	
Polarization	HH	VV
Swath width	1400 km	1840 km
Scanning circle radius	700 km	920 km
Elevation angle (look angle)	42.62°	49.38°
Incidence angle	48.90°	57.60°
Footprint	26 × 46 Km	31 × 65 Km
Scanning rate	20.5 rpm	

TABLE A1.5 Specification of the OSCAT Instrument

Due to the very low receiver bandwidth of ±800 kHz, a single channel digital I/Q demodulation scheme has been implemented instead of the conventional analog I/Q demodulator. This approach results in compact RF and digitizer hardware and offers better signal fidelity in handling low bandwidth signals.

For the OSCAT instrument, real-time onboard signal processing involving range compression is mandatory considering the global mode of sensor operation, as it reduces the effective output data rate of the sensor by a large factor (~50). Also, the Doppler shift computation and subsequent Doppler compensation (within ±550 kHz) will be carried out in the signal processor itself, prior to range compression. To extract the range information, FFT (fast Fourier transforms) are performed on the deramped signal and an average periodogram is formed by applying magnitude-squaring operations. The DSP (digital signal processor) hardware is based on an FPGA implementation.

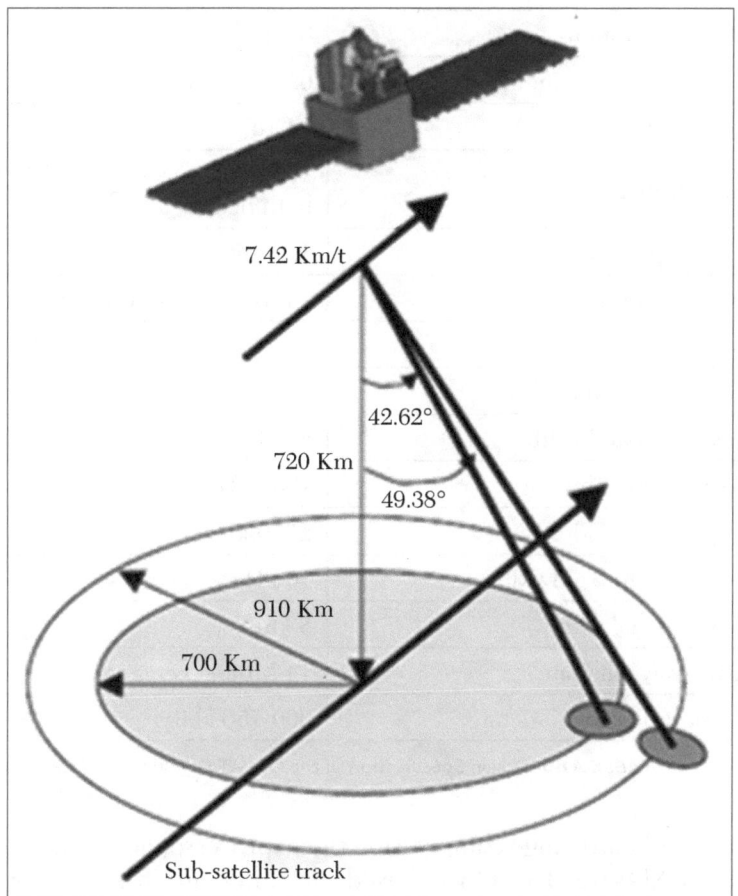

7.42 Km/t

42.62°

720 Km

49.38°

910 Km

700 Km

Sub-satellite track

FIGURE A1.6 Observation geometry of the OSCAT instrument.

The OSCAT instrument employs linear chirp transmission and digital deramping receiver techniques to measure surface backscatter with better accuracy and without compromising on range resolution. It uses digitally generated linear frequency modulation (LFM) transmit signal, having 400 KHz bandwidth and 1.35 ms pulse duration. The return echo signal is subsequently processed by a digital IF Receiver and signal processor based on a high-speed digitizer and FPGA (field programmable gate array).

ADC quantization	8 bit
ADC sampling frequency	62.5 MHz
Decimation factor	14/16
Transmit PRF	100 Hz for each beam, 200 Hz for both beams
Transmit pulse width	1.35 ms
Transmit modulation	LFM (linear frequency modulation)
Receive window	1.835 ms (4,096 samples)
Chirp signal bandwidth	400 kHz
Processing bandwidth	250 kHz
Noise bandwidth	1350 kHz
Doppler bandwidth	±550 kHz
Measurement bandwidth	10 kHz
Measured cell width	8 km
Output raw data rate	14 Mbit/s
Output processed data rate	300–750 kbit/s

TABLE A1.6 Design Specification of the OSCAT Digital Signal Processor

The onboard range compression signal processor algorithm implemented in Xilinx XQVR600 FPGA is based on periodogram estimation approach, where multiple 1K FFT of contiguous (50% overlapped) data partitions are averaged and binning is performed on the averaged spectrum to obtain signal+noise and noise-only energy estimates.

The digital receiver-signal processor hardware consists of a multi-layered PC board called a DACS (data acquisition and range compression system) module. It is based on Atmel's (Now E2V) TS8388B ADC and TS81102G0 Demux and Xilinx Virtex XQVR600 FPGA. The DACS unit has a mass of 2.25 kg and consumes 22.5 W of bus power.

ROSA (radio occultation sounder for atmospheric studies): ROSA is a new GPS occultation receiver provided by ASI (Italian Space Agency). A MOU (memorandum of understanding) between ASI and ISRO was signed in Fukuoka, Japan, in October 2005. The objective of ROSA is to characterize the lower atmosphere and the ionosphere, opening the possibilities for

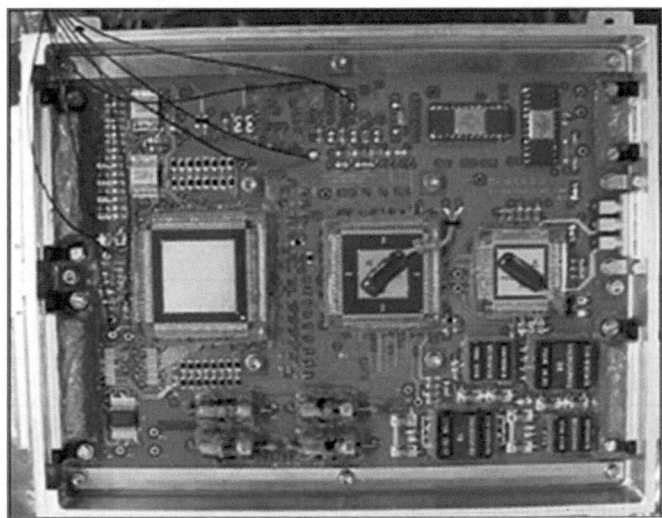

FIGURE A1.7 Digital receiver/processor of OSCAT.

the development of several scientific activities exploiting these new radio occultation data sets.

The ROSA instrument has been developed by TAS-I (Thales Alenia Space-Italia), formerly Laben. The ROSA payload is a dual channel GPS receiver with two antennae and a receiver package. The accommodation of the ROSA instrument on board the OceanSat-2 spacecraft has been driven by the satellite configuration that allows the possibility to install only one radio occultation antenna in the flight direction of the spacecraft. This means that only rising occultation events can be detected by the ROSA instrument in this mission.

The radio occultation antenna, looking along the satellite velocity vector, receives signals from the rising GPS satellites near the Earth horizon. These signals get refracted by the atmosphere and from the bending angle, the temperature and humidity profiles are derived. The POD (precise orbit determination) antenna, looking at the zenith of the satellite, gives precise position of the receiver.

The ROSA instrument, in its complete configuration, is composed of the following parts:

- One hemispherical-coverage navigation and POD (precise orbit determination) antenna to acquire GPS signals to determine position, velocity and time of the LEO satellite.

■ Velocity and anti-velocity radio occultation antennas to acquire GPS signals used in the calculus of all parameters used in the atmospheric sounding (for complete instrument).

■ The receiver unit which processes L1CA and L2P(Y) codeless GPS signals from all the antennas. 16 dual-frequency channels

Frequencies of operation	L1: 1560 - 1590 MHz; L2: 1212 - 1242 MHz
GPS codes used	C/A and P code
Dual-frequency channels (navigation)	16 max (without observation mode) 8 max (in observation mode)
Dual-frequency channels (observation)	8
Doppler shift Doppler rate Doppler acceleration	± 50 kHz ± 110 Hz/s $+ 0.5$ Hz/s^2
Antenna gain	$+ 5$ dBi for navigation antenna $+ 12$ dBi for RO (Radio Occultation) antenna
Polarization	RHCP
Horizontal resolution	< 300 km for temperature and humidity
Vertical resolution	0.3 km (low troposphere), 1–3 km (high troposphere)
Accuracy	< 1.0 K temperature 10% or 0.2 g/kg humidity
Input signal range	127 to -133 dBm POD antenna 130 to -148 dBm RO antenna
Instrument mass, power	17 kg, 36 W (standby) and 38 W (operation)
Data interface	Data/Command: MIL-STD-1553B PPS output: RS-422 (dedicated receiver connector)
Dimensions	Receiver: 287 mm × 250 mm × 206 mm Radio occultation antenna: 1035 mm × 500 mm × 165 mm Navigation antenna: 127 mm × 49 mm

TABLE A1.7 Major Parameters of the ROSA Instrument

(implemented in 4 AGGA-2 chips) are available in the ROSA receiver and can be assigned to different combinations of GPS satellites and POD or RO antennas. A MIL-STD-1553 communication interface is used to exchange telecommand, telemetry, and measurement data with the satellite on-board computer.

- The receiver processing is performed by an ADSP-21020 (analog devices signal processor-21020).

The radio occultation data processing for the ROSA receiver, which is called ROSA-ROSSA (ROSA-research and operational satellite and software activities), has been supported by the Italian Space Agency and has been developed by a pool of Italian universities and research centers.

The ROSA data will be downlinked to the Indian and the Italian receiving stations, to be processed by the ROSA ground segment, completely developed by Italian universities and research centers. This ground segment will be implemented at first level in an integrated computing infrastructure installed in Matera (Italy) and mirrored at Hyderabad, India and, at a second level, on a distributed software and hardware infrastructure. This second infrastructure will perform the rapid POD (precise orbit determination) and prediction, the unambiguous bending and impact parameters profiles extraction, the ionospheric correction and the stratospheric initialization, the refractivity, pressure, temperature, and humidity profile retrieval, the value-added services for meteorology, climate, and space weather applications. This will identify a prototype of distributed and multimission ground processing center distributed through the various research centers and universities involved, connected through a Web-based GRID computing infrastructure.

The ROA (ROSA occultation antenna) accommodation has been driven by the presence of the scatterometer by tilting of 15° on the satellite yaw axis. Moreover, in January 2010 ISRO decided to rotate on yaw axis in the same direction of other 20° for mission operation reasons. This final accommodation of the ROA antenna affects the number and type of occultation events that can be detected with respect to the optimal, velocity pointing, configuration. The quantity and quality of the occultations have been affected by the tilting of the ROA antenna of 35° on yaw axis that moreover introduces multipath effect due to the structure of the scatterometer and solar panel.

ROSA is flown on-board of the following three missions:

1. OceanSat-2 of ISRO, launched on September 23, 2009.

2. SAC-D of CONAE and NASA, launched on June 10, 2011.

- Megha Tropiques of ISRO and CNES, launched on October 12, 2011.

The OceanSat-2 tailoring features only one occultation antenna (the velocity antenna) is shown in Figure A1.8.

Ground segment:

The existing ISTRAC stations at Lucknow, Bearslake, Mauritius and Biak will be used for TT&C (telemetry tracking & command) support under the control of SCC, which will carry our mission operations, satellite health monitoring and analysis, and payload operations scheduling/programming.

The NRSA (National Remote Sensing Agency) DRS (Data Reception Station) located at Shadnagar, Hyderabad, with minor augmentation will receive the payload data both in real time as well as in playback from the onboard

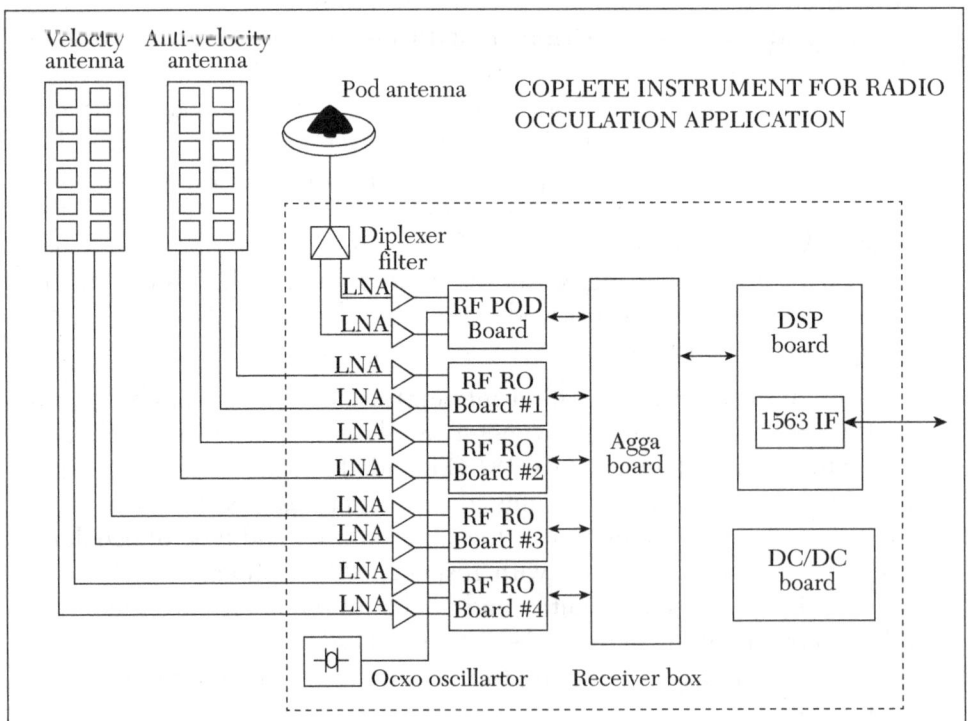

FIGURE A1.8 Block diagram of the ROSA instrument.

memory. The received data will be separated instrument-wise (OCM-2, OSCAT, ROSA) and recorded on the RAIDS (redundant array of independent disk storage) memory of the DAQLB (data acquisition and quicklook browsing) system. Quicklook display, browse generation, calibration analysis, and ADIF (auxiliary data file) generation will be carried out here. The data will be transferred to Balanagar on a high-speed data link in offline mode.

Data processing and products generation: The DPS (data processing system) at Balanagar will be the nodal center for processing the data from OceanSat-2, with support from SAC (Space Applications Center), Ahmedabad. The DPS is in charge to process the raw science data and generates the ocean color data products with in-built work order generation, online quality control, output media preparations, data quality evaluation, and feedback to mission operations. Associated development of mathematical formulations, geometric and radiometric look-up tables, and their updates, associated software tools, and geophysical model functions for wind vector derivation from the OSCAT data are part of the overall data products generation at different levels.

OceanSat-2 will provide two types of science data: LAC (local area coverage) at 360 meter resolution, and GAC (global area coverage) at 1 km and at 4 km resolutions.

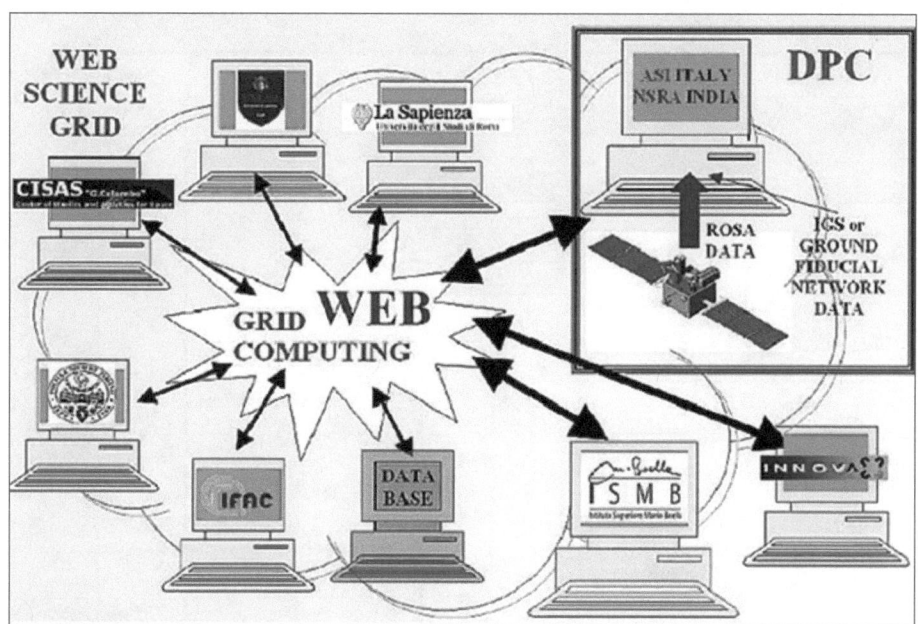

FIGURE A1.9 Schematic view of the ROSA ground segment.

On Sept. 12, 2011, a new state-of-the-art OceanSat-2 ground station (data reception and processing system) was inaugurated in Hyderabad. The OceanSat-2 ground station, fitted with a 7.5 meter diameter antenna, is capable of covering an area of 5,000 km in diameter, covering the Bay of Bengal on the east and the Arabian Sea in the west. The ground station was established to receive and process data from Ocean Color Monitor on-board the Indian Remote Sensing Satellite Oceansat-2 in real time.

Research requirements in India related to the ocean: India should develop capabilities to fully monitor the Indian Ocean region in a complete and 3-dimensional manner. India should develop 80–110 satellites to cover the region.

U.S. OCEAN MONITORING[1]

When you watch the news and see images of weather from around the United States or the world, you are seeing data from NOAA's environmental satellites.

NOAA's environmental satellites provide data from space to monitor the earth to analyze coastal waters, relay life-saving emergency beacons, and predict and track tropical storms and hurricanes.

NOAA operates three types of satellite systems for the United States – **polar-orbiting satellites, geostationary satellites,** and our **deep space satellite**. Polar-orbiting satellites circle the earth and provide global information from 540 miles above the earth. Geostationary satellites constantly monitor the Western Hemisphere from around 22,240 miles above the earth. And our deep space satellite orbits one million miles from earth, providing space weather alerts and forecasts while also monitoring the amounts of solar energy absorbed by Earth every day.

Satellites enable us to provide consistent, long-term observations, 24 hours a day, 7 days a week. By remotely sensing from their orbits high above the earth, they provide us much more information than would be possible to obtain solely from the surface. Over 90 % of the data that goes into our weather models is from satellites.

---

[1] (Source : National Oceanic and Atmospheric Administration/U.S. Dept. of Commerce http://www. noaa.gov/)

They track fast breaking storms across "Tornado Alley" as well as tropical storms in the Atlantic and Pacific oceans.

Using satellites, NOAA researchers can also more closely study the ocean. Information gathered by these satellites can tell us about ocean bathymetry, sea surface temperature, ocean color, coral reefs, and sea and lake ice. NOAA's satellite data improve the Nation's resilience to climate variability, maintain our economic vitality, and improve the security and well-being of the public.

In addition to operating our own satellites, NOAA helps promote and enable commercial uses of satellites and space to benefit the U.S. economy.

Satellites provide other services beyond just imaging the earth. Monitoring conditions in space and solar flares from the sun help us understand how conditions in space affect the earth.

Satellites also relay position information from emergency beacons to help save lives when people are in distress on boats, airplanes, or in remote areas. Each year, thousands of people are rescued through SARSAT - Search And Rescue Satellite Aided Tracking.

Scientists also use a data collection system on the satellites to relay data from transmitters on the ground to researchers in the field – such as measuring tidal heights or the migration of whales.

Monitoring the earth from space helps us understand how the earth works and affects much of our daily lives.

USING SATELLITES FOR FORECASTING (U.S.)[1]

U sing environmental satellites to observe the Earth from space is one of the key tools in forecasting weather, analyzing climate, and monitoring hazards worldwide. This 24-hour global coverage provides us with a never-ending stream of information critical for making decisions affecting everything from what you are going to wear today to governments making decisions about how to deal with climate change. The National Oceanic and Atmospheric Administration (NOAA) Satellite Information Service https://www.nesdis.noaa.gov/ in collaboration with the National Aeronautics and Space Administration (NASA) and the U.S. Air Force, manages and operates fleets of weather and environmental monitoring satellites. There are two main types of environmental satellites: geostationary and polar-orbiting.

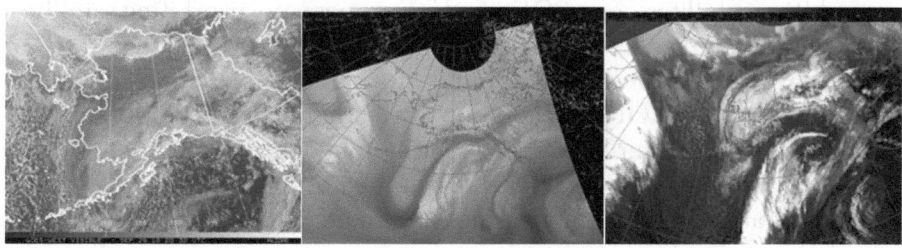

[1] SOURCE: National Weather Service/NOAA http://www.weather.gov/ajk/OurOffice-Sat

Geostationary Operational Environmental Satellites: When you watch your local newscaster present the weather forecast, and they show an image of weather over the whole United States, you are seeing imagery from NOAA Geostationary Operational Environmental Satellites, or GOES. GOES orbit 35,800 km (22,300 miles) above the Earth's equator at speeds equal to Earth's rotation, which maintain their positions relative to Earth. The GOES provide constant monitoring of various areas of the planet. To fully cover Alaska, Hawaii, the entire continental United States, and the Pacific and Atlantic Ocean (for tropical storms), NOAA operates two GOES satellites simultaneously - GOES-East and GOES-West. The satellites provide constant coverage of the western hemisphere by taking photographic images every 15 minutes. These "constant eyes" are critical for identifying severe weather, snow storms, tropical storms and hurricanes. GOES protect our lives and property every day - constantly watching for new storms and severe weather.

Polar Operational Environmental Satellites: When you wonder on Wednesday what the weather will be like over the weekend you turn to weather forecasters, who rely on NOAA Polar Operational Environmental Satellites, or POES, to help make their predictions. POES make regular 360° orbits around the Earth's poles from about 833 km (517 miles) above the Earth's surface. The Earth constantly rotates counterclockwise underneath the path of the satellite making for a different view with each orbit. It takes the satellite approximately 1.5 hours complete a full orbit. In a 24-hour period, the 14 orbits of each polar satellite provide two complete views of weather around the world. By having imagery of the whole globe, meteorologists are able to develop models to predict the weather out to five to ten days. NOAA partners with EUMETSAT to constantly operate two polar-orbiting satellites - one POES and a European polar-orbiting satellite called Metop. When polar-orbiting satellites fly over severe weather, they can also give us very detailed pictures of the storms given how much closer they are to storms than GOES. In addition to weather analysis and forecasting, data from the POES series support a broad range of environmental monitoring applications including climate research and prediction, global sea surface temperature measurements, measurements of temperature and humidity of the atmosphere, ocean dynamics research, volcanic eruption monitoring, forest fire detection, and global vegetation analysis. Instruments on POES are critical for providing long-term, sustained observations used for determining the long term changes in climate conditions around the world.

Finally, POES assist in search and rescue by locating people, planes and ships, who have activated emergency locator beacons.

In addition to basic imagery, on-board sensors detect cloud, land, and ocean temperatures, as well as monitor activities of the sun. NOAA GOES are also used in identifying when satellite emergency locator beacons have been activated to help with Search and Rescue activities.

MetOp is a series of polar orbiting meteorological satellites operated by the European Organisation for the Exploitation of Meteorological Satellites. The satellites are all part of the EUMETSAT Polar System.

Ocean Surface Topography Mission (OSTM)/JASON-2: One aspect of climate change is sea level rise, which affects much of the world's population that live in coastal areas. To measure the height of the ocean around the world, NOAA participates in a joint Ocean Surface Topography Mission (OSTM) program between NOAA, NASA, France's Centre National d'Etudes Spatiales (CNES), and European Organisation for the Exploitation of Meteorological Satellites (EUMETSAT). This is a joint effort by the four organizations to measure sea surface height by using a radar altimeter mounted on a low-earth orbiting satellite called Jason-2. Satellite altimetry data provides sea surface heights for determining ocean circulation, climate change and sea-level rise. These sea surface height measurements are necessary for ocean modeling, forecasting El NiÑo/La NiÑa events, and hurricane intensity prediction.

TRACKING ASH CLOUDS

Satellite imagery is also used to track the ash cloud after a volcanic eruption.

Alaska contains over 130 volcanoes and volcanic fields, which have been active within the last two million years. These volcanoes are catalogued on the National Volcano Observatory's web site.

Of these volcanoes, about 90 have been active within the last 10,000 years (and might be expected to erupt again), and more than 50 have been active within historical time (since about 1760, for Alaska).

The volcanoes in Alaska make up well over three-quarters of U.S. volcanoes that have erupted in the last two hundred years.

Alaska's volcanoes are potentially hazardous to passenger and freight aircraft as jet engines sometimes fail after ingesting volcanic ash. Based on information provided by the Federal Aviation Administration, that more than 80,000 large aircraft per year, and 30,000 people per day, are in the skies over and potentially downwind of Aleutian volcanoes, mostly on the heavily traveled great-circle routes between Europe, North America, and Asia. Volcanic eruptions from Cook Inlet volcanoes (Spurr, Redoubt, Iliamna, and Augustine) can have severe impacts, as these volcanoes are nearest to Anchorage, Alaska's largest population center.

EXPLORING THE OCEAN BASINS WITH SATELLITE ALTIMETER DATA[1]

David T. Sandwell - Scripps Institution of Oceanography
and
Walter H. F. Smith - Geosciences Laboratory, NOAA

The surface of the ocean bulges outward and inward mimicking the topography of the ocean floor. The bumps, too small to be seen, can be measured by a radar altimeter aboard a satellite. Over the past year, data collected by the European Space Agency ERS-1 altimeter along with recently declassified data from the US Navy Geosat altimeter have provided detailed measurements of sea surface height over the oceans. These data provide the first view of the ocean floor structures in many remote areas of the Earth. For scientific applications, the Geosat and ERS-1 altimeter data are comparable in value to the radar altimeter data recently collected by the Magellan spacecraft during its systematic mapping of Venus.

---

[1] SOURCE: National Centers for Environmental Information/NOAA https://www.ngdc.noaa.gov/mgg/bathymetry/predicted/explore.HTML#sat_alt

INTRODUCTION

The geologic and topographic structures of the ocean floor primarily reflect plate tectonic activity that has occurred over the past 150 million years of the 4.5 billion year age of the Earth. Seafloor geology is far simpler than the geology of the continents because erosion rates are lower and also because the continents have suffered multiple collisions associated the opening and closing of ocean basins (Wilson Cycle). Despite their youth and geologic simplicity, most of this deep seafloor has remained poorly understood because it is masked by 3-5 km of seawater. For example, the Pacific-Antarctic rise, which has an area about equal to South America, is a broad rise of the ocean floor caused by sea floor spreading between two major tectonic plates (see Poster southeast of New Zealand). To the west of the ridge lies the Louisville seamount chain which is a chain of large undersea volcanoes having a length equal to the distance between New York and Los Angeles. These features are unfamiliar because they were discovered less than 20 years ago. The Louisville seamount chain was first detected in 1972 using depth soundings collected along random ship crossings of the South Pacific. Six years later the full extent of this chain was revealed by a radar altimeter aboard the Seasat (NASA) spacecraft. Recently, high density data collected by the Geosat (US Navy) and ERS-1 (European Space Agency) spacecraft data show the Pacific-Antarctic Rise and the Louisville Ridge in unprecedented detail. In an age when we are mapping the surfaces of Venus and Mars, it is difficult to believe that so little is known about our own planet.

The reason that the ocean floor, especially the southern hemisphere oceans, is so poorly charted is that electromagnetic waves cannot penetrate the deep ocean (3-5 km = 2-3 mi). Instead, depths are commonly measured by timing the two-way travel time of an acoustic pulse. However because research vessels travel quite slowly (6m/s = 12 knots) it would take approximately 125 years to chart the ocean basins using the latest swath-mapping tools. To date, only a small fraction of the sea floor has been charted by ships.

Fortunately, such a major mapping program is largely unnecessary because the ocean surface has broad bumps and dips which mimic the topography of the ocean floor. These bumps and dips can be mapped using a very accurate radar altimeter mounted on a satellite. In this brief report we attempt to answer some basic questions related to satellite measurements of the ocean basins. What causes the surface of the ocean to bulge outward and inward mimicking the topography of the ocean floor? How big are these bumps? How can they be measured in the presence of waves and tides? What are some of the non-military applications of these data? What has been discovered from the new Geosat and ERS-1 data?

SATELLITE ALTIMETRY

According to the laws of physics, the surface of the ocean is an "equipotential surface" of the earth's gravity field. (Let's ignore waves, winds, tides and currents for the moment.) Basically this means that if one could place balls everywhere on the surface of the ocean, none of the balls would roll downhill because they are all on the same "level". To a first approximation, this equipotential surface of the earth is a sphere. However because the earth is rotating, the equipotential ocean surface is more nearly matched by an ellipsoid of revolution where the polar diameter is 43 km less than the equatorial diameter. While this ellipsoidal shape fits the earth remarkably well, the actual ocean surface deviates by up to 100 meters from this ideal ellipsoid. These bumps and dips in the ocean surface are caused by minute variations in the earth's gravitational field. For example the extra gravitational attraction due to a massive mountain on the ocean floor attracts water toward it causing a local bump in the ocean surface; a typical undersea volcano is 2000 m tall and has a radius of about 20 km. This bump cannot be seen with the naked eye because the slope of the ocean surface is very low.

These tiny bumps and dips in the geoid height can be measured using a very accurate radar mounted on a satellite (Figure). For example, the Geosat satellite was launched by the US Navy in 1985 to map the geoid height at a horizontal resolution of 10-15 km (6 - 10 mi) and a vertical resolution of 0.03 m (1 in). Geosat was placed in a nearly polar orbit to obtain high latitude coverage (+- 72 deg latitude). The Geosat altimeter orbits the earth 14.3 times per day resulting in an ocean track speed of about 7 km per second (4 mi/sec). The earth rotates beneath the fixed plane of the satellite orbit, so over a period of 1.5 years, the satellite maps the topography of the surface of the earth with an ground track spacing of about 6 km (4 mi).

Two very precise distance measurements must be made in order to establish the topography of the ocean surface to an accuracy of 0.03 m

(1 in) (Figure). First, the height of the satellite above the ellipsoid h° is measured by tracking the satellite from a globally-distributed network of lasers and/or doppler stations. The trajectory and height of the satellite are further refined by using orbit dynamic calculations. Second, the height of the satellite above the closest ocean surface h is measured with a microwave radar operating in a pulse-limited mode on a carrier frequency of 13 GHz. (The ocean surface is a good reflector at this frequency.) The radar illuminates a rather large spot on the ocean surface about 45 km (28 mi) in diameter. A smaller effective footprint (1-5 km in diameter = 0.6 - 3 mi)) is achieved by forming a sharp radar pulse and accurately recording its 2-way travel time. The footprint of the pulse must be large enough to average out the local irregularities in the surface due to ocean waves. The spherical wave front of the pulse ensures that the altitude is measured to the closest ocean surface. A high repetition rate (1000 pulses per second) is used to improve the signal to noise ratio, especially when the ocean surface is rough. Corrections to the travel time of the pulse are made for ionospheric and atmospheric delays and known tidal corrections are applied as well. The difference between the height above the ellipsoid and the altitude above the ocean surface is approximately equal to the geoid height N = h° - h.

GRAVITY ANOMALY

As the spacecraft orbits the earth it collects a continuous profile of geoid height across an ocean basin. Profiles from many satellites, collected over many years, are combined to make high resolution images. The Poster shows gravity anomaly derived from geoid height measurements from 4.5 years of Geosat measurements and 2 years of ERS-1 measurements. We have developed a new method to convert these raw geoid height measurements, which have a variety of accuracies, track spacings and data densities, into images (or grids) of gravity anomaly. This conversion is done to enhance the small-scale features of the seafloor. Moreover, after the conversion, the satellite-derived gravity measurements can be compared and combined with gravity anomaly measurements made by ships. The algorithms of the conversion are based on laws of physics, geometry and statistics. Since the data sets are large, diverse, and contaminated with errors, many sophisticated computer operations are required. The ultimate test of the accuracy of our methods is through comparisons with shipboard gravity measurements. Our latest grids show agreement with ship data at a level of 5 milligal (mgal). One mgal is about one millionth the normal pull of gravity (9.8 m/s2). Typical variations in the pull of gravity are 20 milligal although over the deep ocean trenches they exceed 300 mgal.

APPLICATIONS OF SATELLITE ALTIMETRY

Navigation

The Geosat data were collected by the US Navy to fulfill their navigational and mapping requirements. Consider measuring accelerations in a moving submarine or aircraft in order to determine your position as a function of time. (Of course your starting position and velocity must also be known.) If the windows of your vehicle are closed, a true acceleration cannot be distinguished from a variation in the pull of gravity. Thus the gravity data are needed for correction of inertial navigation/guidance systems. The military applications are obvious and provided the rationale for the 80 million dollar cost of the Geosat mission as well as the classification of these data, especially during the cold war when nuclear submarines were more active than they are today. On the commercial side, Honeywell Inc. is using these data to update their inertial navigation systems in commercial aircraft. In particular, when this correction is not applied, they have found large navigational errors along Pacific Ocean flight paths which follow the major ocean trenches.

Prediction of Seafloor Depth

We are using these dense satellite altimeter measurements in combination with sparse measurements of seafloor depth to construct a uniform resolution map of the seafloor topography. These maps do not have sufficient accuracy and resolution to be used to assess navigational hazards but they are useful for such diverse applications as locating the obstructions/constrictions to the major ocean currents and locating shallow seamounts where fish and lobster are abundant.

On a broad scale the topography of the ocean floor reflects the cooling and subsidence of the plates as they move away from the spreading center. While this process is fairly well understood, there are interruptions in this normal subsidence caused by mantle plumes and other types of solid-state convection in the mantle of the Earth that are current topics of research. As the seafloor ages it also becomes covered by a slow rain of sediments. The analysis of the gravity data along with measured can be used to map the thickness of the sedimentary layers.

PLANNING SHIPBOARD SURVEYS

The satellite-derived gravity grids reveal all of the major structures of the ocean floor having widths greater than 10-15 km (6-9 mi). This resolution matches the total swath width of the much higher multibeam mapping system on a ship (100 m resolution) so the gravity maps are the perfect reconnaissance tool for planning the more detailed shipboard surveys. Scientists aboard research vessels use the gravity grids along with other measurements to optimize their survey strategy; in many cases this is done in real time. The cost to operate a research vessel is typically $20,000 per day so these gravity data have become an essential item.

PLATE TECTONICS

These satellite altimeter data provide an important and definitive confirmation of the theory of plate tectonics. Indeed, almost everything apparent in the marine gravity field was created by the formation and motion of the plates. The Indian Ocean Triple junction (27 deg S latitude, 70 deg E longitude) is a textbook example of seafloor spreading. Spreading ridges are characterized by an orthogonal pattern of ridges and transform faults. The scar produced in the active transform valley is carried by seafloor spreading out onto older seafloor leaving evidence of the past plate motions. At this Indian Ocean site, three spreading ridges intersect forming a triple junction as described by plate tectonic theory. The theory predicts that the ridges would intersect at 120° angles if the three ridges were spreading at exactly the same rate. In this case, one can measure the intersection angles and infer the relative spreading rates of each ridge.

Plates are created at spreading ridges and destroyed (subducted) at the deep ocean trenches. All of the major ocean trenches are evident in the gravity map as linear troughs. The deep ocean basins away from the trenches are characterized by fracture zone gravity signatures inherited at the spreading ridge axis. This pattern is sometimes overprinted by linear volcanic chains which are believed to be formed as the plate moves over a stationary mantle plume. The hot plume head melts the mantle rocks which erupt on the surface as a hot spot. Because all of these major features are evident in the gravity maps, the geologic history of the ocean basins can now be established in great detail.

UNDERSEA VOLCANOES

The global gravity grids reveal all volcanoes on the seafloor greater than about 1000 m tall. Approximately 1/2 of these volcanoes were not charted previously. One of the more important aspects of these new data will be to locate all of these volcanoes and identify spatial patterns that may help determine how they formed. Many volcanoes appear in chains, perhaps associated with mantle plumes, there are many more that do not fit this simple model. Moreover, numerous undersea volcanoes are long linear ridges with aspect ratios of 20 or more. These features suggest that the plates are not exactly ridged as predicted by the simple plate tectonic theory. Using these data we are exploring the internal deformations of the plates, especially outboard of trenches where the forces generated by the slab-pull force of the subducted plates is greatest.

PETROLEUM EXPLORATION

All of the major petroleum exploration companies use satellite altimeter gravity data from Geosat and ERS-1 to locate offshore sedimentary basins in remote areas. This information is combined with other reconnaissance survey information to determine where to collect or purchase multi-channel seismic survey data. Currently, the regions of most intense exploration interest are the continental shelves of Australia and the former Soviet Union; recently companies have expressed interest in the Caspian Sea. Developments in offshore drilling technology now make it economical to recover oil from continental slope areas in water once thought prohibitively deep.

While we are not directly involved in this activity, we fill data requests from many companies including UNOCAL. Dr. Mark Odegard of UNOCAL Inc. says "We routinely use satellite gravity data in any exploration effort in the oceans outside of the Gulf of Mexico. We consider the current data quality to be better than the standard regional type of survey that was run in the previous twenty five years or so. . . . If the conventional gravity data were collected over the continental shelves of the World, where we have interest in exploration, the cost by my estimate, would be in the range of $200-$400 million dollars, maybe more. This obviously will not be done, but we are beginning to collect high-resolution (shipboard) data over selected targets outside the US The other companies that use the satellite data, that I know of, are: Exxon, Mobil, and Texaco. UNOCAL has been a recognized leader in potential fields in the oil industry, so we have probably been quicker to utilize the data than many companies. I would suspect BP, Total, and AGIP are foreign companies that probably utilize the data."

LITHOSPHERIC STRUCTURE

There are numerous other scientific applications that cannot be described in a short report. One of the traditional uses of marine gravity measurements is to estimate the thickness of the elastic portion of the tectonic plates. When a volcano forms on the ocean floor it provides a large downward load on the plate causing it to deform. This deformation is appears in the gravity field as a donut-shaped gravity low surrounding the gravity high associated with the volcano itself. By measuring the amplitude and width of the gravity low and relating this to the size of the volcano as measured my a ship with an echo sounder, one can establish the thickness and strength of the elastic plate. The new satellite-derived gravity data enable researchers to perform this type of analysis everywhere in the oceans. Thus scientists can now probe the outermost part of the earth using these and other methods.

ACRONYMS

AOP	Apparent Optical Properties
ANN	Artificial Neural Network
CDOM	Colored Dissolved Organic Matter
CNES	Centre National d'Etudes Spatiales
CNSA	China National Space Administration
COASTLOOC	Coastal Surveillance Through Observation of Ocean Color
COCTS	China Ocean Color and Temperature Scanner
CZCS	Coastal Zone Color Scanner
DLR	German Aerospace Center
DOM	Dissolved Organic Matter
ESA	European Space Agency
GLI	Global Imager
IOP	Inherent Optical Properties
ISRO	Indian Space Research Organization
KARI	Korean Aerospace Research Institute

MERIS	Medium Resolution Imaging Spectrometer (ESA)
MLP	Multi-Layer Perceptron
MISR	Multi-angle Imaging SpectroRadiometer
MODIS	Moderate Resolution Imaging Spectrometer (NASA)
MOMO	Matrix Operator Model
MOS	Modular Optoelectric Scanner
NASA	National Aeronautics and Space Administration
NASDA	National Space Development Agency of Japan
OCM	Ocean Color Monitor
OCTS	Ocean Color and Temperature Scanner
OSMI	Ocean Scanning Multispectral Imager
PMNS	Particulate Matter North Sea
POLDER	Polarization and Directionality of the Earth's Reflectances
POM	Particulate Organic Matter
RMSE	Root Mean Square Error
RT	Radiative Transfer
RTC	Radiative Transfer Calculation
S-GLI	Second generation GLobal Imager
SeaBAM	SeaWiFS Bio-optical Algorithm Mini-Workshop
SeaBASS	SeaWiFS Bio-optical Archive and Storage System
SeaWiFS	Sea-viewing Wide Field-of-view Sensor (NASA)
SPM	Suspended Particulate Matter
TOA	Top of Atmosphere
VIIRS	Visible Infrared Imager Radiometer Suite

INDEX